New Trends in Algebraic Geometry and Its Applications

New Trends in Algebraic Geometry and Its Applications

Editor

Sonia Pérez-Díaz

MDPI • Basel • Beijing • Wuhan • Barcelona • Belgrade • Manchester • Tokyo • Cluj • Tianjin

Editor
Sonia Pérez-Díaz
University of Alcalá
Spain

Editorial Office
MDPI
St. Alban-Anlage 66
4052 Basel, Switzerland

This is a reprint of articles from the Special Issue published online in the open access journal *Mathematics* (ISSN 2227-7390) (available at: https://www.mdpi.com/journal/mathematics/special_issues/new_trends_in_algebraic_geometry_and_its_applications).

For citation purposes, cite each article independently as indicated on the article page online and as indicated below:

LastName, A.A.; LastName, B.B.; LastName, C.C. Article Title. *Journal Name* **Year**, *Volume Number*, Page Range.

ISBN 978-3-0365-2307-1 (Hbk)
ISBN 978-3-0365-2308-8 (PDF)

© 2021 by the authors. Articles in this book are Open Access and distributed under the Creative Commons Attribution (CC BY) license, which allows users to download, copy and build upon published articles, as long as the author and publisher are properly credited, which ensures maximum dissemination and a wider impact of our publications.

The book as a whole is distributed by MDPI under the terms and conditions of the Creative Commons license CC BY-NC-ND.

Contents

About the Editor vii

Preface to "New Trends in Algebraic Geometry and Its Applications" ix

Sonia Pérez-Díaz and J. Rafael Sendra
Computing Birational Polynomial Surface Parametrizations without Base Points
Reprinted from: *Mathematics* **2020**, *8*, 2224, doi:10.3390/math8122224 1

Jorge Caravantes, J. Rafael Sendra, David Sevilla and Carlos Villarino
Covering Rational Surfaces with Rational Parametrization Images
Reprinted from: *Mathematics* **2021**, *9*, 338, doi:10.3390/math9040338 29

Sonia Pérez-Díaz and Li-Yong Shen
The μ-Basis of Improper Rational Parametric Surface and Its Application
Reprinted from: *Mathematics* **2021**, *9*, 640, doi:10.3390/math9060640 45

Francisco G. Montoya, Raúl Baños, Alfredo Alcayde, Francisco M. Arrabal-Campos and Javier Roldán-Pérez
Vector Geometric Algebra in Power Systems: An Updated Formulation of Apparent Power under Non-Sinusoidal Conditions
Reprinted from: *Mathematics* **2021**, *9*, 1295, doi:10.3390/math9111295 67

Alberto Besana and Cristina Martínez
A Topological View of Reed–Solomon Codes
Reprinted from: *Mathematics* **2021**, *9*, 578, doi:10.3390/math9050578 85

Juan G. Alcázar
On the Affine Image of a Rational Surface of Revolution
Reprinted from: *Mathematics* **2020**, *8*, 2061, doi:10.3390/math8112061 101

Francisco G. Montoya, Raúl Baños, Alfredo Alcayde, Francisco M. Arrabal-Campos and Javier Roldán Pérez
Geometric Algebra Framework Applied to Symmetrical Balanced Three-Phase Systems for Sinusoidal and Non-Sinusoidal Voltage Supply
Reprinted from: *Mathematics* **2021**, *9*, 1259, doi:10.3390/math9111259 119

Anastasis Kratsios
Lower-Estimates on the Hochschild (Co)Homological Dimension of Commutative Algebras and Applications to Smooth Affine Schemes and Quasi-Free Algebras
Reprinted from: *Mathematics* **2021**, *9*, 251, doi:10.3390/math9030251 137

Muhammad Imran Qureshi
Polarized Rigid Del Pezzo Surfaces in Low Codimension
Reprinted from: *Mathematics* **2020**, *8*, 1567, doi:10.3390/math8091567 159

About the Editor

Sonia Pérez-Díaz is an Associate Professor ("Profesor Titular de Universidad") of the Department of Physic and Mathematics at the University of Alcalá, in the area of Applied Mathematics. She has published many papers in journals and conferences, both national and international, and the book Rational Algebraic Curves: A Computer Algebra Approach (2007) in Springer Verlag. Her areas of expertise include the study and analysis of symbolic, numeric and approximate algorithms, as well as its applications in the theory of curves and surfaces (https://orcid.org/0000-0002-0174-5325).

Preface to "New Trends in Algebraic Geometry and Its Applications"

Algebraic geometry is an old and important research topic. Its applications include architectural designs, number theoretic problems, models of biological shapes, error-correcting codes, cryptographic algorithms, or as central items in computer-aided geometric design.

In this Special Issue, we explore the interplay between geometry, algebra, and numerical computation when designing algorithms for different varieties, and provide a complexity analysis of the running time of such algorithms. More precisely, we focus on recent problems concerning algebraic geometry and its applications.

This Special Issue is mainly intended for graduate students specializing in constructive algebraic curve geometry. However, we are sure that researchers wanting an overview of where algebraic geometry can be applied to symbolic or numeric algebraic computation will find this book helpful.

Sonia Pérez-Díaz
Editor

Article

Computing Birational Polynomial Surface Parametrizations without Base Points

Sonia Pérez-Díaz *,† **and Juan Rafael Sendra** †

Departamento de Física y Matemáticas, Universidad de Alcalá, E-28871 Madrid, Spain; rafael.sendra@uah.es
* Correspondence: sonia.perez@uah.es
† These authors contributed equally to this work.

Received: 17 November 2020; Accepted: 10 December 2020; Published: 14 December 2020

Abstract: In this paper, we present an algorithm for reparametrizing birational surface parametrizations into birational polynomial surface parametrizations without base points, if they exist. For this purpose, we impose a transversality condition to the base points of the input parametrization.

Keywords: proper (i.e., birational) parametrization; polynomial parametrization; base point

1. Introduction

Algebraic surfaces are mainly studied from three different, but related, points of view, namely: pure theoretical, algorithmic and because of their applications. In this paper, we deal with some computational problems of algebraic surfaces taking into account the potential applicability.

In many different applications, as for instance in geometric design (see e.g., [1]) parametric representations of surfaces are more suitable than implicit representations. Among the different types of parametric representations, one may distinguish radical parametrizations (see [2]) and rational parametrizations (see e.g., [3]), the first being tuples of fractions of nested radical of bivariate polynomials, and the second being tuples of fractions of bivariate polynomials; in both cases the tuples are with generic Jacobian of rank 2. Other parametric representations by means of series can be introduced, but this is not within the scope of this paper. One may observe that the set of rational parametrizations is a subclass of the class of radical parametrizations. Indeed, in [4], one can find an algorithm to decide whether a radical parametrization can be transformed by means of a change of the parameters into a rational parametrization; in this case, we say that a reparametrization has been performed.

Now, we consider a third type of parametric representation of the surface, namely, the polynomial parametrization. That is, tuple of bivariate polynomials with generic Jacobian of rank 2. Clearly the class of polynomial parametrizations is a subclass of the class of the rational parametrizations, and the natural question of deciding whether a given rational parametrization can be reparametrized into a polynomial parametrization appears. This is, indeed, the problem we deal with in the paper. Unfortunately the inclusion of each of these classes into the next one is strict, and hence the corresponding reparametrizations are not always feasible. In some practical applications, the alternative is to use piecewise parametrizations with the desired property (see e.g., [5,6]).

Before commenting the details of our approach to the problem, let us look at some reasons why polynomial parametrizations may be more interesting than rational ones. In general, rational parametrizations are dominant over the surface (i.e., the Zariski closure of its image is the surface), but not necessarily surjective. This may introduce difficulties when applying the parametric representation to a problem, since the answer might be within the non-covered area of the surface. For the curve case, polynomial parametrizations are always surjective (see [7]). For the surface case, the result is not so direct but there are some interesting results for polynomial parametrizations to be surjective (see [8]) as well as subfamilies of polynomial parametrizations that are surjective (see [9]). Another issue that could be mention is the numerical instability when the values, substituted in the parameters of the parametrizations, get close to the poles of the rational functions; note that, in this case, the denominators define algebraic curves which points are all poles of the parametrization. One may also think on the advantages of providing a polynomial parametrization instead of a rational parametrization, when facing surface integrals. Let us mention a last example of motivation: the algebra-geometric technique for solving non autonomous ordinary differential equations (see [10,11]). In these cases, the differential equation is seen algebraically and hence representing a surface. Then, under the assumption that this surface is rational (resp. radical) the general rational (resp. radical) solution, if it exists, of the differential equation is determined from a rational parametrization of the surface. This process may be simplified if the associated algebraic parametrization admits a polynomial parametrization.

Next, let us introduce, and briefly comment on, the notion of base points of a rational parametrization. A base point of a given rational parametrization is a common solution of all numerators and denominators of the parametrization (see e.g., [12,13]). The presence of this type of points is a serious obstacle when approaching many theoretical, algorithmic or applied questions related to the surface represented by the parametrizations; examples of this phenomenon can be found in, e.g., [14–17]. In addition, it happens that rational surface may admit, both, birational parametrizations with empty base locus and with non-empty base locus. Moreover, the behavior of the base locus is not controlled, at least to our knowledge, by the existing parametrization algorithms or when the resulting parametrization appears as the consequence of the intersection of higher dimension varieties, or as the consequence of cissoid, conchoid, offsetting, or any other geometric design process applied to a surface parametrization (see e.g., [18–21]).

In this paper, we solve the problem, by means of reparametrizations, of computing a birational polynomial parametrization without base points of a rational surface, if it exists. For this purpose, we assume that we are given a birational parametrization of the surface that has the property of being transversal (this is a notion introduced in the paper, see Section 3 for the precise definition). Essentially, the idea of transversality is to assume that the multiplicity of the base points is minimal. Since, by definition (see Section 3) this multiplicity is introduced as a multiplicity of intersection of two algebraic curves, one indeed is requiring the transversality of the corresponding tangents. In this paper, we have not approached the problem of eliminating this hypothesis, and we leave it as future work in case it exists.

The general idea to solve the problem is as follows. We are given a birational parametrization $\mathcal{P}$ and let $\mathcal{Q}$ be the searched birational polynomial parametrization without base points; let us say, first of all, that throughout the paper we work projectively. Then, there exists a birational map, say $\mathcal{S}_\mathcal{P}$, that relates both parametrizations as $\mathcal{Q} = \mathcal{P} \circ \mathcal{S}_\mathcal{P}$. Then, taking into account that the base locus of $\mathcal{S}_\mathcal{P}$ and $\mathcal{P}$ are the same, that they coincide also in multiplicity, and applying some additional properties on base points stated in Section 3.1, we introduce a 2-dimensional linear system of curves, associated to an effective divisor generated by the base points of $\mathcal{P}$. Then, using the transversality we prove that every basis of the linear system, composed with a suitable birational transformation, provides a reparametrization of $\mathcal{P}$ that yields to a polynomial parametrization with empty base locus.

To give a better picture of these ideas, let us briefly illustrate them here by means of an example. We consider the projective surface $\mathscr{S}$ defined by the polynomial

$$-2w^2y^2 + 2w^2yz + 2w^2z^2 - wxy^2 - wxyz - wxz^2 - 8wy^3 + 5wy^2z + 5wyz^2 - 3wz^3 + x^2y^2 - 2xy^3$$
$$+ 4xy^2z - 2xyz^2 + y^4 - 4y^3z + 6y^2z^2 - 4yz^3 + z^4.$$

$\mathscr{S}$ is rational and can be birationally parametrized as

$$\mathcal{P} = (t_2^2 t_3^2 + t_3 t_1^3 + t_3 t_2^2 t_1 - t_1^4 - 2t_1^2 t_2^2 - t_2^4 : -t_3(t_1 + t_2)(t_1^2 - t_1 t_3 + t_2^2 - t_2 t_3)$$
$$: t_3^2(t_1 + 2t_2)(t_1 + t_2) : (t_1^2 + t_2^2)^2).$$

$\mathcal{P}$ provides the affine non-polynomial parametrization

$$\left(\frac{-t_1^4 - 2t_1^2 t_2^2 - t_2^4 + t_1^3 + t_1 t_2^2 + t_2^2}{(t_1^2 + t_2^2)^2}, -\frac{(t_1 + t_2)(t_1^2 + t_2^2 - t_1 - t_2)}{(t_1^2 + t_2^2)^2}, \frac{(t_1 + 2t_2)(t_1 + t_2)}{(t_1^2 + t_2^2)^2} \right).$$

On the other hand, $\mathscr{S}$ can also be parametrized as

$$\mathcal{Q} = (t_1^2 + t_2 t_3 - t_1 t_3 - t_3^2 : t_2^2 - t_2 t_3 : t_2^2 + t_2 t_1 : t_3^2),$$

that provides the affine polynomial parametrization

$$(t_1^2 + t_2 - t_1 - 1, t_2^2 - t_2, t_2^2 + t_2 t_1).$$

The question is how to compute $\mathcal{Q}$ from $\mathcal{P}$. Since both parametrizations are birational, there exists a birational change of parameters $\mathcal{S}_\mathcal{P}$ such that $\mathcal{Q} = \mathcal{P} \circ \mathcal{S}_\mathcal{P}$. Furthermore, it holds that the base locus of $\mathcal{S}_\mathcal{P}$ and $\mathcal{P}$ are the same. So, the problem of finding $\mathcal{Q}$ is reduced to the problem of determining a birational map $\mathcal{S}_\mathcal{P}$ satisfying that the base locus of $\mathcal{S}_\mathcal{P}$ and $\mathcal{P}$ are the same. For this purpose, we introduce a 2-dimensional linear system of curves, associated to an effective divisor generated by the base points of $\mathcal{P}$ and, using the transversality, we prove that every basis of the linear system provides a polynomial parametrization of $\mathcal{P}$ with empty base locus.

The structure of the paper is as follows. In Section 2, we introduce the notation and we recall some definitions and properties on base points, essentially taken from [12]. In Section 3 we state some additional required properties on base points, we introduce the notion of transversality of a base locus, both for birational maps of the projective plane and for rational surface projective parametrizations. Moreover, we establish some fundamental properties that require the transversality. Section 4 is devoted to state the theoretical frame for solving the central problem treated in the paper. In Section 5, we derive the algorithm that is illustrated by means of some examples. We finish the paper with a section on conclusions.

2. Preliminary on Basic Points and Notation

In this section, we briefly recall some of the notions related to base points and we introduce some notation; for further results on this topic we refer to [12]. We distinguish three subsections. In Section 2.1, the notation that will be used throughout the paper is introduced. The next subsection focuses on birational surface parametrizations, and the third subsection on birational maps of the projective plane.

2.1. Notation

Let, first of all, start fixing some notation. Throughout this paper, $\mathbb{K}$ is an algebraically closed field of characteristic zero. $\overline{x} = (x_1, \ldots, x_4)$, $\overline{y} = (y_1, \ldots, y_4)$ and $\overline{t} = (t_1, t_2, t_3)$. $\mathbb{F}$ is the algebraic closure of $\mathbb{K}(\overline{x}, \overline{y})$. In addition, $\mathbb{P}^k(\mathbb{K})$ denotes the k–dimensional projective space, and $\mathscr{G}(\mathbb{P}^k(\mathbb{K}))$ is the set of all projective transformations of $\mathbb{P}^k(\mathbb{K})$.

Furthermore, for a rational map

$$\mathcal{M}: \quad \mathbb{P}^{k_1}(\mathbb{K}) \quad \dashrightarrow \quad \mathbb{P}^{k_2}(\mathbb{K})$$
$$\overline{h} = (h_1 : \cdots : h_{k_1+1}) \quad \longmapsto \quad (m_1(\overline{h}) : \cdots : m_{k_2+1}(\overline{h})),$$

where the non-zero m_i are homogenous polynomial in $\overline{h}$ of the same degree, we denote by $\deg(\mathcal{M})$ the degree $\deg_{\overline{h}}(m_i)$, for m_i non-zero, and by $\operatorname{degMap}(\mathcal{M})$ the degree of the map $\mathcal{M}$; that is, the cardinality of the generic fiber of $\mathcal{M}$ (see e.g., [22]).

For $L \in \mathscr{G}(\mathbb{P}^{k_2}(\mathbb{K}))$, and $M \in \mathscr{G}(\mathbb{P}^{k_1}(\mathbb{K}))$ we denote the left composition and the right composition, respectively, by
$$^L\mathcal{M} := L \circ \mathcal{M}, \quad \mathcal{M}^M := \mathcal{M} \circ M.$$

Let $f \in \mathbb{L}[t_1, t_2, t_3]$ be homogeneous and non-zero, where $\mathbb{L}$ is a field extension of $\mathbb{K}$. Then $\mathscr{C}(f)$ denotes the projective plane curve defined by f over the algebraic closure of $\mathbb{L}$.

Let $\mathscr{C}(f), \mathscr{C}(g)$ be two curves in $\mathbb{P}^2(\mathbb{K})$. For $A \in \mathbb{P}^2(\mathbb{K})$, we represent by $\text{mult}_A(\mathscr{C}(f), \mathscr{C}(g))$ the multiplicity of intersection of $\mathscr{C}(f)$ and $\mathscr{C}(g)$ at A. In addition, we denote by $\text{mult}(A, \mathscr{C}(f))$ the multiplicity of $\mathscr{C}(f)$ at A.

Finally, $\mathscr{S} \subset \mathbb{P}^3(\mathbb{K})$ represents a rational projective surface.

2.2. Case of Surface Parametrizations

In this subsection, we consider a rational parametrization of the projective rational surface $\mathscr{S}$, namely,

$$\mathcal{P}: \begin{array}{ccc} \mathbb{P}^2(\mathbb{K}) & \dashrightarrow & \mathscr{S} \subset \mathbb{P}^3(\mathbb{K}) \\ \bar{t} & \longmapsto & (p_1(\bar{t}) : \cdots : p_4(\bar{t})), \end{array} \quad (1)$$

where $\bar{t} = (t_1, t_2, t_3)$ and the p_i are homogenous polynomials of the same degree such that $\gcd(p_1, \ldots, p_4) = 1$.

Definition 1. *A base point of $\mathcal{P}$ is an element $A \in \mathbb{P}^2(\mathbb{K})$ such that $p_i(A) = 0$ for every $i \in \{1, 2, 3, 4\}$. We denote by $\mathscr{B}(\mathcal{P})$ the set of base points of $\mathcal{P}$. That is $\mathscr{B}(\mathcal{P}) = \mathscr{C}(p_1) \cap \cdots \cap \mathscr{C}(p_4)$.*

In order to deal with the base points of the parametrization, we introduce the following auxiliary polynomials:

$$\begin{array}{l} W_1(\bar{x}, \bar{t}) := \sum_{i=1}^{4} x_i \, p_i(t_1, t_2, t_3) \\ W_2(\bar{y}, \bar{t}) := \sum_{i=1}^{4} y_i \, p_i(t_1, t_2, t_3), \end{array} \quad (2)$$

where x_i, y_i are new variables. We will work with the projective plane curves $\mathscr{C}(W_i)$ in $\mathbb{P}^3(\mathbb{F})$. Similarly, for $M = (M_1 : M_2 : M_3) \in \mathscr{G}(\mathbb{P}^3(\mathbb{K}))$, we define,

$$\begin{array}{l} W_1^M(\bar{x}, \bar{t}) := \sum_{i=1}^{4} x_i \, M_i(\mathcal{P}(\bar{t})) \\ W_2^M(\bar{y}, \bar{t}) := \sum_{i=1}^{4} y_i \, M_i(\mathcal{P}(\bar{t})). \end{array} \quad (3)$$

Remark 1. *Sometimes, we will need to specify the parametrization in the polynomials above. In those cases, we will write $W_i^{\mathcal{P}}$ or $W_i^{M,\mathcal{P}}$ instead of W_i or W_i^M; similarly, we may write $\mathscr{C}(W_1^{\mathcal{P}})$ and $\mathscr{C}(W_1^{M,\mathcal{P}})$.*

Using the multiplicity of intersection of these two curves, we define the multiplicity of a base point as follows.

Definition 2. *The multiplicity of a base point $A \in \mathscr{B}(\mathcal{P})$ is $\text{mult}_A(\mathscr{C}(W_1), \mathscr{C}(W_2))$, that is, is the multiplicity of intersection at A of $\mathscr{C}(W_1)$ and $\mathscr{C}(W_2)$; we denote it by*

$$\text{mult}(A, \mathscr{B}(\mathcal{P})) := \text{mult}_A(\mathscr{C}(W_1), \mathscr{C}(W_2)) \quad (4)$$

In addition, we define the **multiplicity of the base points locus** of $\mathcal{P}$, denoted $\text{mult}(\mathscr{B}(\mathcal{P}))$, as

$$\text{mult}(\mathscr{B}(\mathcal{P})) := \sum_{A \in \mathscr{B}(\mathcal{P})} \text{mult}(A, \mathscr{B}(\mathcal{P})) = \sum_{A \in \mathscr{B}(\mathcal{P})} \text{mult}_A(\mathscr{C}(W_1), \mathscr{C}(W_2)). \quad (5)$$

Note that, since $\gcd(p_1, \ldots, p_4) = 1$, the set $\mathscr{B}(\mathcal{P})$ is either empty of finite.

For the convenience of the reader we recall here some parts of Proposition 2 in [12].

Lemma 1. *If $L \in \mathscr{G}(\mathbb{P}^3(\mathbb{K}))$, then:*

1. *If $A \in \mathscr{B}(\mathcal{P})$, then*

$$\mathrm{mult}(A, \mathscr{C}(W_1^L)) = \mathrm{mult}(A, \mathscr{C}(W_2^L)) = \min\{\mathrm{mult}(A, \mathscr{C}(p_i)) \mid i = 1, \ldots, 4\}.$$

2. *If $A \in \mathscr{B}(\mathcal{P})$, then the tangents to $\mathscr{C}(W_1^L)$ at A (similarly to $\mathscr{C}(W_2^L)$), with the corresponding multiplicities, are the factors in $\mathbb{K}[\overline{x}, \overline{t}] \setminus \mathbb{K}[\overline{x}]$ of*

$$\epsilon_1 x_1 T_1 + \epsilon_2 x_2 T_2 + \epsilon_3 x_3 T_3 + \epsilon_4 x_4 T_4,$$

where T_i is the product of the tangents, counted with multiplicities, of $\mathscr{C}(L_i(\mathcal{P}))$ at A, and where $\epsilon_i = 1$ if $\mathrm{mult}(A, \mathscr{C}(L_i(\mathcal{P}))) = \min\{\mathrm{mult}(A, \mathscr{C}(L_i(\mathcal{P}))) \mid i = 1, \ldots, 4\}$ and 0 otherwise.

2.3. Case of rational maps of $\mathbb{P}^2(\mathbb{K})$

In this subsection, let

$$\mathcal{S}: \begin{array}{ccc} \mathbb{P}^2(\mathbb{K}) & \dashrightarrow & \mathbb{P}^2(\mathbb{K}) \\ \overline{t} = (t_1 : t_2 : t_3) & \longmapsto & \mathcal{S}(\overline{t}) = (s_1(\overline{t}) : s_2(\overline{t}) : s_3(\overline{t})), \end{array} \quad (6)$$

where $\gcd(s_1, s_2, s_3) = 1$, is a dominant rational transformation of $\mathbb{P}^2(\mathbb{K})$.

Definition 3. *$A \in \mathbb{P}^2(\mathbb{K})$ is a base point of $\mathcal{S}(\overline{t})$ if $s_1(A) = s_2(A) = s_3(A) = 0$. That is, the base points of $\mathcal{S}$ are the intersection points of the projective plane curves, $\mathscr{C}(s_i)$, defined over $\mathbb{K}$ by $s_i(\overline{t})$, $i = 1, 2, 3$. We denote by $\mathscr{B}(\mathcal{S})$ the set of base points of $\mathcal{S}$.*

We introduce the polynomials

$$\begin{array}{l} V_1 = \sum_{i=1}^3 x_i s_i(\overline{t}) \in \mathbb{K}(\overline{x}, \overline{y})[\overline{t}] \\ V_2 = \sum_{i=1}^3 y_i s_i(\overline{t}) \in \mathbb{K}(\overline{x}, \overline{y})[\overline{t}], \end{array} \quad (7)$$

where x_i, y_j are new variables and we consider the curves $\mathscr{C}(V_i)$ over the field $\mathbb{F}$; compare with (3). Similarly, for every $L \in \mathscr{G}(\mathbb{P}^2(\mathbb{K}))$ we introduce the polynomials

$$\begin{array}{l} V_1^L = \sum_{i=1}^3 x_i L_i(\mathcal{S}) \in \mathbb{K}(\overline{x}, \overline{y})[\overline{t}] \\ V_2^L = \sum_{i=1}^3 y_i L_i(\mathcal{S}) \in \mathbb{K}(\overline{x}, \overline{y})[\overline{t}], \end{array} \quad (8)$$

Remark 2. *Sometimes, we will need to specify the rational map in the polynomials above. In those cases, we will write $V_i^{\mathcal{S}}$ or $V_i^{L,\mathcal{S}}$ instead of V_i or V_i^L; similarly, we may write $\mathscr{C}(V_1^{\mathcal{S}})$ and $\mathscr{C}(V_1^{L,\mathcal{S}})$.*

As we did in Section 2.2, we have the following notion of multiplicity.

Definition 4. *For $A \in \mathscr{B}(\mathcal{S})$, we define the multiplicity of intersection of A, and we denote it by $\mathrm{mult}(A, \mathscr{B}(\mathcal{S}))$, as*

$$\mathrm{mult}(A, \mathscr{B}(\mathcal{S})) := \mathrm{mult}_A(\mathscr{C}(V_1), \mathscr{C}(V_2)). \quad (9)$$

In addition, we define the multiplicity of the base points locus of $\mathcal{S}$, denoted $\mathrm{mult}(\mathscr{B}(\mathcal{S}))$, as (note that, since $\gcd(s_1, s_2, s_3) = 1$, $\mathscr{B}(\mathcal{S})$ is either finite or empty)

$$\mathrm{mult}(\mathscr{B}(\mathcal{S})) := \sum_{A \in \mathscr{B}(\mathcal{S})} \mathrm{mult}(A, \mathscr{B}(\mathcal{S})) = \sum_{A \in \mathscr{B}(\mathcal{S})} \mathrm{mult}_A(\mathscr{C}(V_1), \mathscr{C}(V_2)) \quad (10)$$

The next result is a direct extension of Proposition 2 in [12] to the case of birational transformation of $\mathbb{P}^2(\mathbb{K})$.

Lemma 2. *If $L \in \mathscr{G}(\mathbb{P}^2(\mathbb{K}))$ then*

1. $\mathscr{B}(\mathcal{S}) = \mathscr{C}(V_1^L) \cap \mathscr{C}(V_2^L) \cap \mathbb{P}^2(\mathbb{K})$.
2. *Let $A \in \mathscr{B}(\mathcal{S})$ then*

$$\mathrm{mult}(A, \mathscr{C}(V_1^L)) = \mathrm{mult}(A, \mathscr{C}(V_2^L)) = \min\{\mathrm{mult}(A, \mathscr{C}(s_i)) \mid i = 1, 2, 3\}.$$

3. *Let $A \in \mathscr{B}(\mathcal{S})$. The tangents to $\mathscr{C}(V_1^L)$ at A (similarly to $\mathscr{C}(V_2^L)$), with the corresponding multiplicities, are the factors in $\mathbb{K}[\bar{x}, \bar{t}] \setminus \mathbb{K}[\bar{x}]$ of*

$$\epsilon_1 x_1 T_1 + \epsilon_2 x_2 T_2 + \epsilon_3 x_3 T_3,$$

where T_i is the product of the tangents, counted with multiplicities, of $\mathscr{C}(L_i(\mathcal{S}))$ at A, and where $\epsilon_i = 1$ if $\mathrm{mult}(A, \mathscr{C}(L_i(\mathcal{S}))) = \min\{\mathrm{mult}(A, \mathscr{C}(L_i(\mathcal{S}))) \mid i = 1, 2, 3\}$ and 0 otherwise.

3. Transversal Base Locus

In this section, we present some new results on base points that complement those in [12] and we introduce and analyze the notion of transversality in conexion with the base locus.

Throughout this section, let $\mathcal{S} = (s_1 : s_2 : s_3)$, with $\gcd(s_1, s_2, s_3) = 1$, be as in (6). In the sequel, we assume that $\mathcal{S}$ is birational. Let the inverse of $\mathcal{S}$ be denoted by $\mathcal{R} = (r_1 : r_2 : r_3)$; that is $\mathcal{R} := \mathcal{S}^{-1}$. In addition, we consider a rational surface parametrization $\mathcal{P} = (p_1 : \cdots : p_4)$, with $\gcd(p_1, \ldots, p_4) = 1$, be as in (1). We assume that $\mathcal{P}$ is birational.

3.1. Further Results on Base Points

We start analyzing the rationality of the curve $\mathscr{C}(V_i^L)$ (see (8)).

Lemma 3. *There exists a non-empty open subset Ω_1 of $\mathscr{G}(\mathbb{P}^2(\mathbb{K}))$ such that if $L \in \Omega_1$ then $\mathscr{C}(V_1^L)$ is a rational curve. Furthermore,*

$$\mathcal{V}_1(\bar{x}, h_1, h_2) = \mathcal{R}^{L^{-1}}(h_1 x_3, h_2 x_3, -(h_1 x_1 + x_2 h_2))$$

is a birational parametrization of $\mathscr{C}(V_1^L)$.

Proof. We start proving that for every $L \in \mathscr{G}(\mathbb{P}^2(\mathbb{K}))$, V_1^L is irreducible. Indeed, let $L = (\sum \lambda_i t_i : \sum \mu_i t_i : \sum \gamma_i t_i) \in \mathscr{G}(\mathbb{P}^2(\mathbb{K}))$. Then $V_1^L = (\lambda_1 x_1 + \mu_1 x_2 + \gamma_1 x_3) s_1 + (\lambda_2 x_1 + \mu_2 x_2 + \gamma_2 x_3) s_2 + (\lambda_3 x_1 + \mu_3 x_2 + \gamma_3 x_3) s_3$. $\gcd(s_1, s_2, s_3) = 1$ and $\gcd(\lambda_1 x_1 + \mu_1 x_2 + \gamma_1 x_3, \lambda_2 x_1 + \mu_2 x_2 + \gamma_2 x_3, \lambda_3 x_1 + \mu_3 x_2 + \gamma_3 x_3) = 1$ because the determinant of the matrix associated to L is non-zero. Therefore, V_1^L is irreducible.

In the following, to define the open set Ω_1, let $\mathcal{L}(t_1, t_2, t_3) = (\mathcal{L}_1 : \mathcal{L}_2 : \mathcal{L}_3)$ be a generic element of $\mathscr{G}(\mathbb{P}^2(\mathbb{K}))$; that is, $\mathcal{L}_i = z_{i,1} t_1 + z_{i,2} t_2 + z_{i,3} t_3$, where $z_{i,j}$ are undetermined coefficients satisfying that the determinant of the corresponding matrix is not zero. Furthermore, for $L \in \mathscr{G}(\mathbb{P}^2(\mathbb{K}))$, we denote by $\bar{z}^L$ the coefficient list of L. We also introduce the polynomial $R^{\mathcal{L}} = x_1 \mathcal{L}^1 + x_2 \mathcal{L}^2 + x_3 \mathcal{L}^3 = (\sum z_{i,1} x_i) t_1 + (\sum z_{i,2} x_i) t_2 + (\sum z_{i,3} x_i) t_3$. Similarly, for $L \in \mathscr{G}(\mathbb{P}^2(\mathbb{K}))$, we denote $R^L = R_{\mathcal{L}}(\bar{z}^L, \bar{x}, \bar{t})$.

We consider the birational extension $\mathcal{R}_{\bar{x}} : \mathbb{P}^2(\mathbb{F}) \dashrightarrow \mathbb{P}^2(\mathbb{F})$ of $\mathcal{R}$ from $\mathbb{P}^2(\mathbb{K})$ to $\mathbb{P}^2(\mathbb{F})$. Let $\mathcal{U}_{\bar{x}} \subset \mathbb{P}^2(\mathbb{F})$ be the open set where the $\mathcal{R}_{\bar{x}}$ is bijective; say that $\mathcal{U}_{\bar{x}} = \mathbb{P}^2(\mathbb{F}) \setminus \Delta$. We express the close set Δ as $\Delta = \Delta_1 \cup \Delta_2$ where Δ_1 is either empty or it is a union of finitely many curves, and Δ_2 is either empty or finite many points. We fix our attention in Δ_1. Let $f(\bar{t})$ be the defining polynomial of Δ_1. Let $Z(\bar{z}, t_1, t_2)$ be the remainder of f when diving by $V_1^\mathcal{L}$ w.r.t t_3. Note that $R^\mathcal{L}$ does not divide f since $R^\mathcal{L}$ is irreducible and depends on $\bar{z}$. Hence Z is no zero. Let $\alpha(\bar{z})$ be the numerator of a non-zero coefficient of Z w.r.t. $\{t_1, t_2\}$ and let $\beta(\bar{z})$ the l.c.m. of the denominators of all coefficients of Z w.r.t. $\{t_1, t_2\}$. Then, we define Ω_1 as

$$\Omega_1 = \{L \in \mathcal{G}(\mathbb{P}^2(\mathbb{K})) \mid \alpha(\bar{z}^L)\beta(\bar{z}^L) \neq 0\}$$

We observe that, by construction, if $L \in \Omega_1$ then $\mathscr{C}(R^L) \cap \mathcal{U}_{\bar{x}}$ is dense in $\mathscr{C}(R^L)$.

Let $\bar{a}, \bar{b} \in \mathscr{C}(R^L) \cap \mathcal{U}_{\bar{x}}$ be two different points, then by injectivity $\mathcal{R}_{\bar{x}}(\mathscr{C}(R^L))$ contains at least two points, namely $\mathcal{R}_{\bar{x}}(\bar{a})$ and $\mathcal{R}_{\bar{x}}(\bar{b})$. In this situation, since $\mathcal{R}_{\bar{x}}(\mathscr{C}(R^L))$ and $\mathscr{C}(V_1)$ are irreducible we get that $\overline{\mathcal{R}_{\bar{x}}(\mathscr{C}(R^L))} = \mathscr{C}(V_1)$, and hence $\mathscr{C}(V_1)$ is a rational curve $\mathbb{P}^2(\mathbb{F})$. Furthermore, one easily may check that $\mathcal{V}_1$ parametrizes $\mathscr{C}(V_1)$ and it is proper since $\mathcal{R}$ is birational. □

Remark 3. Note that $\mathcal{V}_1(t_3, 0, -t_1, t_1, t_2) = \mathcal{R}^{L^{-1}}(-t_1 t_1, -t_1 t_2, -t_1 t_3) = \mathcal{R}^{L^{-1}}(\bar{t})$. Hence

$$^L\mathcal{S}(\mathcal{V}_1(t_3, 0, -t_1, t_1, t_2)) = {}^L\mathcal{S}(\mathcal{R}^{L^{-1}}(\bar{t})) = (t_1 : t_2 : t_3).$$

Therefore,

$$L_i(\mathcal{S}(\mathcal{V}_1(t_3, 0, -t_1, t_1, t_2))) = t_i \cdot \wp(\bar{t}),\ i = 1, 2, 3.$$

Next lemma analyzes the rationality of the curves $\mathscr{C}(L_i(\mathcal{S}))$ where $L = (L_1 : L_2 : L_3) \in \mathcal{G}(\mathbb{P}^2(\mathbb{K}))$.

Lemma 4. *There exists a non-empty Zariski open subset Ω_2 of $\mathcal{G}(\mathbb{P}^2(\mathbb{K}))$ such that if $L \in \mathcal{G}(\mathbb{P}^2(\mathbb{K}))$ then $\mathscr{C}(L_i(\mathcal{S}))$, where $i \in \{1, 2, 3\}$, is rational.*

Proof. Let $\mathcal{U}$ be the open subset where $\mathcal{R}$ is a bijective map, and let $\{\rho_{j,1} t_1 + \rho_{j,2} t_2 + \rho_{j,3} t_3\}_{j=1,\ldots,n}$ be the linear forms defining the lines, if any, included in $\mathbb{P}^2(\mathbb{K}) \setminus \mathcal{U}$. Then, we take $\Omega_2 = \cap_{j=1}^n \Sigma_j$ where

$$\Sigma_j = \left\{ (\sum \lambda_i t_i : \sum \mu_i t_i : \sum \gamma_i t_i) \in \mathcal{G}(\mathbb{P}^2(\mathbb{K})) \;\middle|\; \begin{array}{l} (\lambda_1 : \lambda_2 : \lambda_3) \neq (\rho_{j,1} : \rho_{j,2} : \rho_{j,3}), \\ (\mu_1 : \mu_2 : \mu_3) \neq (\rho_{j,1} : \rho_{j,2} : \rho_{j,3}), \\ (\gamma_1 : \gamma_2 : \gamma_3) \neq (\rho_{j,1} : \rho_{j,2} : \rho_{j,3}) \end{array} \right\}.$$

Now, let $L = (L_1 : L_2 : L_3) \in \Omega_2$. By construction, $\mathscr{C}(L_i) \cap \mathcal{U}$ is dense in $\mathscr{C}(L_i)$, for $i \in \{1, 2, 3\}$. In this situation, reasoning as in the last part of the proof of Lemma 3, we get that $\overline{\mathcal{R}(\mathscr{C}(L_i))} = \mathscr{C}(L_i(\mathcal{S}))$. Therefore, $\mathscr{C}(L_i(\mathcal{S}))$ is rational. □

The following lemma follows from Lemma 2.

Lemma 5. *If $L \in \mathcal{G}(\mathbb{P}^2(\mathbb{K}))$ then*

1. *$\mathscr{B}(\mathcal{S}) = \mathscr{B}({}^L\mathcal{S})$.*
2. *For $A \in \mathscr{B}(\mathcal{S})$ it holds that $\mathrm{mult}(A, \mathscr{B}(\mathcal{S})) = \mathrm{mult}(A, \mathscr{B}({}^L\mathcal{S}))$.*
3. *$\mathrm{mult}(\mathscr{B}(\mathcal{S})) = \mathrm{mult}(\mathscr{B}({}^L\mathcal{S}))$.*

Proof. (1) Let $A \in \mathscr{B}(\mathcal{S})$ then $s_1(A) = s_2(A) = s_3(A) = 0$. Thus, $L_1(\mathcal{S})(A) = L_2(\mathcal{S})(A) = L_3(\mathcal{S})(A) = 0$. So, $A \in \mathscr{B}({}^L\mathcal{S})$. Conversely, let $A \in \mathscr{B}({}^L\mathcal{S})$. Then expressing $L(\mathcal{S}(A)) = \bar{0}$ in terms of matrices, since L is invertible, we have that $\mathcal{S}(A) = L^{-1}(0, 0, 0) = (0, 0, 0)$. Thus, $A \in \mathscr{B}(\mathcal{S})$. (2) and (3) follows from Theorem 5 in [12]. □

Lemma 6. *There exists a non-empty Zariski open subset Ω_3 of $\mathscr{G}(\mathbb{P}^2(\mathbb{K}))$ such that if $L \in \Omega_3$ then for every $A \in \mathscr{B}(\mathcal{S})$ it holds that*

$$\text{mult}(A, \mathscr{C}(V_1^L)) = \text{mult}(A, \mathscr{C}(L_1(\mathcal{S}))) = \text{mult}(A, \mathscr{C}(L_2(\mathcal{S}))) = \text{mult}(A, \mathscr{C}(L_3(\mathcal{S}))).$$

Proof. Let $A \in \mathscr{B}(\mathcal{S})$. Then, by Lemma 2(2), we have that

$$m_A := \text{mult}(A, \mathscr{C}(V_1^L)) = \min\{\text{mult}(A, \mathscr{C}(s_i)) \mid i \in \{1,2,3\}\}, \; \forall \; L \in \mathscr{G}(\mathbb{P}^2(\mathbb{K})). \tag{11}$$

Let us assume w.l.o.g. that the minimum above is reached for $i = 1$. Then all $(m_A - 1)$—order derivatives of the forms s_i vanish at A, and there exists an m_A-order partial derivative of s_1 not vanishing at A. Let us denote this partial derivative as ∂^{m_A}.

Now, let $\mathcal{L}$ be as in the proof of Lemma 3. Then,

$$g_i(\bar{z}) := \partial^{m_A} \mathcal{L}_i(\mathcal{S})(A) = z_{i,1} \partial^{m_A} s_1(A) + z_{i,2} \partial^{m_A} s_2(A) + z_{i,3} \partial^{m_A} s_3(A) \in \mathbb{K}[\bar{z}]$$

is a non-zero polynomial because $\partial^{m_A} s_1(A) \neq 0$. We then consider the open subset (see proof of Lemma 3 for the notation $\bar{z}^L$)

$$\Omega_A = \{L \in \mathscr{G}(\mathbb{P}^2(\mathbb{K})) \mid g_1(\bar{z}^L) g_2(\bar{z}^L) g_3(\bar{z}^L) \neq 0\} \neq \emptyset.$$

In this situation, we take

$$\Omega_3 = \bigcap_{A \in \mathscr{B}(\mathcal{S})} \Omega_A$$

Note that, since $\mathscr{B}(\mathcal{S})$ is finite then Ω_3 is open. Moreover, since $\mathscr{G}(\mathbb{P}^2(\mathbb{K}))$ is irreducible then Ω_3 is not empty.

Let us prove that Ω_3 satisfies the property in the statement of the lemma. Let $L \in \Omega_3$ and $A \in \mathscr{B}(\mathcal{S})$. Let m_A be as in (11). Then all partial derivatives of $L_i(\mathcal{S})$, of any order smaller than m_A, vanishes at A. Moreover, since $L \in \Omega_3 \subset \Omega_A$, it holds that $\partial^{m_A} L_i(\mathcal{S})(A) \neq 0$ for $i = 1, 2, 3$. Therefore,

$$\text{mult}(A, \mathscr{C}(V_1^L)) = m_A = \text{mult}(A, \mathscr{C}(L_1(\mathcal{S}))) = \text{mult}(A, \mathscr{C}(L_2(\mathcal{S}))) = \text{mult}(A, \mathscr{C}(L_3(\mathcal{S})))$$

□

Remark 4. *We note that the proofs of Lemmas 5 and 6 are directly adaptable to the case of birational surface parametrizations. So, both lemmas hold if $M \in \mathscr{G}(\mathbb{P}^3(\mathbb{K}))$ and we replace $\mathcal{S}$ by the birational surface parametrizaion $\mathcal{P}$ and ${}^L\mathcal{S}$ by ${}^M\mathcal{P} := M \circ \mathcal{P}$.*

In the following, we denote by $\text{Sing}(\mathcal{D})$, the set of singularities of an algebraic plane curve $\mathcal{D}$.

Corollary 1. *Let Ω_3 be the open subset in Lemma 6 and $L \in \Omega_3$. It holds that*

1. $\cap_{i=1}^{3} \text{Sing}(\mathscr{C}(L_i(\mathcal{S}))) \cap \mathscr{B}(\mathcal{S}) \subset \text{Sing}(\mathscr{C}(V_1^L))$.
2. *Let $A \in \mathscr{B}(\mathcal{S})$. The tangents to $\mathscr{C}(V_1^L)$ at A, with the corresponding multiplicities, are the factors in $\mathbb{K}[\bar{x}, \bar{t}] \setminus \mathbb{K}[\bar{x}]$ of*

$$x_1 T_1 + x_2 T_2 + x_3 T_3,$$

where T_i is the product of the tangents, counted with multiplicities, to $\mathscr{C}(L_i(\mathcal{S}))$ at A.

3. *Let $A \in \mathscr{B}(\mathcal{S})$, and let T_i be the product of the tangents, counted with multiplicities, to $\mathscr{C}(L_i(\mathcal{S}))$, at A. If $\gcd(T_1, T_2, T_3) = 1$, then*

$$\text{mult}_A(\mathscr{C}(V_1^L), \mathscr{C}(V_2^L)) = \text{mult}(A, \mathscr{C}(L_i(\mathcal{S})))^2, \; i \in \{1,2,3\}$$

Proof.

(1) Let $A \in \cap_{i=1}^{3} \text{Sing}(\mathscr{C}(L_i(\mathcal{S}))) \cap \mathscr{B}(\mathcal{S})$. By Lemma 6, $m := \text{mult}(A, \mathscr{C}(L_i(\mathcal{S}))) > 0$, for $i \in \{1,2,3\}$, and $\text{mult}(A, \mathscr{C}(V_1^L)) = m > 0$. So, $A \in \text{Sing}(\mathscr{C}(V_1^L))$.

(2) follows from Lemmas 2 and 6.

(3) By (2) the tangents to $\mathscr{C}(V_1^L)$ and to $\mathscr{C}(V_2^L)$ at A are $\mathcal{T}_1 := \sum x_i T_i$ and $\mathcal{T}_2 := \sum y_i T_i$, respectively. Since $\gcd(T_1, T_2, T_3) = 1$, then $\mathcal{T}_i$ is primitive, and hence $\gcd(\mathcal{T}_1, \mathcal{T}_2) = 1$. That is, $\mathscr{C}(V_1^L)$ and $\mathscr{C}(V_2^L)$ intersect transversally at A. From here, the results follows. □

3.2. Transversality

We start introducing the notion of transversality for birational maps of $\mathbb{P}^2(\mathbb{K})$.

Definition 5. *We say that $\mathcal{S}$ is transversal if either $\mathscr{B}(\mathcal{S}) = \emptyset$ or for every $A \in \mathscr{B}(\mathcal{S})$ it holds that (see (7))*

$$\text{mult}(A, \mathscr{B}(\mathcal{S})) = \text{mult}(A, \mathscr{C}(V_1))^2$$

In this case, we also say that the base locus of $\mathcal{S}$ is transversal.

In the following lemma, we see that the transversality is invariant under left composition with elements in $\mathscr{G}(\mathbb{P}^2(\mathbb{K}))$.

Lemma 7. *If $\mathcal{S}$ is transversal, then for every $L \in \mathscr{G}(\mathbb{P}^2(\mathbb{K}))$ it holds that $^L\mathcal{S}$ is transversal.*

Proof. By Lemma 5(1), $\mathscr{B}(\mathcal{S}) = \mathscr{B}(^L\mathcal{S})$. So, if $\mathscr{B}(\mathcal{S}) = \emptyset$, there is nothing to prove. Let $A \in \mathscr{B}(\mathcal{S}) \neq \emptyset$, and let $L := (L_1 : L_2 : L_3)$. Then

$$\begin{aligned}
\text{mult}(A, \mathscr{B}(^L\mathcal{S})) &= \text{mult}(A, \mathscr{B}(\mathcal{S})) && \text{(see Lemma 5(2))} \\
&= \text{mult}(A, \mathscr{C}(V_1))^2 && (\mathcal{S} \text{ is transversal}) \\
&= \text{mult}(A, \mathscr{C}(V_1^{L,\mathcal{S}}))^2 && \text{(see Lemma 2(2) and Remark 2)}
\end{aligned}$$

Therefore, $^L\mathcal{S}$ is transversal. □

The next lemma characterizes the transversality by means of the tangents of $\mathscr{C}(s_i)$ at the base points. A direct generalization of this lemma to the case of surface parametrizations appears in Lemma 10, and will be used in Algorithm 1 for checking the transversality.

Lemma 8. *The following statements are equivalent*

1. *$\mathcal{S}$ is transversal.*
2. *For every $A \in \mathscr{B}(\mathcal{S})$ it holds that $\gcd(T_1, T_2, T_3) = 1$, where T_i is the product of the tangents, counted with multiplicities, to $\mathscr{C}(s_i)$ at A.*

Proof. If $\mathscr{B}(\mathcal{S}) = \emptyset$, the result if trivial. Let $\mathscr{B}(\mathcal{S}) \neq \emptyset$. First of all, we observe that, because of Lemma 7, we may assume w.l.o.g. that Lemma 6 applies to $\mathcal{S}$. So, by Definition 5, $\mathcal{S}$ is transversal if and only if for every $A \in \mathscr{B}(\mathcal{S})$ it holds that

$$\text{mult}(A, \mathscr{B}(\mathcal{S})) = \text{mult}(A, \mathscr{C}(V_1))^2,$$

and, by Definition 2, if and only if

$$\text{mult}(A, \mathscr{C}(V_1))^2 = \text{mult}_A(\mathscr{C}(V_1), \mathscr{C}(V_2)).$$

Furthermore, using Theorem 2.3.3 in [23], we have that

$$\text{mult}_A(\mathscr{C}(V_1), \mathscr{C}(V_2)) = \text{mult}(A, \mathscr{C}(V_1))\,\text{mult}(A, \mathscr{C}(V_2))$$

if and only V_1 and V_2 intersect transversally at A i.e., if the curves have no common tangents at A which is equivalent to $\gcd(T_1, T_2, T_3) = 1$. The proof finishes taking into account that, by Lemma 6 $\text{mult}(A, \mathscr{C}(V_1)) = \text{mult}(A, \mathscr{C}(V_2))$. □

In the last part of this section, we analyze the relationship of the transversality of a birational map of the projective plane and the transversality of a birational projective surface parametrization. For this purpose, first we introduce the notion of transversality for parametrizations.

Definition 6. *Let $\mathcal{P}$ be a birational surface parametrization of $\mathbb{P}^3(\mathbb{K})$. We say that $\mathcal{P}$ is* transversal *if either $\mathscr{B}(\mathcal{P}) = \emptyset$ or for every $A \in \mathscr{B}(\mathcal{P})$ it holds that (see (3))*

$$\text{mult}(A, \mathscr{B}(\mathcal{P})) = \text{mult}(A, \mathscr{C}(W_1))^2$$

In this case, we say that the base locus of $\mathcal{P}$ is transversal.

We start with some technical lemmas. The next lemma states that transversality does not change under projective transformations of the cartesian coordinates, i.e., under left composition. This has to be taken into account when extending the results of this paper to the case of non-transversality.

Lemma 9. *If $\mathcal{P}$ is transversal, then for every $M \in \mathscr{G}(\mathbb{P}^2(\mathbb{K}))$ it holds that $^M\mathcal{P}$ is transversal.*

Proof. The proof is analogous to the proof of Lemma 7. □

The next lemma provides a characterization of the transversality of a parametrization by means of the tangents that will be used in Algorithm 1. The proof of this lemma is a direct generalization of the proof of Lemma 8.

Lemma 10. *The following statements are equivalent*

1. *$\mathcal{P}$ is transversal.*
2. *For every $A \in \mathscr{B}(\mathcal{P})$ it holds that $\gcd(T_1, \ldots, T_4) = 1$, where T_i is the product of the tangents, counted with multiplicities, to $\mathscr{C}(p_i)$ at A.*

The following lemma focusses on the behavior of the base points of $\mathcal{P}$ when right composing with elements in $\mathscr{G}(\mathbb{P}^2(\mathbb{K}))$.

Lemma 11. *Let $L \in \mathscr{G}(\mathbb{P}^2(\mathbb{K}))$. It holds that*

1. *$\mathscr{B}(\mathcal{P}) = L(\mathscr{B}(\mathcal{P}^L))$. Furthermore, $A \in \mathscr{B}(\mathcal{P})$ if and only $L^{-1}(A) \in \mathscr{B}(\mathcal{P}^L)$.*
2. *For $A \in \mathscr{B}(\mathcal{P})$, $\text{mult}(A, \mathscr{B}(\mathcal{P})) = \text{mult}(L^{-1}(A), \mathscr{B}(\mathcal{P}^L))$*
3. *$\text{mult}(\mathscr{B}(\mathcal{P})) = \text{mult}(\mathscr{B}(\mathcal{P}^L))$.*
4. *If $\mathcal{P}$ is transversal then $\mathcal{P}^L$ is also transversal.*

Proof. (1) $A \in \mathscr{B}(\mathcal{P})$ iff $p_i(A) = 0$ for $i \in \{1, \ldots, 4\}$ iff $p_i(L(L^{-1}(A))) = 0$ for $i \in \{1, \ldots, 4\}$ iff $L^{-1}(A) \in \mathscr{B}(\mathcal{P}^L)$ iff $A \in L(\mathscr{B}(\mathcal{P}^L))$. So (1) follows.

We consider the curves $\mathscr{C}(W_i^{\mathcal{P}})$ and $\mathscr{C}(W_i^{\mathcal{P}^L})$ (see Remark 1), and we note that $\mathscr{C}(W_i^{\mathcal{P}^L})$ is the transformation of $\mathscr{C}(W_i)$ under the birational transformation L^{-1} of $\mathbb{P}^2(\mathbb{K})$. Now, (2) and (3) follow from Definition 2, and (4) from Lemma 10. □

The next results analyze the base loci of birational surface parametrizations assuming that there exists one of them with empty base locus.

Lemma 12. *Let $\mathcal{P}$ and $\mathcal{Q}$ be two birational projective parametrizations of the same surface $\mathcal{S}$ such that $\mathcal{Q}(\mathcal{S}) = \mathcal{P}$ and $\mathcal{B}(\mathcal{Q}) = \emptyset$. It holds that*

1. $\mathcal{B}(\mathcal{S}) = \mathcal{B}(\mathcal{P})$.
2. *If $A \in \mathcal{B}(\mathcal{S})$ then $\deg(\mathcal{S})\, \text{mult}(A, \mathcal{B}(\mathcal{S})) = \text{mult}(A, \mathcal{B}(\mathcal{P}))$.*

Proof. Since $\mathcal{Q} = \mathcal{P}(\mathcal{S})$ and $\mathcal{B}(\mathcal{Q}) = \emptyset$, by Theorem 11 in [12] we get that $\mathcal{B}(^{L_\mathcal{S}}\mathcal{S}) = \mathcal{B}(^{L_\mathcal{P}}\mathcal{P})$ for $L_\mathcal{S}$ in a certain open subset of $\mathscr{G}(\mathbb{P}^2(\mathbb{K}))$ and $L_\mathcal{P}$ in a certain open subset of $\mathscr{G}(\mathbb{P}^3(\mathbb{K}))$. Now, using Lemma 5, and Remark 4 one concludes the proof of statement (1). Statement (2) follows from Theorem 11 in [12], taking into account that $\mathcal{Q}$ is birational. □

Lemma 13. *Let $\mathcal{P}$ and $\mathcal{Q}$ be two birational projective parametrizations of the same surface $\mathcal{S}$ such that $\mathcal{Q}(\mathcal{S}) = \mathcal{P}$ and $\mathcal{B}(\mathcal{Q}) = \emptyset$. Then, for every $A \in \mathcal{B}(\mathcal{S})$ it holds that (see (3) and (7))*

$$\text{mult}(A, \mathscr{C}(W_1)) = \text{mult}(A, \mathscr{C}(V_1)) \deg(\mathcal{Q}).$$

Proof. Let $\mathcal{P} = (p_1 : \cdots : p_4)$, and $\mathcal{Q} = (q_1 : \cdots : q_4)$, where $\gcd(q_1, \ldots, q_4) = 1$. We know that $p_i = q_i(\mathcal{S})$. Moreover, since $\mathcal{B}(\mathcal{Q}) = \emptyset$, by Theorem 10 in [12], we have that $\gcd(p_1, \ldots, p_4) = 1$.

We start observing that because of Lemma 12 one has that $\mathcal{B}(\mathcal{S}) = \mathcal{B}(\mathcal{P})$. Now, let us consider $L \in \mathscr{G}(\mathbb{P}^2(\mathbb{K}))$ and $M \in \mathscr{G}(\mathbb{P}^3(\mathbb{K}))$. Let $\mathcal{Q}^* = {}^MQ^{L^{-1}}$, $\mathcal{S}^* = {}^L\mathcal{S}$ and $\mathcal{P}^* = {}^M\mathcal{P}$. Note that $\mathcal{Q}^*(\mathcal{S}^*) = \mathcal{P}^*$. Moreover, $\mathcal{B}(\mathcal{Q}^*) = \emptyset$. Indeed: if $A \in \mathbb{P}^3(\mathbb{K})$ then $B := L^{-1}(A) \in \mathbb{P}^3(\mathbb{K})$ and, since $\mathcal{B}(\mathcal{Q}) = \emptyset$, $C := \mathcal{Q}(B) \in \mathbb{P}^3(\mathbb{K})$. Therefore $\mathcal{Q}^*(A) = M(B) \in \mathbb{P}^3(\mathbb{K})$ and, in consequence, $\mathcal{B}(\mathcal{Q}^*) = \emptyset$. Moreover, $\mathcal{Q}^*$ and $\mathcal{P}^*$ parametrize the same surface. Furthermore, by Lemma 7, $\mathcal{S}^*$ is transversal. Thus, $\mathcal{S}^*, \mathcal{P}^*, \mathcal{Q}^*$ satisfy the hypotheses of the lemma. On the other hand, by Lemma 5 and Remark 4, we have that $\mathcal{B}(\mathcal{S}^*) = \mathcal{B}(\mathcal{S}) = \mathcal{B}(\mathcal{P}) = \mathcal{B}(\mathcal{P}^*)$. Furthermore, by Lemmas 1 and 2 we have that $\text{mult}(A, \mathscr{C}(V_1)) = \text{mult}(A, \mathscr{C}(V_1^L))$ and $\text{mult}(A, \mathscr{C}(W_1)) = \text{mult}(A, \mathscr{C}(W_1^M))$. Therefore, by Lemma 6 and Remark 4, we can assume w.l.o.g. that for every $A \in \mathcal{B}(\mathcal{S}) = \mathcal{B}(\mathcal{P})$ it holds that

$$\begin{aligned} \text{mult}(A, \mathscr{C}(V_1)) &= \text{mult}(A, \mathscr{C}(s_i)) \text{ for } i \in \{1,2,3\} \\ \text{mult}(A, \mathscr{C}(W_1)) &= \text{mult}(A, \mathscr{C}(p_i)) \text{ for } i \in \{1,2,3,4\} \end{aligned} \quad (12)$$

Now, let $A \in \mathcal{B}(\mathcal{P})$ and let $m := \text{mult}(A, \mathscr{C}(V_1))$. We can assume w.l.o.g that $A = (0:0:1)$. Let T_i denote the product of the tangents to s_i at A. Additionally, let $\deg(\mathcal{S}) = \mathfrak{s}$, $\deg(\mathcal{P}) = \mathfrak{p}$, and $\deg(\mathcal{Q}) = \mathfrak{q}$. Then, by (12), we may write:

$$s_i = T_i t_3^{\mathfrak{s}-m} + g_{m+1,i} t_3^{\mathfrak{s}-m-1} + \cdots + g_{\mathfrak{s},i} \quad (13)$$

where $g_{j,i}(t_1, t_2)$ are homogeneous forms of degree j. In addition, let q_i be expressed as

$$q_i(\bar{t}) = F_{\mathfrak{q},i} + F_{\mathfrak{q}-1,i} t_3 + \cdots + F_{\ell_i,i} t_3^{\mathfrak{q}-\ell_i}, \quad (14)$$

where $F_{j,i}(t_1, t_2)$ are homogeneous forms of degree j. Then

$$p_i(\bar{t}) = F_{\mathfrak{q},i}(s_1, s_2) + F_{\mathfrak{q}-1,i}(s_1, s_2) s_3 + \cdots + F_{\ell_i,i}(s_1, s_2) s_3^{\mathfrak{q}-\ell_i}.$$

Using this expression and (13) it can be expressed as

$$p_i(\bar{t}) = \left(F_{\mathfrak{q},i}(T_1, T_2) + F_{\mathfrak{q}-1,i}(T_1, T_2) T_3 + \cdots + F_{\ell_i,i}(T_1, T_2) T_3^{\mathfrak{q}-\ell_i} \right) t_3^{\mathfrak{q}(\mathfrak{s}-m)} \quad (15)$$
$$+ (\text{terms of degree in } t_3 \text{ strictly smaller than } \mathfrak{q}(\mathfrak{s}-m)).$$

Let
$$H_i := F_{q,i}(T_1, T_2) + F_{q-1,i}(T_1, T_2)T_3 + \cdots + F_{\ell_i,i}(T_1, T_2)T_3^{q-\ell_i}.$$

Now, let us prove that H_i is not identically zero. We first observe that $H_i = q_i(T_1, T_2, T_3)$ for $i \in \{1, 2, 3, 4\}$. We also note that if there exists $i \in \{1, 2, 3, 4\}$ such that $H_i = 0$, by (12), it must happen that for all $i \in \{1, 2, 3, 4\}$ it holds that $H_i = 0$. Let H_1 be zero. Then, $\mathcal{T} = (T_1 : T_2 : T_3) \notin \mathbb{P}^2(\mathbb{K})$, because otherwise $\mathcal{T} \in \mathcal{B}(\mathcal{Q})$ and $\mathcal{B}(\mathcal{Q}) = \emptyset$. Therefore, $\mathcal{T}$ is a curve parametrization. Thus, if $H_i = 0$, then $\mathcal{T}$ parametrizes a common component of the four curves $\mathscr{C}(p_i)$. However, this implies that $\gcd(q_1, q_2, q_3, q_4) \neq 1$ which is a contradiction.

Thus, by (12), $\mathrm{mult}(A, \mathscr{C}(p_1)) = \mathrm{mult}(A, \mathscr{C}(W_i)) = q\, m = \deg(\mathcal{Q})\, \mathrm{mult}(A, \mathscr{C}(V_i))$. □

We finish this section stating the relationship between the transversality of $\mathcal{S}$ and $\mathcal{P}$ under the assumption that $\mathcal{P}(\mathcal{S}^{-1})$ does not have base points.

Theorem 1. *Let $\mathcal{P}$ and $\mathcal{Q}$ be two birational projective parametrizations of the same surface $\mathscr{S}$ such that $\mathcal{Q}(\mathcal{S}) = \mathcal{P}$ and $\mathcal{B}(\mathcal{Q}) = \emptyset$. Then, $\mathcal{S}$ is transversal if and only if $\mathcal{P}$ is transversal.*

Proof. Let $A \in \mathcal{B}(\mathcal{S}) = \mathcal{B}(\mathcal{P})$. First we note that from Lemma 13, and Corollary 5 in [12], it holds that
$$\mathrm{mult}(A, \mathscr{C}(W_1))^2 = \mathrm{mult}(A, \mathscr{C}(V_1))^2 \deg(\mathcal{Q})^2 = \mathrm{mult}(A, \mathscr{C}(V_1))^2 \deg(\mathscr{S})$$

Using Corollary 9 in [12], we have that
$$\mathrm{mult}(A, \mathcal{B}(\mathcal{P})) = \deg(\mathscr{S}) \mathrm{mult}(A, \mathcal{B}(\mathcal{S})).$$

Therefore,
$$\mathrm{mult}(A, \mathcal{B}(\mathcal{S}))\, \mathrm{mult}(A, \mathscr{C}(W_1))^2 = \mathrm{mult}(A, \mathcal{B}(\mathcal{P}))\, \mathrm{mult}(A, \mathscr{C}(V_1))^2.$$

Thus, $\mathcal{S}$ is transversal if and only if $\mathcal{P}$ is transversal. □

4. Proper Polynomial Reparametrization

In this section, we deal with the central problem of the paper, namely, the determination, if they exist, of proper (i.e., birational) polynomial parametrizations of rational surfaces. For this purpose, we distinguish several subsections. In the first subsection, we fix the general assumptions and we propose our strategy. In the second subsection, we perform the theoretical analysis, and in the last subsection we prove the existence of a linear subspace, computable from the input data, and containing the solution to the problem.

We start recalling what we mean with a polynomial projective parametrization. We say that a projective parametrization is polynomial if its dehomogenization w.r.t. the fourth component, taking $t_i = 1$ for some $i \in \{1, 2, 3\}$, is polynomial; note that the fourth component of a polynomial projective parametrization has to be a power of t_i for some $i \in \{1, 2, 3\}$. Clearly, a similar reasoning is applicable w.r.t. other dehomogenizations. On the other hand, we say that a parametrization is *almost polynomial* if its fourth component is the power of a linear form.

The important fact is that a rational surface admits a birational polynomial parametrization if and only if it admits a birational almost polynomial parametrization. Furthermore, if we have an almost polynomial parametrization, and its fourth component is a power of the linear form $L_3^*(\bar{t})$, we may consider two additional linear forms L_1^*, L_2^* such that $L^* = (L_1^* : L_2^* : L_3^*) \in \mathscr{G}(\mathbb{P}^2(\mathbb{K}))$ and then the composition of the almost polynomial parametrization with $(L^*)^{-1}$ is a polynomial parametrization of the same surface.

4.1. General Assumptions and Strategy

In our analysis we have two main assumptions. We assume that the rational surface $\mathcal{S}$ admits a polynomial birational parametrization with empty base locus. Throughout the rest of the paper, let us fix one of these parametrizations and denote it by $\mathcal{Q}$; that is,

$$\mathcal{Q}(\bar{t}) = (q_1(\bar{t}) : q_2(\bar{t}) : q_3(\bar{t}) : q_4(\bar{t})), \tag{16}$$

with q_i homogenous polynomials of the same degree such that $\gcd(q_1, \ldots, q_4) = 1$, is a proper polynomial parametrization of $\mathcal{S}$ satisfying that $\mathcal{B}(\mathcal{Q}) = \emptyset$. Note that, by Corollary 6 in [12], the degree of $\mathcal{S}$ is then the square of a natural number. Moreover, we introduce a second assumption. We assume that we are given a transversal birational parametrization of $\mathcal{S}$. Note that, because Lemmas 9 and 11, this hypothesis is invariant under right and/or left projective transformations. Throughout the rest of the paper, let us fix $\mathcal{P}$ as a transversal proper parametrization of $\mathcal{S}$, and let $\mathcal{P}$ be expressed as in (1).

Our goal is to reach $\mathcal{Q}$, or more precisely an almost polynomial parametrization of $\mathcal{S}$, from $\mathcal{P}$. For this purpose, first we observe that, since both $\mathcal{P}$ and $\mathcal{Q}$ are birational, they are related by means of a birational map of $\mathbb{P}^2(\mathbb{K})$, say $\mathcal{S}_\mathcal{P}$. More precisely, $\mathcal{S}_\mathcal{P} := \mathcal{Q}^{-1} \circ \mathcal{P}$. In the following, we represent $\mathcal{S}_\mathcal{P}$ as

$$\mathcal{S}_\mathcal{P}(\bar{t}) = (s_1(\bar{t}) : s_2(\bar{t}) : s_3(\bar{t})), \tag{17}$$

where $\gcd(s_1, s_2, s_3) = 1$. Note that, because of Theorem 1, since $\mathcal{P}$ is transversal, then $\mathcal{S}_\mathcal{P}$ is transversal. In addition, let $\mathcal{R}_\mathcal{P} := \mathcal{S}_\mathcal{P}^{-1}(\bar{t}) = \mathcal{P}^{-1} \circ \mathcal{Q}$. In the sequel, we represent $\mathcal{R}_\mathcal{P}$ as

$$\mathcal{R}_\mathcal{P}(\bar{t}) = (r_1(\bar{t}) : r_2(\bar{t}) : r_3(\bar{t})), \tag{18}$$

where $\gcd(r_1, r_2, r_3) = 1$.

So, in order to derive $\mathcal{Q}$ from $\mathcal{P}$ it would be sufficient to determine $\mathcal{S}_\mathcal{P}$, and hence $\mathcal{R}_\mathcal{P}$, because $\mathcal{Q} = \mathcal{P}(\mathcal{R}_\mathcal{P})$. Furthermore if, instead of determining $\mathcal{S}_\mathcal{P}$, we obtain ${}^L\mathcal{S}_\mathcal{P} := L \circ \mathcal{S}_\mathcal{P}$, for some $L \in \mathcal{G}(\mathbb{P}^2(\mathbb{K}))$, then instead of $\mathcal{Q}$ we get

$$\mathcal{P}(({}^L\mathcal{S}_\mathcal{P})^{-1}) = \mathcal{P}(\mathcal{R}_\mathcal{P}^{L^{-1}}) = \mathcal{Q}(L^{-1}) = \mathcal{Q}^{L^{-1}},$$

which is almost polynomial, and hence solves the problem. Taking into account this fact we make the following two considerations:

1. We can assume w.l.o.g. that $\mathcal{B}(\mathcal{P}) \neq \emptyset$. Indeed, if $\mathcal{B}(\mathcal{P}) = \emptyset$, by Theorem 10 and Corollary 9 in [12], we get that $\mathcal{B}(\mathcal{S}_\mathcal{P}) = \emptyset$. Furthermore, by Corollary 7 in [12], we obtained that $\deg(\mathcal{S}_\mathcal{P}) = 1$. Thus, using that $\mathcal{Q}$ is indeed polynomial, we get that the fourth component of $\mathcal{P}$ is the power of a linear form, and therefore the input parametrization $\mathcal{P}$ would be already almost polynomial, and hence the problem would be solved.

2. We can assume w.l.o.g. that $\mathcal{S}_\mathcal{P}$ satisfies whatever property reachable by means of a left composition with elements in $\mathcal{G}(\mathbb{P}^2(\mathbb{K}))$, as for instance those stated in Lemma 3, or Lemma 4, or Lemma 6. In particular, by Lemma 7, the transversality is preserved. In other words, in the set $\mathcal{R}$ of all birational transformations of $\mathbb{P}^2(\mathbb{K})$, we consider the equivalence relation $\sim$, defined as $\mathcal{S} \sim \mathcal{S}^*$ if there exists $L \in \mathcal{G}(\mathbb{P}^2(\mathbb{K}))$ such that $L \circ \mathcal{S} = \mathcal{S}^*$, and we work with the equivalence classes in $\mathcal{R}/\sim$.

Therefore, our strategy will be to find a birational map $\mathcal{M}$ of $\mathbb{P}^2(\mathbb{K})$ such that $\mathcal{P}(\mathcal{M}^{-1})$ is almost polynomial. For this purpose, we will see that it is enough to determine a dominant rational transformation $\mathcal{M}$ of $\mathbb{P}^2(\mathbb{K})$ (later, we will prove that such a transformation is indeed birational) such that

1. $\deg(\mathcal{M}) = \deg(\mathcal{S}_\mathcal{P})$.
2. $\mathcal{B}(\mathcal{M}) = \mathcal{B}(\mathcal{S}_\mathcal{P})$.

3. $\forall A \in \mathcal{B}(\mathcal{M})$ it holds that $\text{mult}(A, \mathcal{B}(\mathcal{M})) = \text{mult}(A, \mathcal{B}(\mathcal{S}_\mathcal{P}))$.

The difficulty is that both $\mathcal{M}$ and $\mathcal{S}_\mathcal{P}$ are unknown. Nevertheless, by Corollary 10 and Theorem 3 in [12], we have that

$$\deg(\mathcal{S}_\mathcal{P}) = \frac{\deg(\mathcal{P})}{\sqrt{\deg(\mathcal{S})}}.$$

Note that $\deg(\mathcal{P})$ is given and $\deg(\mathcal{S})$ can be determined by applying, for instance, the formulas in [24] (see also [25]). On the other hand, taking into account Lemma 12, we can achieve our goal by focusing on $\mathcal{P}$. More precisely, we reformulate the above conditions into the equivalent following conditions.

Conditions 1. *We say that a rational dominant map $\mathcal{M}$ of $\mathbb{P}^2(\mathbb{K})$ satisfies Conditions 1 if*

1. $\deg(\mathcal{M}) = \frac{\deg(\mathcal{P})}{\sqrt{\deg(\mathcal{S})}}$.
2. $\mathcal{B}(\mathcal{M}) = \mathcal{B}(\mathcal{P})$,
3. $\text{mult}(A, \mathcal{B}(\mathcal{M})) = \frac{\text{mult}(A, \mathcal{B}(\mathcal{P}))}{\deg(\mathcal{S})}$ for all $A \in \mathcal{B}(\mathcal{P})$.

In the following subsections, we will see that rational dominant maps satisfying Conditions 1 provide an answer to the polynomiality problem.

4.2. Theoretical Analysis

We start this analysis with some technical lemmas. For this purpose, $\mathcal{S}, \mathcal{Q}, \mathcal{P}, \mathcal{S}_\mathcal{P}, \mathcal{R}_\mathcal{P}$ are as in the previous subsection. We recall that $\mathcal{Q}(\mathcal{S}_\mathcal{P}) = \mathcal{P}, \mathcal{B}(\mathcal{Q}) = \emptyset, \mathcal{R}_\mathcal{P} = \mathcal{S}_\mathcal{P}^{-1}, \mathcal{P}$ is transversal, and hence $\mathcal{S}_\mathcal{P}$ is also transversal. Moreover, by Lemma 12, $\mathcal{S}_\mathcal{P}$ satisfies Conditions 1. Furthermore, in the sequel, let

$$\overline{S}(\overline{t}) = (\overline{s}_1(\overline{t}) : \overline{s}_2(\overline{t}) : \overline{s}_3(\overline{t})), \tag{19}$$

with $\gcd(\overline{s}_1, \overline{s}_2, \overline{s}_3) = 1$, be dominant rational map of $\mathbb{P}^2(\mathbb{K})$ satisfying Conditions 1.

Lemma 14. *Let $\mathcal{M}$ be a birational map of $\mathbb{P}^2(\mathbb{K})$. Then, $\deg(\mathcal{M}) = \deg(\mathcal{M}^{-1})$.*

Proof. We use the notation introduced in Lemma 3. We take $L \in \mathscr{G}(\mathbb{P}^3(\mathbb{K}))$ such that

1. $\mathscr{C}(V_1^L)$ is rational (see (8) for the definition of V_1^L constructed from $\mathcal{M}$, and Lemma 3 for the existence of L).
2. $\gcd(\eta_1^L, \eta_3^L) = 1$, where $\mathcal{M}^{-1} = (\eta_1 : \eta_2 : \eta_3)$.

In addition, we consider a projective transformation $N(\overline{t})$ in the parameters $\overline{t}$ such that $\deg_{\overline{t}}(V_1^L(\overline{x}, N(\overline{t}))) = \deg_{t_2}(V_1^L(\overline{x}, N(\overline{t})))$ and $\deg_{\overline{t}}(\eta_1^L(N(\overline{t}))) = \deg_{t_2}(\eta_1^L(N(\overline{t}))) = \deg_{t_2}(\eta_3^L(N(\overline{t})))$. Then, it holds

$$\begin{aligned}
\deg(\mathcal{M}) &= \deg_{\overline{t}}(V_1^L(\overline{x}, \overline{t})) \\
&= \deg_{\overline{t}}(V_1^L(\overline{x}, N(\overline{t}))) && (\mathcal{M} \text{ is a proj. transf.}) \\
&= \deg_{t_2}(V_1^L(\overline{x}, N(\overline{t}))) && \text{(see above)} \\
&= \deg_{t_2}(\eta_1^L(N(x_1, h_1, x_3))/\eta_3^L(N(x_1, h_1, x_3))) && (*) \\
&= \deg_{t_2}(\eta_1^L((N(x_1, h_1, x_3)))) && (\gcd(\eta_1^L, \eta_3^L) = 1) \\
&= \deg(\mathcal{M}^{-1})
\end{aligned}$$

* See Theorem 4.21 in [26]. □

Our goal will be to compute birational transformations satisfying Conditions 1. Next lemma shows that the birationality will be derived from Conditions 1, and hence we will not have to check it computationally.

Lemma 15. Let $\mathcal{M}$ be a rational dominant map of $\mathbb{P}^2(\mathbb{K})$. If $\mathcal{M}$ satisfies Conditions 1, then $\mathcal{M}$ is birational.

Proof. Since $\deg(\mathcal{M}) = \deg(\mathcal{S}_\mathcal{P})$, and $\text{mult}(\mathcal{B}(\mathcal{M})) = \text{mult}(\mathcal{B}(\mathcal{S}_\mathcal{P}))$, by Theorem 7(a) in [12], we have that $\text{degMap}(\mathcal{M}) = \text{degMap}(\mathcal{S}_\mathcal{P})$. So $\overline{\mathcal{S}}$ is birational. □

Therefore, since we have assume above (see (19)) that $\overline{\mathcal{S}}$ satisfies Conditions 1, $\overline{\mathcal{S}}$ is birational (see Lemma 15). Let

$$\overline{\mathcal{R}}(\overline{t}) = \overline{\mathcal{S}}^{-1}(\overline{t}) = (\overline{r}_1(\overline{t}) : \overline{r}_2(\overline{t}) : \overline{r}_3(\overline{t})) \tag{20}$$

be its inverse. Clearly, $\overline{\mathcal{S}}(\overline{\mathcal{R}}) = (t_1 : t_2 : t_3)$, which implies that $\overline{s}_i(\overline{\mathcal{R}}(\overline{t})) = t_i \wp(\overline{t})$, for $i \in \{1,2,3\}$, and where $\deg(\wp) = \deg(\overline{\mathcal{S}})^2 - 1$ (see Lemma 14) and hence $\deg(\wp) = \text{mult}(\mathcal{B}(\overline{\mathcal{S}})) = \text{mult}(\mathcal{B}(\mathcal{S}_\mathcal{P})) = \text{mult}(\mathcal{B}(\mathcal{P}))$. In the next result we prove that $\wp$ is directly related to $\mathcal{B}(\overline{\mathcal{S}})$, and using that $\mathcal{B}(\overline{\mathcal{S}}) = \mathcal{B}(\mathcal{S}_\mathcal{P})$, we study the common factor appearing in the composition $\mathcal{S}(\overline{\mathcal{R}})$. We start with a technical lemma.

Lemma 16. Let $L \in \mathcal{G}(\mathbb{P}^2(\mathbb{K}))$. It holds that

1. $\mathcal{B}(\mathcal{S}_\mathcal{P}^L) = \mathcal{B}(\mathcal{P}^L) = \mathcal{B}(\overline{\mathcal{S}}^L)$.
2. If $A \in \mathcal{B}(\mathcal{S}_\mathcal{P}^L)$ then $\text{mult}(A, \mathcal{B}(\mathcal{S}_\mathcal{P}^L)) = \text{mult}(A, \mathcal{P}^L)/\deg(\mathcal{S})$
3. $\overline{\mathcal{S}}^L$ satisfies Conditions 1.
4. $\mathcal{S}_\mathcal{P}^L$ is transversal
5. If $\overline{\mathcal{S}}$ is transversal, then $\overline{\mathcal{S}}^L$ is transversal.

Proof.

(1) $A \in \mathcal{B}(\mathcal{S}_\mathcal{P}^L)$ iff $\mathcal{S}_\mathcal{P}^L(A) = \overline{0}$ iff $\mathcal{S}_\mathcal{P}(L(A)) = \overline{0}$ iff $L(A) \in \mathcal{B}(\mathcal{S}_\mathcal{P}) = \mathcal{B}(\mathcal{P})$ iff $\mathcal{P}^L(A) = \mathcal{P}(L(A)) = \overline{0}$ iff $A \in \mathcal{B}(\mathcal{P}^L)$. Moreover, the second equality follows as in the previous reasoning, taking into account that $\overline{\mathcal{S}}$ satisfies Condition 1, and hence $\mathcal{B}(\mathcal{S}_\mathcal{P}) = \mathcal{B}(\overline{\mathcal{S}}) = \mathcal{B}(\mathcal{P})$.

(2) follows taking into account that the multiplicity of a point on a curve, as well as the multiplicity of intersection, does not change under projective transformations.

(3) Condition (1) follows taking into account that $\deg(L) = 1$. Statement (1) implies condition (2). For condition (3), we apply statement (2) and that $\text{mult}(A, \mathcal{B}(\overline{\mathcal{S}}^L)) = \text{mult}(A, \mathcal{B}(\mathcal{S}))$ because the multiplicity of intersection does not change with L.

(4) and (5) follow arguing as in the proof of Lemma 11(4). □

Theorem 2. Let $\overline{\mathcal{S}}$ be transversal. If $i \in \{1,2,3\}$ then $\overline{s}_i(\overline{\mathcal{R}}) = t_i \wp(\overline{t})$ where $\deg(\wp(\overline{t})) = \text{mult}(\mathcal{B}(\mathcal{P}))$ and such that $\wp$ is uniquely determined by $\mathcal{B}(\mathcal{P})$.

Proof. We first observe that we can assume w.l.o.g. that no base point of $\mathcal{P}$ is on the line at infinity $x_3 = 0$. Indeed, let $L \in \mathcal{G}(\mathbb{P}^2(\mathbb{K}))$ be such that $\mathcal{B}(\mathcal{P})$ is contained in the affine plane $x_3 = 1$. We consider $\overline{\mathcal{S}}^* := \overline{\mathcal{S}}^L = (\overline{s}_1^* : \overline{s}_2^* : \overline{s}_3^*)$ and $\overline{\mathcal{R}}^* := (\overline{\mathcal{S}}^L)^{-1}$, then $\overline{s}_i^*(\overline{\mathcal{R}}^*) = \overline{s}_i^*(L^{-1}(\overline{\mathcal{R}}))$, and $\overline{s}_i^* = \overline{s}_i(L)$; hence $\overline{s}_i^*(\overline{\mathcal{R}}^*) = \overline{s}_i(\overline{\mathcal{R}})$. In addition, because of Lemma 16, $\overline{\mathcal{S}}^*$ satisfies the hypothesis of the theorem.

Let $\mathcal{C}(\overline{V}_1)$ denote the curve associated to $\overline{\mathcal{S}}$ as in (7). By Lemma 3, taking L in the corresponding open subset of $\mathcal{G}(\mathbb{P}^2(\mathbb{K}))$, we have that $\mathcal{C}(\overline{V}_1^L)$ is a rational curve. So, we assume w.l.o.g. that $\mathcal{C}(\overline{V}_1)$ is rational. Let $\overline{\mathcal{V}}(\overline{x}, h_1, h_2)$ be the rational parametrization of $\mathcal{C}(\overline{V}_1)$ provided by Lemma 3. We apply a Möbius transformation $\phi \in \mathcal{G}(\mathbb{P}^1(\mathbb{K}))$ such that if $\overline{\mathcal{W}}(\overline{x}, h_1, h_2) = (\overline{w}_1(\overline{x}, h_1, h_2) : \overline{w}_2(\overline{x}, h_1, h_2) : \overline{w}_3(\overline{x}, h_1, h_2)) := \overline{\mathcal{V}}(\overline{x}, \phi(h_1, h_2))$ then the affine parametrization $\overline{p}(\overline{x}, h_1) := (\overline{w}_1(\overline{x}, h_1, 1)/\overline{w}_3(\overline{x}, h_1, 1), \overline{w}_2(\overline{x}, h_1, 1)/\overline{w}_3(\overline{x}, h_1, 1))$ is affinely surjective (see [7,23]).

Now, let $A = (a_1 : a_2 : 1) \in \mathcal{B}(\mathcal{P})$. By Lemma 2, $P \in \mathscr{C}(\overline{V}_1)$. We observe that, by taking L in the open subset of Lemma 6, we may assume that

$$m_A := \text{mult}(A, \mathscr{C}(\overline{V}_1)) = \text{mult}(A, \mathscr{C}(\overline{s}_i)), \ i \in \{1,2,3\}. \tag{21}$$

We consider the polynomial

$$g_A = \gcd(\overline{w}_1(\overline{x}, h_1, h_2) - a_1\overline{w}_3(\overline{x}, h_1, h_2), \overline{w}_2(\overline{x}, h_1, h_2) - a_2\overline{w}_3(\overline{x}, h_1, h_2)).$$

Since the affine parametrization has been taken surjective, we have that

$$\deg_{\overline{h}}(g_A) = m_A \tag{22}$$

and that for every root t_0 of g_A it holds that $\overline{\rho}(t_0) = (a_1, a_2)$. We write $\overline{w}_i$ as

$$\overline{w}_i = g_A \cdot \overline{w}_i^* + a_i \overline{w}_3, \ i = 1, 2.$$

On the other hand, we express $\overline{s}_i$ as

$$\overline{s}_i(\overline{t}) = \overline{T}_{i,m_A}(\overline{t}) t_3^{\deg(\overline{S}) - m_A} + \cdots + \overline{T}_{i,\deg(\overline{S})}(\overline{t}),$$

where $\deg(\overline{T}_{i,j}) = j$, $j \in \{m_A, \ldots, \deg(\overline{S})\}$, and $\overline{T}_{i,j}(\overline{t}) = \sum_{k_1 + k_2 = j}(t_1 - a_1 t_3)^{k_1}(t_2 - a_2 t_3)^{k_2}$. Therefore

$$\overline{s}_i(\overline{\mathcal{W}}) = g_A^{m_A} \cdot \left(\overline{T}_{i,m_A}(\overline{w}_1^*, \overline{w}_2^*) \overline{w}_3^{\deg(\overline{S}) - m_A} + \cdots + g_A^{\deg(\overline{S}) - m_A} \overline{T}_{i,\deg(\overline{S})}(\overline{w}_1^*, \overline{w}_2^*) \right).$$

In other words, g_A divides $\overline{s}_i(\overline{\mathcal{W}})$. Now, for $B = (b_1 : b_2 : 1) \in \mathcal{B}(\mathcal{P})$, with $A \neq B$, it holds that $\gcd(g_A, g_B) = 1$, since otherwise there would exist a root t_0 of $\gcd(g_A, g_B)$, and this implies that $\overline{\rho}(t_0) = (a_1, a_2) = (b_1, b_2) = \overline{\rho}(t_0)$ which is a contradiction. Therefore, we have that

$$\overline{s}_i(\overline{\mathcal{W}}) = \prod_{A \in \mathcal{B}(\mathcal{P})} g_A(\overline{x}, h_1, h_2)^{m_A} f_i(\overline{x}, h_1, h_2) \tag{23}$$

We observe that the factor defined by the base points does not depend on i. Thus, since $\overline{s}_i(\overline{\mathcal{W}})$ does depend on i, we get that f_i is the factor depending on i. Furthermore,

$$\begin{aligned}
\deg_{\overline{h}}\left(\prod_{A \in \mathcal{B}(\mathcal{P})} g_A^{m_A}\right) &= \sum_{A \in \mathcal{B}(\mathcal{P})} \deg_{\overline{h}}(g_A)^{m_A} & \\
&= \sum_{A \in \mathcal{B}(\mathcal{P})} m_A^2 & \text{(see (22))} \\
&= \sum_{A \in \mathcal{B}(\mathcal{P})} \text{mult}(A, \mathscr{C}(\overline{V}_1))^2 & \text{(see (21))} \\
&= \sum_{A \in \mathcal{B}(\mathcal{P})} \text{mult}(A, \mathcal{B}(\overline{S})) & (\overline{S} \text{ is transversal}) \\
&= \sum_{A \in \mathcal{B}(\mathcal{P})} \text{mult}(A, \mathcal{B}(\overline{\mathcal{P}})) & \text{(See Conditions 1)} \\
&= \text{mult}(\mathcal{B}(\overline{\mathcal{P}})) & \text{(See Definition 2)}
\end{aligned}$$

Moreover, by Theorem 4.21 in [23], since $\overline{\mathcal{W}}$ is birational it holds that $\deg(\overline{\mathcal{W}}) = \deg(\mathscr{C}(\overline{V}_1)) = \deg(\overline{S})$. Hence, $\deg(\overline{s}_i(\overline{\mathcal{W}})) = \deg(\overline{S})^2 = \text{mult}(\mathcal{B}(\mathcal{P})) + 1$. Therefore, f_i in (28) is a linear form.

In this situation, let us introduce the notation $\overline{t}^* := (t_3, 0, -t_1, t_1, t_2)$ and $\overline{t}^{**} = (t_3, 0, -t_1, \phi^{-1}(t_1, t_2))$. Then, for $i \in \{1, 2, 3\}$, we have that

$$\begin{aligned}
t_i \wp &= \overline{s}_i(\overline{\mathcal{R}}) & \\
&= \overline{s}_i(\overline{V}(\overline{t}^*)) & \text{(see Remark 3)} \\
&= \overline{s}_i(\overline{\mathcal{W}}(\overline{t}^{**})) & \text{(see definition of } \overline{\mathcal{W}}) \\
&= \prod_{A \in \mathcal{B}(\mathcal{P})} g_A(\overline{t}^{**})^{m_A} f_i(\overline{t}^{**}) & \text{(see (28))}
\end{aligned}$$

Taking into account that $\prod_{A\in\mathcal{B}(\mathcal{P})} g_A(\bar{t}^{**})^{m_A}$ does not depend on i, we get that $t_1 f_2(\bar{t}^{**}) = t_2 f_1(\bar{t}^{**})$. This implies that t_1 divides $f_1(\bar{t}^{**})$, and since $f_1(\bar{t}^{**})$ is linear we get that $t_1 = \lambda f_1(\bar{t}^{**})$ for $\lambda \in \mathbb{K} \setminus \{0\}$. Then, substituting above, we get $\lambda f_1(\bar{t}^{**}) f_2(\bar{t}^{**}) = t_2 f_1(\bar{t}^{**})$, which implies that $t_2 = \lambda f_2(\bar{t}^{**})$. Similarly, for $t_3 = \lambda f_3(\bar{t}^{**})$. Therefore, we get that

$$\bar{s}_i(\overline{\mathcal{R}}) = t_i \lambda \prod_{A\in\mathcal{B}(\mathcal{P})} g_A(\bar{t}^{**})^{m_A}, \text{ with } \lambda \in \mathbb{K} \setminus \{0\}$$

This concludes the proof. □

For the next theorem, we recall that $\mathcal{S}_\mathcal{P} = (s_1 : s_2 : s_3)$ with $\gcd(s_1, s_2, s_3) = 1$; see (17).

Theorem 3. *Let $\overline{\mathcal{S}}$ be transversal. If $i \in \{1,2,3\}$ then $s_i(\overline{\mathcal{R}}) = Z_i(\bar{t})\wp(\bar{t})$ where Z_i is a linear form, $\deg(\wp(\bar{t})) = \mathrm{mult}(\mathcal{B}(\mathcal{P}))$ and such that $\wp$ is uniquely determined by $\mathcal{B}(\mathcal{P})$.*

Proof. We first observe that we can assume w.l.o.g. that no base point of $\mathcal{P}$ is on the line at infinity $x_3 = 0$. Indeed, let $L \in \mathcal{G}(\mathbb{P}^2(\mathbb{K}))$ be such that $\mathcal{B}(\mathcal{P})$ is contained in the affine plane $x_3 = 1$. We consider $\mathcal{S}^* := \mathcal{S}_\mathcal{P}^L = (s_1^* : s_2^* : s_3^*)$, $\overline{\mathcal{S}}^* := \overline{\mathcal{S}}^L$, and $\overline{\mathcal{R}}^* := (\overline{\mathcal{S}}^*)^{-1}$. Then $s_i^*(\overline{\mathcal{R}}^*) = s_i(L(L^{-1}(\overline{\mathcal{R}}))) = s_i(\overline{\mathcal{R}})$. In addition, because of Lemma 16, $\overline{\mathcal{S}}^*$ and $\mathcal{S}^*$ satisfy the hypotheses of the theorem.

Let $\mathscr{C}(\overline{V}_1), \overline{\mathcal{V}}(\bar{x}, h_1, h_2), \overline{\mathcal{W}}(\bar{x}, h_1, h_2) = (\overline{w}_1 : \overline{w}_2 : \overline{w}_3)$ and $\bar{\rho}$ be as in the proof of Theorem 2.

Now, let $A = (a_1 : a_2 : 1) \in \mathcal{B}(\mathcal{P})$. By Lemma 2, $P \in \mathscr{C}(\overline{V}_1)$. We recall that $\mathcal{B}(\mathcal{P}) = \mathcal{B}(\overline{\mathcal{S}}) = \mathcal{B}(\mathcal{S}_\mathcal{P})$. Let $\Omega_3^{\overline{\mathcal{S}}}$ and $\Omega_3^{\mathcal{S}_\mathcal{P}}$ be the open subset in Lemma 6 applied to $\overline{\mathcal{S}}$ and $\mathcal{S}_\mathcal{P}$, respectively. Taking $L \in \Omega_3^{\overline{\mathcal{S}}} \cap \Omega_3^{\mathcal{S}_\mathcal{P}}$ (note that $\mathcal{G}(\mathbb{P}^2(\mathbb{K}))$ is irreducible and hence the previous intersection is non-empty), we may assume that

$$m_A := \mathrm{mult}(A, \mathscr{C}(\overline{V}_1)) = \mathrm{mult}(A, \mathscr{C}(\bar{s}_i)), \ i \in \{1,2,3\}. \tag{24}$$

and

$$\mathrm{mult}(A, \mathscr{C}(V_1)) = \mathrm{mult}(A, \mathscr{C}(s_i)), \ i \in \{1,2,3\}. \tag{25}$$

Since $\mathcal{S}_\mathcal{P}$ and $\overline{\mathcal{S}}$ are transversal, and taking into account Condition 1, it holds that

$$\mathrm{mult}(A, \mathscr{C}(V_1))^2 = \mathrm{mult}(A, \mathcal{B}(\mathcal{S}_\mathcal{P})) = \mathrm{mult}(A, \mathcal{B}(\overline{\mathcal{S}})) = \mathrm{mult}(A, \mathscr{C}(\overline{V}_1))^2$$

Therefore,

$$\mathrm{mult}(A, \mathscr{C}(V_1)) = m_A = \mathrm{mult}(A, \mathscr{C}(\overline{V}_1)). \tag{26}$$

We consider the polynomial $g_A = \gcd(\overline{w}_1 - a_1\overline{w}_3, \overline{w}_2 - a_2\overline{w}_3)$. Reasoning as in the Proof of Theorem 2 we get that

$$\deg_{\bar{h}}(g_A) = m_A \tag{27}$$

and that for every root t_0 of g_A it holds that $\bar{\rho}(t_0) = (a_1, a_2)$. We write $\overline{w}_i$ as $\overline{w}_i = g_A \cdot \overline{w}_i^* + a_i\overline{w}_3$ for $i = \in \{1,2\}$.

On the other hand, by (25) and (26), we have that $\mathrm{mult}(A, \mathscr{C}(s_i)) = m_A$. Therefore, we can express s_i as

$$s_i(\bar{t}) = T_{i,m_A}(\bar{t}) t_3^{\deg(\mathcal{S}_\mathcal{P}) - m_A} + \cdots + T_{i,\deg(\mathcal{S}_\mathcal{P})}(\bar{t}),$$

where $\deg(T_{i,j}) = j, j \in \{m_A, \ldots, \deg(\mathcal{S}_\mathcal{P})\}$, and $T_{i,j}(\bar{t}) = \sum_{k_1+k_2=j}(t_1 - a_1 t_3)^{k_1}(t_2 - a_2 t_3)^{k_2}$. Therefore

$$s_i(\overline{\mathcal{W}}) = g_A^{m_A} \cdot \left(T_{i,m_A}(\overline{w}_1^*, \overline{w}_2^*) \overline{w}_3^{\deg(\mathcal{S}_\mathcal{P}) - m_A} + \cdots + g_A^{\deg(\mathcal{S}_\mathcal{P}) - m_A} T_{i,\deg(\mathcal{S})}(\overline{w}_1^*, \overline{w}_2^*)\right).$$

In other words, g_A divides $s_i(\overline{\mathcal{W}})$. Now, for $B = (b_1 : b_2 : 1) \in \mathcal{B}(\mathcal{P})$, with $A \neq B$, reasoning as in the proof of Theorem 2, it holds that $\gcd(g_A, g_B) = 1$. Therefore, we have that

$$s_i(\overline{\mathcal{W}}) = \prod_{A \in \mathcal{B}(\mathcal{P})} g_A(\overline{x}, h_1, h_2)^{m_A} f_i(\overline{x}, h_1, h_2) \tag{28}$$

Furthermore,

$$\begin{aligned}
\deg_{\overline{h}}\left(\prod_{A \in \mathcal{B}(\mathcal{P})} g_A^{m_A}\right) &= \sum_{A \in \mathcal{B}(\mathcal{P})} \deg_{\overline{h}}(g_A)^{m_A} \\
&= \sum_{A \in \mathcal{B}(\mathcal{P})} m_A^2 & \text{(see (27))} \\
&= \sum_{A \in \mathcal{B}(\mathcal{P})} \mathrm{mult}(A, \mathcal{C}(\overline{V}_1))^2 & \text{(see (24))} \\
&= \sum_{A \in \mathcal{B}(\mathcal{P})} \mathrm{mult}(A, \mathcal{B}(\overline{S})) & (\overline{S} \text{ is transversal}) \\
&= \sum_{A \in \mathcal{B}(\mathcal{P})} \mathrm{mult}(A, \mathcal{B}(\overline{\mathcal{P}})) & \text{(See Conditions 1)} \\
&= \mathrm{mult}(\mathcal{B}(\overline{\mathcal{P}})) & \text{(See Definition 2)}
\end{aligned}$$

Moreover, by Theorem 4.21 in [23], since $\overline{\mathcal{W}}$ is birational it holds that $\deg(\overline{\mathcal{W}}) = \deg(\mathcal{C}(\overline{V}_1)) = \deg(\overline{S})$. Hence, by Condition 1, $\deg(s_i(\overline{\mathcal{W}})) = \deg(\mathcal{S}_\mathcal{P}) \deg(\overline{S}) = \deg(\mathcal{S}_\mathcal{P})^2 = \mathrm{mult}(\mathcal{B}(\mathcal{P})) + 1$. Therefore, f_i in (28) is a linear form.

In this situation, let us introduce the notation $\overline{t}^* := (t_3, 0, -t_1, t_1, t_2)$ and $\overline{t}^{**} = (t_3, 0, -t_1, \phi^{-1}(t_1, t_2))$. Then, for $i \in \{1, 2, 3\}$, we have that

$$\begin{aligned}
s_i(\overline{\mathcal{R}}) &= s_i(\overline{\mathcal{V}}(\overline{t}^*)) & \text{(see Remark 3)} \\
&= s_i(\overline{\mathcal{W}}(\overline{t}^{**})) & \text{(see definition of } \overline{\mathcal{W}}) \\
&= \prod_{A \in \mathcal{B}(\mathcal{P})} g_A(\overline{t}^{**})^{m_A} f_i(\overline{t}^{**}) & \text{(see (28))}
\end{aligned}$$

This concludes the proof. $\square$

Corollary 2. *If $\overline{S}$ is transversal, there exists $L \in \mathcal{G}(\mathbb{P}^2(\mathbb{K}))$ such that $\overline{S} = {}^L\mathcal{S}_\mathcal{P}$.*

Proof. From Theorem 3, we get that $\mathcal{S}_\mathcal{P}(\overline{\mathcal{R}}) = (Z_1(\overline{t}) : Z_2(\overline{t}) : Z_3(\overline{t}))$, where Z_i is a linear form. Thus, ${}^L\mathcal{S}_\mathcal{P} = \overline{S}$, where $L \in \mathcal{G}(\mathbb{P}^2(\mathbb{K}))$ is the inverse of (Z_1, Z_2, Z_3). $\square$

Corollary 3. *The following statements are equivalent*

1. *$\overline{S}$ is transversal.*
2. *There exists $L \in \mathcal{G}(\mathbb{P}^2(\mathbb{K}))$ such that $\overline{S} = {}^L\mathcal{S}_\mathcal{P}$.*

Proof. If (1) holds, then (2) follows from Corollary 2. Conversely, if (2) holds, then (1) follows from Lemma 7. $\square$

4.3. The Solution Space

In this subsection we introduce a linear projective variety containing the solution to our problem and we show how to compute it. We start identifying the set of all projective curves, including multiple component curves, of a fixed degree d, with the projective space (see [23,27,28] for further details)

$$\mathcal{V}_d := \mathbb{P}^{\frac{d(d+3)}{2}}(\mathbb{K}).$$

More precisely, we identify the projective curves of degree d with the forms in $\mathbb{K}[\overline{t}]$ of degree d, up to multiplication by non-zero $\mathbb{K}$-elements. Now, these forms are identified with the elements in $\mathcal{V}_d$ corresponding to their coefficients, after fixing an order of the monomials. By abuse of notation, we will refer to the elements in $\mathcal{V}_d$ by either their tuple of coefficients, or the associated form, or the corresponding curve.

Let $\mathcal{M} = (m_1(\bar{t}) : m_2(\bar{t}) : m_3(\bar{t}))$, $\gcd(m_1, m_2, m_3) = 1$, be a birational transformation of $\mathbb{P}^2(\mathbb{K})$. We consider $\mathcal{V}_{\deg(\mathcal{M})}$. Then, $m_1, m_2, m_3 \in \mathcal{V}_{\deg(\mathcal{M})}$. Moreover, in $\mathcal{V}_{\deg(\mathcal{M})}$, we introduce the projective linear subspace

$$\mathcal{L}(\mathcal{M}) := \{a_1 m_1(\bar{t}) + a_2 m_2(\bar{t}) + a_3 m_3(\bar{t}) \mid (a_1 : a_2 : a_3) \in \mathbb{P}^2(\mathbb{K})\}.$$

We observe that if $\{m_1, m_2, m_3\}$ are linearly dependent, then the image of $\mathbb{P}^2(\mathbb{K})$ via $\mathcal{M}^{-1}$ would be a line in $\mathbb{P}^2(\mathbb{K})$ which is impossible because $\mathcal{M}$ is birational on $\mathbb{P}^2(\mathbb{K})$. Therefore, the following holds.

Lemma 17. *If $\mathcal{M}$ is a birational transformation of $\mathbb{P}^2(\mathbb{K})$, then $\dim(\mathcal{L}(\mathcal{M})) = 2$.*

Similarly, one has the next lemma.

Lemma 18. *If $\mathcal{M}$ is a birational transformation of $\mathbb{P}^2(\mathbb{K})$ and $L \in \mathscr{G}(\mathbb{P}^2(\mathbb{K}))$ then $\mathcal{L}(\mathcal{M}) = \mathcal{L}(^L\mathcal{M})$.*

Furthermore, the following theorem holds

Theorem 4. *Let $\mathcal{M}$ be a birational transformation of $\mathbb{P}^2(\mathbb{K})$ and let $\{n_1, n_2, n_3\}$ be a basis of $\mathcal{L}(\mathcal{M})$ and $\mathcal{N} := (n_1 : n_2 : n_3)$. There exists $L \in \mathscr{G}(\mathbb{P}^2(\mathbb{K}))$ such that $^L\mathcal{M} = \mathcal{N}$.*

Proof. Let $\mathcal{M} = (m_1 : m_2 : m_3)$, with $\gcd(m_1, m_2, m_3) = 1$. Since $m_1, m_2, m_3 \in \mathcal{L}(\mathcal{M})$, and $\{n_1, n_2, n_3\}$ is a basis of $\mathcal{L}(\mathcal{M})$, there exist $(\lambda_{i,1} : \lambda_{i,2} : \lambda_{i,3}) \in \mathbb{P}^2(\mathbb{K})$ such that

$$m_i = \sum \lambda_{i,j} n_j.$$

Since $\{m_1, m_2, m_3\}$ is also a basis of $\mathcal{L}(\mathcal{M})$, one has that $L := (\sum \lambda_{1,j} t_j : \sum \lambda_{2,j} t_j : \sum \lambda_{3,j} t_j) \in \mathscr{G}(\mathbb{P}^2(\mathbb{K}))$ and $\mathcal{N} = L \circ \mathcal{M}$. □

Remark 5. *Observe that, by Theorem 4, all bases of $\mathcal{L}(\mathcal{M})$ generate birational maps of $\mathbb{P}^2(\mathbb{K})$.*

Corollary 4. *Let $\mathcal{M}$ be a birational transformation of $\mathbb{P}^2(\mathbb{K})$. The following statements are equivalent*

1. *$\mathcal{M}$ is transversal.*
2. *There exists a basis $\{n_1, n_2, n_3\}$ of $\mathcal{L}(\mathcal{M})$ such that $(n_1 : n_2 : n_3)$ is transversal.*
3. *For all bases $\{n_1, n_2, n_3\}$ of $\mathcal{L}(\mathcal{M})$ it holds that $(n_1 : n_2 : n_3)$ is transversal.*

Proof. It follows from Theorem 4 and Lemma 7. □

In the following results we analyze the bases of $\mathcal{L}(\mathcal{S}_\mathcal{P})$. So, $\mathcal{S}, \mathcal{R}, \mathcal{P}, \mathcal{Q}$ and $\overline{\mathcal{S}}$ are as the in previous subsections.

Corollary 5. *Let $\{m_1, m_2, m_3\}$ a basis of $\mathcal{L}(\mathcal{S}_\mathcal{P})$. Then, $(m_1 : m_2 : m_3)$ satisfies Conditions 1.*

Proof. It is a direct consequence of Theorem 4. □

Corollary 6. *If $\mathcal{M} := (m_1 : m_2 : m_3)$ is transversal and satisfies Condition 1, then $\{m_1, m_2, m_3\}$ is a basis of $\mathcal{L}(\mathcal{S}_\mathcal{P})$.*

Proof. By Corollary 3, there exists $L \in \mathscr{G}(\mathbb{P}^2(\mathbb{K}))$ such that $\mathcal{M} = {}^L\mathcal{S}_\mathcal{P}$. Now, by Lemma 18, $\mathcal{L}(\mathcal{S}_\mathcal{P}) = \mathcal{L}(\mathcal{M})$. Taking into account that $\{m_1, m_2, m_3\}$ are linearly independent, we get the result. □

The previous results show that the solution to our problem lies in $\mathcal{L}(\mathcal{S}_\mathcal{P})$. However, knowing $\mathcal{L}(\mathcal{S}_\mathcal{P})$ implies knowing $\mathcal{S}_\mathcal{P}$, which is essentially our goal. In the following, we see how to achieve $\mathcal{L}(\mathcal{S}_\mathcal{P})$ by simply knowing $\mathscr{B}(\mathcal{S}_\mathcal{P})$ and the base point multiplicities; note that, under the hypotheses of this section, this information is given by $\mathcal{P}$.

Definition 7. *Let $\mathcal{M}$ be a birational transformation of $\mathbb{P}^2(\mathbb{K})$. We define the linear system of base points of $\mathcal{M}$, and we denote it by $\mathscr{L}(\mathcal{M})$, as the linear system of curves, of degree $\deg(\mathcal{M})$,*

$$\mathscr{L}(\mathcal{M}) = \{f \in \mathscr{V}_{\deg(\mathcal{M})} \mid \operatorname{mult}(A, \mathscr{C}(f)) \geq \sqrt{\operatorname{mult}(A, \mathscr{B}(\mathcal{M}))} \; \forall A \in \mathscr{B}(\mathcal{M})\}$$

Observe that $\mathscr{L}(\mathcal{M})$ is the $\deg(\mathcal{M})$-linear system associated to the effective divisor

$$\sum_{A \in \mathscr{B}(\mathcal{M})} \sqrt{\operatorname{mult}(A, \mathscr{B}(\mathcal{M}))} \cdot A$$

Remark 6. *We observe that if $\mathcal{M}$ satisfies Condition 1, in particular $\mathcal{S}_\mathcal{P}$, then $\mathscr{L}(\mathcal{M})$ is the $\deg(\mathcal{S}_\mathcal{P})$-degree linear system generated by the effective divisor*

$$\sum_{A \in \mathscr{B}(\mathcal{P})} \sqrt{\operatorname{mult}(A, \mathscr{B}(\mathcal{P}))} \cdot A.$$

The following lemma is a direct consequence of the definition above. In Section 4.1, we have mentioned that we will work with equivalence classes. The next lemma states that the $\deg(\mathcal{S}_\mathcal{P})$-degree linear system generated by the effective divisor is invariant within the equivalence class, and hence we may take whatever representant for our computations.

Lemma 19. *Let $\mathcal{M}$ be a birational transformation of $\mathbb{P}^2(\mathbb{K})$. If $L \in \mathscr{G}(\mathbb{P}^2(\mathbb{K}))$ then $\mathcal{L}(\mathcal{M}) = \mathcal{L}(^L\mathcal{M})$ and $\mathscr{L}(\mathcal{M}) = \mathscr{L}(^L\mathcal{M})$.*

The next lemma relates the $\mathcal{L}(\mathcal{M})$ and $\mathscr{L}(\mathcal{M})$.

Lemma 20. *If $\mathcal{M}$ is a transversal birational map of $\mathbb{P}^2(\mathbb{K})$, then $\mathcal{L}(\mathcal{M}) \subset \mathscr{L}(\mathcal{M})$.*

Proof. Let $\mathcal{M} = (m_1 : m_2 : m_3)$, let $f \in \mathcal{L}(\mathcal{M})$ and $A \in \mathscr{B}(\mathcal{M})$. Then, $\deg(f) = \deg(\mathcal{M})$. On the other hand

$$\begin{aligned}
\operatorname{mult}(A, \mathscr{C}(f)) &\geq \min\{\operatorname{mult}(A, \mathscr{C}(m_i)) \mid i \in \{1,2,3\}\} \\
&= \operatorname{mult}(A, \mathscr{C}(V_1)) &\text{(see Lemma 2(2))} \\
&= \sqrt{\operatorname{mult}(A, \mathscr{B}(\mathcal{M}))} &(\mathcal{M} \text{ is transversal}).
\end{aligned}$$

Therefore, $f \in \mathscr{L}(\mathcal{M})$. □

Lemma 21. *If $\mathcal{M}$ is a transversal birational map of $\mathbb{P}^2(\mathbb{K})$, then $\dim(\mathscr{L}(\mathcal{M})) = 2$.*

Proof. Let $\mathcal{M} = (m_1 : m_2 : m_3)$. By Lemmas 17 and 20, we have that $\dim(\mathscr{L}(\mathcal{M})) \geq 2$. Let us assume that $\dim(\mathscr{L}(\mathcal{M})) = k > 2$ and let $\{n_1, \ldots, n_{k+1}\}$ be a basis of $\mathscr{L}(\mathcal{M})$ where $n_1 = m_1, n_2 = m_2, n_3 = m_3$. Then

$$\mathscr{L}(\mathcal{M}) = \{\lambda_1 n_1 + \cdots + \lambda_{k+1} n_{k+1} \mid (\lambda_1 : \cdots : \lambda_{k+1}) \in \mathbb{P}^{k+1}(\mathbb{K})\}.$$

Now, we take three points in $\mathbb{P}^2(\mathbb{K})$ that will be crucial later. For their construction, we first consider an open Zariski subset $\Sigma \subset \mathbb{P}^2(\mathbb{K})$ where $\mathcal{M} : \Sigma \to \mathcal{M}(\Sigma) \subset \mathbb{P}^2(\mathbb{K})$ is a bijective map. Then, $\mathcal{M}(\Sigma)$ is a constructible set of $\mathbb{P}^2(\mathbb{K})$ (see e.g., Theorem 3.16 in [22]). Thus, $\mathbb{P}^2(\mathbb{K}) \setminus \mathcal{M}(\Sigma)$ can only contain finitely many lines. On the other hand, we consider the open subset $\Omega_2 \subset \mathscr{G}(\mathbb{P}^2(\mathbb{K}))$ introduced in Lemma 4, and we take $L = (L_1 : L_2 : L_3) \in \Omega_2$ such that a non-empty open subset of $\mathscr{C}(L_1)$ is included in $\mathcal{M}(\Sigma)$. We take three points $B_1, B_2, B_3 \in \Sigma$ (this, in particular, implies that $B_1, B_2, B_3 \notin \mathscr{B}(\mathcal{M})$) such that:

1. $\mathcal{M}(B_1) \neq \mathcal{M}(B_2)$
2. $\mathcal{M}(B_1), \mathcal{M}(B_2) \in \mathscr{C}(L_1)$,
3. $\mathcal{M}(B_3) \notin \mathscr{C}(L_1)$; note that $\mathcal{M}(B_1), \mathcal{M}(B_2), \mathcal{M}(B_3)$ are not on a line

Since $\mathcal{M}(B_1), \mathcal{M}(B_2), \mathcal{M}(B_3)$ are not collinear, the system

$$\left\{ \sum_{i=1}^{k+1} \lambda_i n_i(B_j) = 0 \right\}_{j \in \{1,2,3\}}$$

has solution. Let $(b_1 : \cdots : b_{k+1})$ be a solution. Then, we consider the polynomials (say that $L_1 := a_1 t_1 + a_2 t_2 + a_3 t_3$)

$$f(\bar{t}) := L_1(\mathcal{M}) = a_1 m_1 + a_2 m_2 + a_3 m_3, \quad g(\bar{t}) := b_1 n_1 + \cdots + b_{k+1} n_{k+1}.$$

We have that $\mathscr{C}(f)$ is irreducible because $L \in \Omega_2$. Moreover, $\deg(\mathscr{C}(f)) = \deg(\mathscr{C}(g))$. In addition, $\mathscr{C}(f) \neq \mathscr{C}(g)$: indeed, $B_3 \in \mathscr{C}(g)$ and $B_3 \notin \mathscr{C}(f)$ because otherwise

$$\begin{pmatrix} m_1(B_1) & m_2(B_1) & m_3(B_1) \\ m_1(B_2) & m_2(B_2) & m_3(B_2) \\ m_1(B_3) & m_2(B_3) & m_3(B_3) \end{pmatrix} \begin{pmatrix} a_1 \\ a_2 \\ a_3 \end{pmatrix} = \begin{pmatrix} 0 \\ 0 \\ 0 \end{pmatrix},$$

and since $\mathcal{M}(B_1), \mathcal{M}(B_2), \mathcal{M}(B_3)$ are not collinear we get that $a_1 = a_2 = a_3 = 0$ that is a contradiction. Therefore, $\mathscr{C}(f)$ and $\mathscr{C}(g)$ do not share components. Thus, by Bézout's theorem the number of intersections of $\mathscr{C}(f)$ and $\mathscr{C}(g)$, properly counted, is $\deg(\mathcal{M})^2$. In addition, we oberve that $f \in \mathcal{L}(\mathcal{M}) \subset \mathscr{L}(\mathcal{M})$ (see Lemma 20) and $g \in \mathscr{L}(\mathcal{M})$. Thus,

$$\mathscr{B}(\mathcal{M}) \cup \{B_1, B_2\} \subset \mathscr{C}(f) \cap \mathscr{C}(g). \tag{29}$$

Therefore

$$\begin{aligned}
\deg(\mathcal{M})^2 &= \sum_{A \in \mathscr{C}(f) \cap \mathscr{C}(g)} \mathrm{mult}_A(\mathscr{C}(f), \mathscr{C}(g)) \\
&\geq \sum_{A \in \mathscr{B}(\mathcal{M})} \mathrm{mult}_A(\mathscr{C}(f), \mathscr{C}(g)) + \sum_{A \in \{B_1, B_2\}} \mathrm{mult}_A(\mathscr{C}(f), \mathscr{C}(g)) \quad &\text{(see (29))} \\
&\geq \sum_{A \in \mathscr{B}(\mathcal{M})} \mathrm{mult}_A(\mathscr{C}(f), \mathscr{C}(g)) + 2 \quad &(B_1, B_2 \notin \mathscr{B}(\mathcal{M})) \\
&\geq \sum_{A \in \mathscr{B}(\mathcal{M})} \mathrm{mult}(A, \mathscr{C}(f)) \, \mathrm{mult}(A, \mathscr{C}(g)) + 2 \\
&\geq \sum_{A \in \mathscr{B}(\mathcal{M})} \mathrm{mult}(A, \mathscr{B}(\mathcal{M})) + 2 \quad &(f, g \in \mathscr{L}(\mathcal{M}))) \\
&= \mathrm{mult}(\mathscr{B}(\mathcal{M})) + 2 \quad &\text{(see Definition 4)} \\
&= \deg(\mathcal{M})^2 + 1 \quad &\text{(see Corollary 7 in [12]).}
\end{aligned}$$

which is a contradiction. □

Theorem 5. *If $\mathcal{M}$ is a transversal birational map of $\mathbb{P}^2(\mathbb{K})$, then $\mathcal{L}(\mathcal{M}) = \mathscr{L}(\mathcal{M})$.*

Proof. By Lemma 20, $\mathcal{L}(\mathcal{M}) \subset \mathscr{L}(\mathcal{M})$. Thus, using Lemmas 17 and 21, we get the result. □

5. Algorithm and Examples

In this section, we use the previous results to derive an algorithm for determining polynomial parametrizations of rational surface, under the conditions stated in Section 4.1. For this purpose we first introduce an auxiliary algorithm for testing the transversality of parametrizations. In addition, we observe that we require to the input parametrization to be proper (i.e., birational). This can be checked for instance using the algorithms in [29].

Observe that Step 2 in Algorithm 1 provides a first direct filter to detect some non-transversal parametrizations, and Step 5 applies the characterization in Lemma 10. This justifies the next theorem.

Algorithm 1 Transversality of a Parametrization

Require: A rational proper projective parametrization $\mathcal{P}(\bar{t})$ of an algebraic surface $\mathcal{S}$.

1: Compute $\mathcal{B}(\mathcal{P}) = \bigcap_{i=1}^{4} \mathscr{C}(p_i)$ and $\text{mult}(A, \mathcal{B}(\mathcal{P})) = \text{mult}_A(\mathscr{C}(W_1), \mathscr{C}(W_2))$ for every $A \in \mathcal{B}(\mathcal{P})$.

2: **if** $\exists A \in \mathcal{B}(\mathcal{P})$, such that $\text{mult}(A, \mathcal{B}(\mathcal{P})) \neq m_A^2$ for some $m_A \in \mathbb{N}$, $m_A \geq 1$ **then**

3: return "$\mathcal{P}$ is not transversal".

4: **end if**

5: **if** $\forall A \in \mathcal{B}(\mathcal{P})$, $\gcd(T_1, T_2, T_3, T_4) = 1$, where T_i is the product of the tangents, counted with multiplicities, to $\mathscr{C}(p_i)$ at A, **then**

6: return "$\mathcal{P}$ is transversal" **else return** "$\mathcal{P}$ is not transversal".

7: **end if**

Theorem 6. *Algorithm 1 is correct.*

The following algorithm is the central algorithm of the paper.

Theorem 7. *Algorithm 2 is correct.*

Proof. For the correctness of the first steps (1-4) we refer to the preamble in Section 4 where the almost polynomial parametrizations are treated. For the rest of the steps, we use the notation introduced in Section 4 and we assume the hypotheses there, namely, $\mathcal{Q}(\mathcal{S_P}) = \mathcal{P}$ and $\mathcal{B}(\mathcal{Q}) = \emptyset$. Since $\mathcal{P}$ is transversal, by Theorem 1, we have that $\mathcal{S_P}$ is transversal. Now, by Theorem 5, $\mathscr{L} = \mathcal{L}(\mathcal{S_P})$. In this situation, by Theorem 4, $\overline{S} = {}^L\mathcal{S_P}$ for some $L \in \mathscr{G}(\mathbb{P}^2(\mathbb{K}))$. Therefore, $\mathcal{P}(\overline{\mathcal{R}})$ has to be almost polynomial, and hence the last step generates a polynomial parametrization. If the fourth component of $\mathcal{Q}$, namely q_4 is not the power of a linear form, then $\mathcal{B}(\mathcal{Q}) \neq \emptyset$. □

Remark 7. *Let us comment some consequences and computational aspects involved in the execution of the previous algorithms.*

1. *We observe that if Algorithm 2 returns a parametrization, then it is polynomial and its base locus is empty.*
2. *In order to check the properness of $\mathcal{P}$, one may apply, for instance, the results in [29] and, for determining $\deg(\mathcal{S})$, one may apply, for instance, the results in [24,25,30]. For the computation of $\overline{\mathcal{R}}$ one may apply well known elimination techniques as resultants or Gröbner basis; see e.g., [31].*
3. *In different steps of both algorithms one need to deal with the base points. Since the base locus is zero-dimensional, one may consider a decomposition of its elements in families of conjugate points, so that all further step can be performed exactly by introducing algebraic extensions of the ground field. For further details on how to deal with conjugate points we refer to Section 3.3 (Chapter 3) in [23]*

We finish this section illustrating the algorithm with some examples. The first two examples provide polynomial parametrizations, while in the third the algorithm detects that the input parametrization, although proper, is not transversal.

Algorithm 2 Birational Polynomial Reparametrization for Surfaces

Require: A rational proper projective parametrization $\mathcal{P}(\bar{t}) = (p_1(\bar{t}) : p_2(\bar{t}) : p_3(\bar{t}) : p_4(\bar{t}))$ of an algebraic surface $\mathscr{S}$, with $\gcd(p_1,\ldots,p_4) = 1$.

1: **if** $p_4(\bar{t})$ is of the form $(a_1 t_1 + a_2 t_2 + a_3 t_3)^{\deg(\mathcal{P})}$, **then**
2:   Consider the projective transformation $L = (t_i, t_j, a_1 t_1 + a_2 t_2 + a_3 t_3)$ where $i, j \in \{1, 2, 3\}$ are different indexes such if $\{k\} = \{1, 2, 3\} \setminus \{i, j\}$, then $a_k \neq 0$.
3:   Compute the inverse L^{-1} of L and **return** "$\mathcal{P}(L^{-1})$ is a rational proper polynomial parametrization of $\mathscr{S}$".
4: **end if**
5: Apply Algorithm 1 to check whether $\mathcal{P}$ is transversal. In the affirmative case go to the next Step. Otherwise **return** "Algorithm 2 is not applicable".
6: Compute $\deg(\mathscr{S})$ and $\deg(\mathcal{S}) = \deg(\mathcal{P})/\sqrt{\deg(\mathscr{S})}$.
7: Compute the $\deg(\mathcal{S})$-linear system associated to the effective divisor

$$\mathscr{L} := \sum_{A \in \mathscr{B}(\mathcal{P})} \sqrt{\mathrm{mult}(A, \mathscr{B}(\mathcal{P}))/\deg(\mathscr{S})} \cdot A$$

8: Determine $\overline{\mathcal{S}}(\bar{t}) = (\bar{s}_1(\bar{t}) : \bar{s}_2(\bar{t}) : \bar{s}_3(\bar{t}))$, where $\{\bar{s}_1, \bar{s}_2, \bar{s}_3\}$ is a basis of $\mathscr{L}$.
9: Compute $\overline{\mathcal{R}}(\bar{t}) = \overline{\mathcal{S}}^{-1}(\bar{t})$.
10: Compute $\mathcal{Q}(\bar{t}) = (q_1 : q_2 : q_3 : q_4)$, where $\mathcal{Q}(\bar{t}) = \mathcal{P}(\overline{\mathcal{R}}(\bar{t}))$.
11: **if** $q_4(\bar{t})$ is of the form $(a_1 t_1 + a_2 t_2 + a_3 t_3)^{\deg(\mathcal{Q})}$, **then**
12:   **return** "$\mathcal{Q}(L^{-1})$ (where L is as in Step 2) is a rational proper polynomial parametrization of $\mathscr{S}$" **else return** "$\mathscr{S}$ does not admit a polynomial proper parametrization with empty base locus".
13: **end if**

Example 1. Let $\mathcal{P}(\bar{t}) = (p_1(\bar{t}) : p_2(\bar{t}) : p_3(\bar{t}) : p_4(\bar{t}))$ be a rational parametrization of an algebraic surface $\mathscr{S}$, where

$$\begin{aligned}
p_1 &= -6t_3^4 t_1 t_2 + 6t_3^2 t_2^2 t_1^2 - t_3 t_2 t_1^4 - 2t_3 t_2^3 t_1^2 + 5t_3^3 t_1^2 t_2 + 3t_3^3 t_1 t_2^2 - t_2^6 + 3t_3^2 t_1^4 + 3t_3^2 t_2^4 - \\
&\quad t_3 t_2^5 - 3t_1^4 t_2^2 - 3t_1^2 t_2^4 + t_3^3 t_1^3 - 6t_3^4 t_1^2 + 3t_3^5 t_1 + 3t_3^3 t_2^3 - 6t_3^4 t_2^2 + 2t_3^5 t_2 - t_1^6. \\
p_2 &= -(t_1 - t_3) t_3 (t_2^2 + t_1^2 - t_1 t_3)(t_2^2 + t_1^2 - 2t_3^2 + t_1 t_3). \\
p_3 &= t_3^2 t_2^3 t_1 + t_3^2 t_1^3 t_2 - 3t_3^4 t_1 t_2 + 39t_3^2 t_2^2 t_1^2 - 8t_3^2 t_2^2 t_1^3 - 4t_3 t_1 t_2^4 - 4t_3 t_2 t_1^4 - 8t_3 t_2^3 t_1^2 + 8t_3^3 t_1^2 t_2 \\
&\quad +6t_3^3 t_1 t_2^2 + 6t_3^6 - 5t_2^6 - 4t_3 t_1^5 + 20t_3^2 t_1^4 + 19t_3^2 t_2^4 - 4t_3 t_2^5 - 15t_1^4 t_2^2 - 15t_1^2 t_2^4 + 8t_3^3 t_1^3 \\
&\quad -29t_3^4 t_1^2 + 4t_3^5 t_1 + 7t_3^3 t_2^3 - 22t_3^4 t_2^2 - 2t_3^5 t_2 - 5t_1^6. \\
p_4 &= (t_1^2 + t_2^2 - t_3^2)^3.
\end{aligned}$$

Applying the results in [29], one deduces that $\mathcal{P}$ is proper. We apply Algorithm 2 in order to compute a rational proper polynomial parametrization $\mathcal{Q}(\bar{t})$ of $\mathscr{S}$, without base points, if it exists. Clearly, $\mathcal{P}$ is not almost polynomial and hence steps 1–4 does not apply. In Step 5, we perform Algorithm 1. The base locus is (we denote by $\pm\imath$ the square roots of -1)

$$\mathscr{B}(\mathcal{P}) = \bigcap_{i=1}^{4} \mathscr{C}(p_i) = \{(1:0:1), (1:\imath:0), (1:-\imath:0)\}.$$

Moreover, it holds that

$$\mathrm{mult}(A, \mathscr{B}(\mathcal{P})) = \mathrm{mult}_A(\mathscr{C}(W_1), \mathscr{C}(W_2)) = 9, \ \forall A \in \mathscr{B}(\mathcal{P}.)$$

Therefore, for every $A \in \mathcal{B}(\mathcal{P})$ we have that $\text{mult}(A, \mathcal{B}(\mathcal{P})) = m_A^2$ for some $m_A \in \mathbb{N}$, $m_A \geq 1$. Thus, the necessary condition in Algorithm 1 is fulfilled. In addition, one may also check that the gcd of the tangents is 1, for each base point. As a consequence, we deduce that $\mathcal{P}$ is transversal.

In Step 6 of Algorithm 2, we get that $\deg(\mathscr{S}) = 9$ (see [30]). Now, using that

$$\deg(\mathcal{S}) = \deg(\mathcal{P})/\sqrt{\deg(\mathscr{S})} = 6/3 = 2,$$

and that

$$\text{mult}(A, \mathcal{B}(\mathcal{S})) = \text{mult}(A, \mathcal{B}(\mathcal{P}))/\deg(\mathscr{S}) = 9/9 = 1 \quad \text{for every } A \in \mathcal{B}(\mathcal{P}),$$

we compute the 2-linear system associated to the effective divisor

$$\sum_{A \in \mathcal{B}(\mathcal{P})} A.$$

For this purpose, one considers a generic polynomial of degree 2 with undetermined coefficients (note that we have 6 undetermined coefficients). We impose the three conditions, i.e., $\{(1:0:1), (1:\pm\imath:0)\}$ should be simple points, and we get

$$\mathscr{L} := \lambda_1(-9t_1^2 - 9t_2^2 + 9t_1t_3 + t_2t_3) + \lambda_2(-10t_1^2 - 10t_2^2 + 9t_1t_3 + t_3^2) + \lambda_3(t_1^2 + t_2^2 - t_3^2).$$

Let

$$\overline{\mathcal{S}}(\bar{t}) = (\bar{s}_1 : \bar{s}_2 : \bar{s}_3) = (-9t_1^2 - 9t_2^2 + 9t_1t_3 + t_2t_3 : -10t_1^2 - 10t_2^2 + 9t_1t_3 + t_3^2 : t_1^2 + t_2^2 - t_3^2),$$

where $\{\bar{s}_1, \bar{s}_2, \bar{s}_3\}$ is a basis of $\mathscr{L}$. Next, we compute

$$\overline{\mathcal{R}}(\bar{t}) = \overline{\mathcal{S}}^{-1}(\bar{t}) = (\bar{r}_1(\bar{t}) : \bar{r}_2(\bar{t}) : \bar{r}_3(\bar{t})) =$$

where

$$\begin{aligned}
\bar{r}_1 &= 81t_1^2 - 162t_1t_2 - 162t_1t_3 + 71t_3^2 + 151t_2t_3 + 80t_2^2, \\
\bar{r}_2 &= -9(2t_2 + 11t_3)(t_1 - t_2 - t_3), \\
\bar{r}_3 &= 181t_3^2 + 82t_2^2 + 81t_1^2 - 162t_1t_2 + 182t_2t_3 - 162t_1t_3.
\end{aligned}$$

In the last step, the algorithm returns

$$\mathcal{Q}(\bar{t}) = \mathcal{P}(\overline{\mathcal{R}}(\bar{t})) = (t_1^3 + t_2t_3^2 - t_1t_3^2 - t_3^3 : t_2(t_2 - t_3)(t_2 + t_3) : t_2^3 + t_1t_2^2 + t_3t_2t_1 - 4t_1t_3^2 - 5t_3^3 : t_3^3)$$

that is a rational proper polynomial parametrization of $\mathscr{S}$ with empty base locus. Note that the affine polynomial parametrization is given as

$$(t_1^3 + t_2 - t_1 - 1,\ t_2(t_2 - 1)(t_2 + 1),\ t_2^3 + t_1t_2^2 + t_2t_1 - 4t_1 - 5).$$

Observe that in this example we have introduced $\pm\imath$. Nevertheless we could have considered conjugate points. More precisely, the base locus decomposes as

$$\{(1:0:1)\} \cup \{(1:s:0) \mid s^2 + 1 = 0\}$$

Then, all remaining computations could have been carried out working in the field extension $\mathbb{Q}(\alpha)$ where $\alpha^2 + 1 = 0$.

Example 2. Let $\mathcal{P}(\bar{t}) = (p_1(\bar{t}) : p_2(\bar{t}) : p_3(\bar{t}) : p_4(\bar{t}))$ be a rational parametrization of an algebraic surface $\mathscr{S}$, where

$$\begin{aligned}
p_1 &= \tfrac{2891876933101}{7056} t_2^4 t_3^2 - \tfrac{94253497}{42} t_2^5 t_3 + \tfrac{79182089}{24} t_1^5 t_3 - \tfrac{15185833}{35} t_2^5 t_1 + \tfrac{230745016769}{19600} t_2^4 t_1^2 \\
&\quad - \tfrac{789948757}{280} t_1^4 t_2^2 - \tfrac{314171}{4} t_1^5 t_2 - \tfrac{3324893202046}{2205} t_2^3 t_3^2 t_1 + \tfrac{17297334852139}{29400} t_2^2 t_3^2 t_1^2 \\
&\quad + \tfrac{835536822991}{5880} t_2^4 t_3 t_1 - \tfrac{3567593339657}{14700} t_2^3 t_3 t_1^2 + \tfrac{8869391921}{420} t_1^4 t_2 t_3 + \tfrac{199437407}{140} t_1^3 t_2^3 \\
&\quad + \tfrac{56021820649}{144} t_1^4 t_3^2 + \tfrac{925548000997}{630} t_1^3 t_2 t_3^2 \\
&\quad - \tfrac{35094007283}{210} t_1^3 t_2^2 t_3 + \tfrac{28561}{4} t_2^6 + \tfrac{3455881}{16} t_1^6,
\end{aligned}$$

$$\begin{aligned}
p_2 &= -\tfrac{1097019300247}{2352} t_2^4 t_3^2 - \tfrac{246980149}{56} t_2^5 t_3 - 2485483 t_1^5 t_3 - \tfrac{32737835}{56} t_2^5 t_1 - \tfrac{35410335273}{3920} t_2^4 t_1^2 \\
&\quad + \tfrac{321945}{4} t_1^4 t_2^2 + \tfrac{314171}{4} t_1^5 t_2 + \tfrac{287134716635}{168} t_2^3 t_3^2 t_1 - \tfrac{52659146973}{80} t_2^2 t_3^2 t_1^2 - \tfrac{35536353385}{294} t_2^4 t_3 t_1 \\
&\quad + \tfrac{60928171523}{280} t_2^3 t_3 t_1^2 + \tfrac{52899535}{3} t_1^4 t_2 t_3 - \tfrac{446331197}{140} t_1^3 t_2^3 - \tfrac{5296771655}{12} t_1^4 t_3^2 - \tfrac{49879553251}{30} t_1^3 t_2 t_3^2 \\
&\quad + \tfrac{23802911463}{140} t_1^3 t_2^2 t_3 - \tfrac{257049}{16} t_2^6,
\end{aligned}$$

$$\begin{aligned}
p_3 &= -\tfrac{2676488123101}{7056} t_2^4 t_3^2 + \tfrac{94253497}{42} t_2^5 t_3 - \tfrac{379182089}{24} t_1^5 t_3 + \tfrac{15185833}{35} t_2^5 t_1 - \tfrac{219513256369}{19600} t_2^4 t_1^2 \\
&\quad + \tfrac{797945837}{280} t_1^4 t_2^2 + \tfrac{314171}{4} t_1^5 t_2 + \tfrac{3079803152296}{2205} t_2^3 t_3^2 t_1 - \tfrac{16059945270739}{29400} t_2^2 t_3^2 t_1^2 \\
&\quad - \tfrac{786351504991}{5880} t_2^4 t_3 t_1 + \tfrac{3371173762457}{14700} t_2^3 t_3 t_1^2 - \tfrac{9628594001}{420} t_1^4 t_2 t_3 - \tfrac{163616167}{140} t_1^3 t_2^3 \\
&\quad - \tfrac{51903261673}{144} t_1^4 t_3^2 - \tfrac{857765630677}{630} t_1^3 t_2 t_3^2 + \tfrac{32679676343}{210} t_1^3 t_2^2 t_3 - \tfrac{28561}{4} t_2^6 - \tfrac{3455881}{16} t_1^6,
\end{aligned}$$

$$p_4 = (-5348 t_1^2 t_3 + 5525 t_2^2 t_3 + 169 t_1^2 t_2 + 757 t_1 t_2^2 - 10059 t_1 t_2 t_3)^2.$$

Applying the results in [29], one deduces that $\mathcal{P}$ *is proper. We apply Algorithm 2. Clearly,* $\mathcal{P}$ *is not almost polynomial and hence steps 1–4 does not apply. In Step 5, we perform Algorithm 1. The base locus is*

$$\mathscr{B}(\mathcal{P}) = \{(0:0:1), (1:2:1), (5:7:1), (1/3:-1/7:1), (-13:7:1)\}.$$

Moreover, it holds that

$$\mathrm{mult}(A, \mathscr{B}(\mathcal{P})) = 4$$

for every $A \in \mathscr{B}(\mathcal{P})$ *except for* $A = (0:0:1)$ *that satisfies that* $\mathrm{mult}(A, \mathscr{B}(\mathcal{P})) = 16$. *Thus, the necessary condition in Algorithm 1 is fulfilled. In addition, one may also check that the gcd of the tangents is 1, for each base point. As a consequence, we deduce that* $\mathcal{P}$ *is transversal. Now, using that*

$$\deg(\mathcal{S}) = \deg(\mathcal{P}) / \sqrt{\deg(\mathscr{S})} = 6/2 = 3,$$

and that

$$\mathrm{mult}(A, \mathscr{B}(\mathcal{S})) = \mathrm{mult}(A, \mathscr{B}(\mathcal{P})) / \deg(\mathscr{S}) = 1,$$

for every $A \in \mathscr{B}(\mathcal{P})$ *except for* $A = (0:0:1)$ *that satisfies that* $\mathrm{mult}(A, \mathscr{B}(\mathcal{S})) = 4$, *we compute the 3-linear system associated to the effective divisor*

$$4(0:0:1) + (1:2:1) + (5:7:1) + (1/3:-1/7:1) + (-13:7:1).$$

We get that $\mathscr{L} = \lambda_1 \bar{s}_1 + \lambda_2 \bar{s}_2 + \lambda_3 \bar{s}_3$ *where*

$$\begin{aligned}
\bar{s}_1 &= \tfrac{203971}{12} t_1^2 t_3 - \tfrac{1463501}{84} t_2^2 t_3 + \tfrac{3373732}{105} t_1 t_2 t_3 + \tfrac{1859}{4} t_1^3 + \tfrac{169}{2} t_2^3 - \tfrac{169}{2} t_1^2 t_2 - \tfrac{438913}{140} t_1 t_2^2, \\
\bar{s}_2 &= \tfrac{37443}{2} t_1^2 t_3 - \tfrac{538707}{28} t_2^2 t_3 + \tfrac{140997}{4} t_1 t_2 t_3 - \tfrac{507}{4} t_3^3 - 507 t_1^2 t_2 - \tfrac{71637}{28} t_1 t_2^2, \\
\bar{s}_2 &= \tfrac{26747}{2} t_1^2 t_3 - \tfrac{384007}{28} t_2^2 t_3 - 338 t_1^2 t_2 - \tfrac{50441}{28} t_1 t_2^2 + \tfrac{100761}{4} t_1 t_2 t_3 - \tfrac{507}{4} t_2^3.
\end{aligned}$$

So, we take, for instance, $\overline{\mathcal{S}}(\bar{t}) = (\bar{s}_1(\bar{t}) : \bar{s}_2(\bar{t}) : \bar{s}_3(\bar{t}))$ *and we compute* $\overline{\mathcal{R}}(\bar{t}) = \overline{\mathcal{S}}^{-1}(\bar{t}) = (\bar{r}_1(\bar{t}) : \bar{r}_2(\bar{t}) : \bar{r}_3(\bar{t}))$ *where*

$$\bar{r}_1 = \frac{1}{11}(-34331t_2 + 7140t_1 + 39091t_3)(-1240370879t_2^2 + 4693319730t_2t_3$$
$$- 957816090t_1t_2 - 4096303731t_3^2 + 26989200t_1^2 + 1272637170t_1t_3),$$

$$\bar{r}_2 = -\frac{7}{3}(-5349t_3 + 3821t_2)(-1240370879t_2^2 + 4693319730t_2t_3 - 957816090t_1\,t_2$$
$$- 4096303731t_3^2 + 26989200t_1^2 + 1272637170t_1t_3),$$

$$\bar{r}_3 = 9122349600t_1^2t_3 - 6081566400t_1^2t_2 + 5962839694227t_3^3 - 13840668860013t_2t_3^2$$
$$+ 10640657052993t_2^2t_3 - 2711599696487t_2^3 + 503701536030t_1t_2^2$$
$$- 1409880894660t_1t_2t_3 + 985048833510t_1t_3^2.$$

Finally, we obtain

$$\mathcal{Q}(\bar{t}) = \mathcal{P}(\overline{\mathcal{R}}(\bar{t})) = (t_1^2 + t_2^2 - t_2t_3 : -t_1t_2 - t_2^2 + t_1t_3 : -t_1^2 + t_3^2 - t_2t_3 : (t_2 - t_3)^2).$$

Since $q_4(\bar{t}) = (t_2 - t_3)^2$, the algorithm returns

$$\mathcal{Q}((t_1, t_2, t_2 - t_3)^{-1}) = (t_1^2 + t_2t_3 : -t_2^2 - t_1t_3 : -t_1^2 + t_3^2 - t_2t_3 : t_3^2)$$

that is a rational proper polynomial parametrization of $\mathscr{S}$ with empty base locus. Note that the affine polynomial parametrization is given as

$$(t_1^2 + t_2, -t_2^2 - t_1, -t_1^2 + 1 - t_2).$$

Example 3. Let $\mathcal{P}(\bar{t}) = (p_1(\bar{t}) : p_2(\bar{t}) : p_3(\bar{t}) : p_4(\bar{t}))$ be a rational parametrization of an algebraic surface $\mathscr{S}$, where

$$p_1 = (-14065142t_1^3t_3 + 29410550t_2^3t_3 - 29410550t_2t_1^2t_3 + 14065142t_2^2t_1t_3 + 27633480t_1^4$$
$$- 46976541t_1t_2^3 + 64760061t_1^3t_2)^2,$$

$$p_2 = 15452942581758441/7\, t_2t_1^6t_3 - 317479084729363299/49\, t_2^6t_1t_3$$
$$- 68267697305871459/7\, t_2^5t_1^2t_3 - 18666824719928010/7\, t_2^5t_1t_3^2$$
$$+ 212684946864036627/49t_2^4t_1^2t_3^2 + 37333649439856020/7t_2^3t_1^2t_3^2$$
$$- 2927680573060371t_2^2t_1^5t_3 - 18666824719928010/7t_2t_1^5t_3^2$$
$$+ 10954535298967494/7t_2^3t_1^5 + 1789545850442280t_2^2t_1^6 - 1587369926524977t_2t_1^7$$
$$- 700212410256675t_1^6t_3^2 + 1255537783884564t_1^7t_3 + 69932525820304176/7t_2^7\, t_1$$
$$- 202783295759585328/49t_2^6t_1^2 - 3042203660729001t_2^5t_1^3 + 48537853394156778/7t_2^7\, t_3$$
$$- 123497677483306851/49t_2^6t_3^2 + 399414081398977842/49t_2^4t_1^3\, t_3$$
$$- 4193865500723721t_2^8 - 987075578994849t_1^8 - 217339297920270t_2^4t_1^4$$
$$- 54876861278152701/49t_2^2t_1^4t_3^2 + 4276901329956240/7t_2^3t_1^4t_3,$$

$$p_3 = 3/7(24511557t_1^4 - 64760061t_1^2t_2^2 + 11755445t_2t_1^2t_3 + 38554704t_1t_2^3 - 1125425t_2^2t_1\, t_3$$
$$- 11755445t_2^3t_3 + 1125425t_1^3t_3)(-151106809t_2^4 + 97487778t_2^3t_3 + 269939512t_1t_2^3$$
$$+ 59811570t_2^2t_1t_3 - 151106809t_1^2t_2^2 - 97487778t_2t_1^2t_3 + 98258706t_1^4 - 59811570t_1^3t_3)$$

$$p_4 = (24511557t_1^4 - 64760061t_1^2t_2^2 + 11755445t_2t_1^2t_3 + 38554704t_1t_2^3 - 1125425t_2^2t_1\, t_3$$
$$- 11755445t_2^3t_3 + 1125425t_1^3t_3)^2.$$

Applying the results in [29], one gets that $\mathcal{P}$ is proper. However, when applying Algorithm 1, we get that

$$\mathcal{B}(\mathcal{P}) = \{(0:0:1),(1:2:1),(5:7:1),(1/3:-1/7:1),(-13:7:1)\}$$

and that $\mathrm{mult}(A, \mathcal{B}(\mathcal{P})) = 4$ for every $A \in \mathcal{B}(\mathcal{P})$ except for $A = (0:0:1)$ where $\mathrm{mult}(A, \mathcal{B}(\mathcal{P})) = 44$. Since $\mathrm{mult}(A, \mathcal{B}(\mathcal{P})) = 44$, which is not the square of a natural number, the algorithm returns that $\mathcal{P}$ is not transversal. Thus, we can not apply Algorithm 2.

6. Conclusions

Some crucial difficulties in many applications, and algorithmic questions, dealing with surface parametrizations are, on one hand, the presence of base points and, on the other, the existence of non-constant denominators of the parametrizations. In this paper, we have seen how to provide a polynomial parametrization with empty base locus, and hence an algorithm to avoid the two complications mentioned above, if it is possible. For this purpose, we have had to introduce, and indeed impose, the notion of transversal base locus. This notion directly affects to the transversality of the tangents at the base points of the algebraic plane curves V_i or W_i (see (3) and (7)). This, somehow, implies that in general one may expect transversality in the input. In any case, we do deal here with the non-transversal case and we leave it as an open problem. We think that using the ideas, pointed out by J. Schicho in [32], on blowing up the base locus, one might transform the given problem (via a finite sequence of Cremone transformations and projective transformations) into the case of transversality.

Author Contributions: The authors contributed equally to this work and they worked together through the whole paper. Both authors have read and agreed to the published version of the manuscript.

Funding: This work has been partially supported by FEDER/Ministerio de Ciencia, Innovación y Universidades-Agencia Estatal de Investigación/MTM2017-88796-P (Symbolic Computation: new challenges in Algebra and Geometry together with its applications).

Acknowledgments: Authors belong to the Research Group ASYNACS (Ref. CT-CE2019/683).

Conflicts of Interest: The authors declare no conflict of interest.

References

1. Hoschek, J.; Lasser, D. *Fundamentals of Computer Aided Geometric Design*; A.K. Peters, Ltd.: Natick, MA, USA, 1993.
2. Sendra, J.R.; Sevilla, D. First Steps Towards Radical Parametrization of Algebraic Surfaces. *Comput. Aided Geom. Des.* **2013**, *30*, 374–388. [CrossRef]
3. Schicho, J. Rational Parametrization of Surfaces. *J. Symb. Comput.* **1998**, *26*, 1–9. [CrossRef]
4. Sendra, J.R.; Sevilla, D.; Villarino, C. Algebraic and algorithmic aspects of radical parametrizations. *Comput. Aided Geom. Des.* **2017**, *55*, 1–14. [CrossRef]
5. Bizzarri, M.; Lávička, M.; Vršek, J. Hermite interpolation by piecewise polynomial surfaces with polynomial area element. *Comput. Aided Geom. Des.* **2017**, *51*, 30–47. [CrossRef]
6. Šír, Z.; Gravesen, J.; Juttler, B. Curves and surfaces represented by polynomial support functions. *Theor. Comput. Sci.* **2008**, *392*, 141–157. [CrossRef]
7. Sendra, J.R. Normal Parametrization of Algebraic Plane Curves. *J. Symb. Comput.* **2002**, *33*, 863–885. [CrossRef]
8. Pérez-Díaz, S.; Sendra, J.R.; Villarino, C. A First Approach Towards Normal Parametrizations of Algebraic Surfaces. *Int. J. Algebra Comput.* **2010**, *20*, 977–990. [CrossRef]
9. Sendra, J.R.; Sevilla, D.; Villarino, C. Some results on the surjectivity of surface parametrizations. In *Lecture Notes in Computer Science 8942*; Schicho, J., Weimann, M., Gutierrez, J., Eds.; Springer International Publishing: Cham, Switzerland, 2015; pp. 192–203.
10. Grasegger, G. Radical solutions of first order autonomous algebraic ordinary differential equations. In Proceedings of the 39th International Symposium on Symbolic and Algebraic Computation, Kobe, Japan, 23–25 July 2014; Nabeshima, K., Ed.; ACM Press: New York, NY, USA, 2014; pp. 217–223.
11. Ngô, L.X.C.; Winkler, F. Rational general solutions of first order non-autonomous parametrizable ODEs. *J. Symb. Comput.* **2010**, *45*, 1426–1441.

12. Cox, D.A.; Pérez-Díaz, S.; Sendra, J.R. On the base point locus of surface parametrizations: Formulas and consequences. *arXiv* **2020**, arXiv:2008.08009.
13. Pérez-Díaz, S.; Sendra, J.R. Behavior of the Fiber and the Base Points of Parametrizations under Projections. *Math. Comput. Sci.* **2013**, *7*, 167–184. [CrossRef]
14. Busé, L.; Cox, D.; D'Andrea, C. Implicitization of surfaces in $\mathbb{P}^3$ in the presence of base points. *J. Algebra Appl.* **2003**, *2*, 189–214 [CrossRef]
15. Cox, D.; Goldman, R.; Zhang, M. On the validity of implicitization by moving quadrics for rational surfaces with no base points. *J. Symb. Comput.* **2000**, *29*, 419–440. [CrossRef]
16. Sendra, J.R.; Sevilla, D.; Villarino, C. Covering of surfaces parametrized without projective base points. In Proceedings of the 39th International Symposium on Symbolic and Algebraic Computation, Kobe, Japan, 23–25 July 2014; ACM Press: New York, NY, USA, 2014; pp. 375–380.
17. Shen, L.Y.; Goldman, R. Strong μ-Bases for Rational Tensor Product Surfaces and Extraneous Factors Associated to Bad Base Points and Anomalies at Infinity. *J. Appl. Algebra Geom.* **2017**, *1*, 328–351. [CrossRef]
18. Arrondo, E.; Sendra, J.; Sendra, J.R. Parametric generalized offsets to hypersurfaces. *J. Symb. Comput.* **1997**, *23*, 267–285. [CrossRef]
19. Sendra, J.R.; Peternell, M.; Sendra, J. Cissoid Constructions of Augmented Rational Ruled Surfaces. *Comput. Aided Geom. Des.* **2018**, *60*, 1–9. [CrossRef]
20. Sendra, J.; Sendra, J.R. An algebraic analysis of conchoids to algebraic curves. *Appl. Algebra Eng. Commun. Comput* **2008**, *19*, 413–428. [CrossRef]
21. Vršek, J. Lávička M. On convolutions of algebraic curves. *J. Symb. Comput.* **2010**, *45*, 657–676. [CrossRef]
22. Harris, J. *Algebraic Geometry. A First Course*; Springer Science and Business Media: New York, NY, USA, 1995.
23. Sendra, J.R.; Winkler, F.; Pérez-Díaz, S. Rational Algebraic Curves: A Computer Algebra Approach. In *Algorithms and Computation in Mathematics*; Springer: Berlin/Heilderbarg, Germany, 2007; Volume 22.
24. Pérez-Díaz, S.; Sendra, J.R. A Univariate Resultant Based Implicitization Algorithm for Surfaces. *J. Symb. Comput.* **2008**, *43*, 118–139. [CrossRef]
25. Pérez-Díaz, S.; Sendra, J.R. Partial Degree Formulae for Rational Algebraic Surfaces. In Proceedings of the International Symposium on Symbolic and Algebraic Computation (ISSAC), Beijing, China, 24–27 July 2005; ACM Press: New York, NY, USA, 2005; pp. 301–308.
26. Pérez-Díaz, S.; Sendra, J.R.; Schicho, J. Properness and Inversion of Rational Parametrizations of Surfaces. *Appl. Algebra Eng. Commun. Comput.* **2002**, *13*, 29–51. [CrossRef]
27. Miranda, R. Linear Systems of Plane Curves. *Notices AMS 46* **1999**, *2*, 192–201.
28. Walker, R.J. *Algebraic Curves*; Princeton University Press: Princeton, NJ, USA, 1950.
29. Pérez-Díaz, S.; Sendra, J.R. Computation of the Degree of Rational Surface Parametrizations. *J. Pure Appl. Algebra* **2004**, *193*, 99–121. [CrossRef]
30. Pérez–Díaz, S.; Sendra, J.R.; Villarino, C. Computing the Singularities of Rational Surfaces. *Math. Comput.* **2015**, *84*, 1991–2021. [CrossRef]
31. Schicho, J. Inversion of birational maps with Gröbner bases. In *Gröbner Bases and Applications*; Buchberger, B., Winkler, F., Eds.; London Mathematical Society Lecture Note Series 251; Cambridge University Press: Cambridge, UK, 1998; pp. 495–503.
32. Schicho, J. Simplification of Surface Parametrizations. In Proceedings of the International Symposium on Symbolic and Algebraic Computation, Lille, France, 7–10 July 2002; ACM Press: New York, NY, USA, 2002; pp. 229–237.

Publisher's Note: MDPI stays neutral with regard to jurisdictional claims in published maps and institutional affiliations.

© 2020 by the authors. Licensee MDPI, Basel, Switzerland. This article is an open access article distributed under the terms and conditions of the Creative Commons Attribution (CC BY) license (http://creativecommons.org/licenses/by/4.0/).

Article

Covering Rational Surfaces with Rational Parametrization Images

Jorge Caravantes [1,*], J. Rafael Sendra [1], David Sevilla [2] and Carlos Villarino [1]

1   Department of Physics and Mathematics, The University of Alcalá, 28801 Alcalá de Henares, Madrid, Spain; rafael.sendra@uah.es (J.R.S.); carlos.villarino@uah.es (C.V.)
2   Department of Mathematics, The University of Extremadura, 06800 Mérida, Badajoz, Spain; sevillad@unex.es
*   Correspondence: jorge.caravantes@uah.es

Abstract: Let S be a rational projective surface given by means of a projective rational parametrization whose base locus satisfies a mild assumption. In this paper we present an algorithm that provides three rational maps $f, g, h : \mathbb{A}^2 \dashrightarrow S \subset \mathbb{P}^n$ such that the union of the three images covers S. As a consequence, we present a second algorithm that generates two rational maps $\bar{f}, \bar{g} : \mathbb{A}^2 \dashrightarrow S$, such that the union of its images covers the affine surface $S \cap \mathbb{A}^n$. In the affine case, the number of rational maps involved in the cover is in general optimal.

Keywords: rational surface; birational parametrization; surjective parametrization; surface cover; base points

Citation: Caravantes, J.; Sendra, J.R.; Sevilla, D.; Villarino, C. Covering Rational Surfaces with Rational Parametrization Images. *Mathematics* **2021**, *9*, 338. https://doi.org/10.3390/math9040338

Academic Editor: Gabriel Eduard Vîlcu

Received: 12 January 2021
Accepted: 2 February 2021
Published: 8 February 2021

Publisher's Note: MDPI stays neutral with regard to jurisdictional claims in published maps and institutional affiliations.

Copyright: © 2021 by the authors. Licensee MDPI, Basel, Switzerland. This article is an open access article distributed under the terms and conditions of the Creative Commons Attribution (CC BY) license (https://creativecommons.org/licenses/by/4.0/).

1. Introduction

Rational parametrizations of algebraic varieties are an important tool in many geometric applications like those in computer aided design (see, e.g., [1,2]) or computer vision (see, e.g., [3,4]). Nevertheless, the applicability of this tool can be negatively affected if the parametrization is missing basic properties: for instance its injectivity, its surjectivity, or the nature of the ground field where the coefficients belong to; see, e.g., the introductions of the papers [5–7] for some illustrating examples of this phenomenon.

Let us mention here some illustrating situations of this phenomenon; for more detailed examples, we refer the reader, e.g., to the introductions of the papers [5–7]. Let us say that we are given a rational parametrization $\mathcal{P}(t)$ of a curve that describes the possible positions that a given robot may achieve. Now, for a position P in the affine space, we want to detect the value t_0 of the parameter t such that $\mathcal{P}(t_0) = P$. If the parametrization $\mathcal{P}$ is not injective, the answer may contain unnecessary information, for instance several parameter values for the same goal, and hence the computation time may not be optimal. On the other hand, if the parametrization is not surjective the situation can be worse because the point P may be on the trajectory curve, but not reachable via the given parametrization for any parameter value. In addition, if the parametrization has complex non-real coefficients, it would be unnecessarily complicated to detect the real point positions of the robot.

These types of problems have been addressed by different authors from the theoretical and computational points of view. For the optimality of the ground field we refer to [8–10] where the notions of hypercircle and ultraquadric were introduced to approach the problem. In the case of the injectivity, the answer is based on Lüroth's theorem for the case of curves (see, e.g., [11]) and on Castelnuovo's theorem for the case of surfaces (see, e.g., [12]). Computationally, the injectivity problem has also been studied and there are different approaches; for the case of curves, we refer to [13–15], and for the case of surfaces, we refer to [14,16,17]. It is worth mentioning that, up to our knowledge, the determination of injective surface re-parametrizations is still an open problem.

In this paper we focus on the third problem, namely, the surjectivity of rational parametrizations. Surjective parametrizations, also called normal parametrizations, have been studied extensively but many important questions, both theoretical and computational, still remain open. The case of curves, over algebraically closed fields of characteristic zero, is comprehensively understood, and one can always find a surjective, indeed also injective, affine parametrization of the affine algebraic curve (see, e.g., [11,18–20]). Furthermore, the case of curves defined over the field of the real numbers is also studied and characterized in [20].

The situation, as usually happens, turns to be much more complicated when dealing with surfaces. In [21,22] the first steps in this direction were given and answers for certain types of surfaces, like quadrics, were provided. Also, in [23] the relation of the polynomiality of parametrizations to the surjectivity was analyzed. Nevertheless, in [24] it is shown that there exist rational surfaces that cannot be parametrized birationally and surjectively. As a consequence of this fact, the question of whether every rational surface can be covered by the union of the images of finitely many birational parametrizations is of interest. The answer is positive and can be deduced from the results in [25]. In previous papers, the second, the third and the fourth authors have studied this problem for special types of surfaces. In [5] the unreachable points of parametrizations of surfaces of revolution are characterized. In [7] it is proved that ruled surfaces can be covered by using two rational parametrizations. In addition, in [6] an algorithm to cover an affine rational surface without based points at infinity with at most three parametrizations is presented.

In this paper we continue the research described above and we present two main algorithmic and theoretical results. Moreover, we provide an algorithm that, for any projective surface parametrization, generates a cover of the projective surface with three parametrizations, assuming that, either the base locus of the input is empty, or the Jacobian of the input parametrization, specialized at each base point, has rank two. As a consequence of this result, we also present an algorithm that, for a given affine parametrization whose projectivization satisfies the condition on the base points mentioned above, returns a cover of the affine surface with two affine parametrizations.

Taking into account the results in [24], the affine cover presented in this paper is, in general, optimal. Furthermore, it improves the results in [6] and extends the results in [7] to a much more general class of surfaces. With respect to the projective cover case, although theoretically interesting, we cannot ensure that the number of parametrizations involved in the cover is optimal for a generically large class of projective surfaces since, for instance, the whole projective plane can be covered with just two maps from the affine plane as the following example shows. We leave this theoretical question open as future reseach.

Example 1. *Consider the following two maps:*

$$f : \mathbb{A}^2 \to \mathbb{P}^2 \qquad g : \mathbb{A}^2 \to \mathbb{P}^2$$
$$(s,t) \mapsto (1:s:t) \ ' \qquad (s,t) \mapsto (s(1-s):t:(1-s)^2)$$

Then for any point $P := (x : y : z) \in \mathbb{P}^2$, *if* $x \neq 0$ *then* $P := f(\frac{y}{x}, \frac{z}{x})$; *if* $x = 0 \neq z$ *then* $P := g(0, \frac{y}{z})$; *and if* $x = z = 0$ *then* $P := g(1,y)$, $y \neq 0$. *This means that we can cover the whole projective plane with just two maps from the affine plane.*

The paper is structured as follows. In Section 2, we introduce some notation and we briefly recall some notions and results that will be used throughout the paper. In Section 3, we present the projective cover algorithm and in Section 4 we illustrate the result by means of some examples. In Section 5, we apply the results in Section 3 to derive the two affine parametrization cover algorithm.

2. Preliminaries

In this section, we briefly recall some concepts and results that will be used in the subsequent sections. We essentially recall some results on the fundamental locus of rational maps and some consequences and the characterization of zero dimensional ideals via

Gröbner bases. Throughout this paper $\mathbb{K}$ is an algebraically closed field of characteristic zero, and $\mathbb{P}^n$ the projective space over $\mathbb{K}$. Moreover, we denote by $\mathbb{A}_i$ the affine space $\{(x_0 : \cdots : x_n) \in \mathbb{P}^n, | x_i \neq 0\}$. In the examples, the field $\mathbb{K}$ will be the field $\mathbb{C}$ of the complex numbers.

Let X be an irreducible projective variety and let $f : X \dashrightarrow \mathbb{P}^n$ be a rational map. The *fundamental locus* of f is the algebraic set $F(f)$ of points to which f cannot be extended regularly. Any $P \in F(f)$ is called a *fundamental point* of f. The following theorem analyzes the dimension of the fundamental locus.

Theorem 1. *(Lemma V.5.1 [12]) Let X be a smooth irreducible projective variety and let $f : X \dashrightarrow \mathbb{P}^n$ be a rational map generically finite. The fundamental locus of f has codimension at least 2 in X.*

Corollary 1. *Let X be a smooth irreducible surface and f as in Theorem 1. $F(f)$ is either empty or zero dimensional.*

The traditional way for solving indeterminacies in algebraic geometry consists in blowing up fundamental points (see, e.g., IV.3.3 [26]) and composing with the corresponding map as the next theorem shows.

Theorem 2. *(Example II.7.17.3 [12] or Theorem II.7 [27]) Let X be a smooth surface. Let $f : X \dashrightarrow \mathbb{P}^n$ be a rational map. Then there exists a commutative diagram*

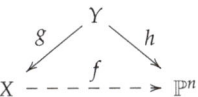

where g is a composite of blowups involving fundamental points of f and h is a morphism.

A first consequence of Theorem 2 is the following.

Corollary 2. *(Corollary 2.5 [24]) Let X and f be as in Theorem 2. For any fundamental point P of f, $h(g^{-1}(P))$ is a connected finite union of rational curves.*

Remark 1. *Let $f : \mathbb{P}^2 \dashrightarrow \mathbb{P}^n$ be as in Theorem 2, and let S be the Zariski closure of $f(\mathbb{P}^2)$ in $\mathbb{P}^n$. The complementary in S of the $f(\mathbb{A}^2)$ is, according to Theorem 2, contained in $h(g^{-1}(F(f) \cup L_\infty))$, where $L_\infty = \mathbb{P}^2 - \mathbb{A}^2$. Such a subset consists of some rational curves and, if f contracts L_∞, a closed point (see Corollary 2.5 [24]).*

We end this section with a well-known result on elimination theory that will be used in Section 3.

Theorem 3. *(Chapter 5, Theorem 6 [28]) Let I be an ideal in $\mathbb{K}[x_1, .., x_n]$. Then, the following statements are equivalent:*

1. *The algebraic subset of $\mathbb{K}^n$ defined by I is a finite set.*
2. *Let B be a Gröbner basis for I with respect to a fixed monomial ordering. Then, for each $1 \leq i \leq n$, there is some $m_i \in \mathbb{N}$ such that $x_i^{m_i}$ is the leading monomial of an element of B.*

3. Covering Projective Surfaces with Three Parametrizations

Throughout this section, let $S \subset \mathbb{P}^n$ be a rational projective surface and let

$$F = (F_0 : \cdots : F_n) : \mathbb{P}^2 \dashrightarrow S \subset \mathbb{P}^n$$

be a (not necessarily birational) parametrization of S, given by $n+1$ homogeneous coprime polynomials $F_0, ..., F_n$ where the nonzero polynomials have degree d. In addition, let the homogeneous ideal $I = (F_0, ..., F_n)\mathbb{K}[x_0, x_1, x_2]$ be called the fundamental ideal associated to F.

Since the polynomials defining F are coprime, by Corollary 1, I defines a closed algebraic subset $\mathcal{A}$ of $\mathbb{P}^2$ that is either empty or consists of a finite amount of points. If $\mathcal{A} = \emptyset$, then F defines a regular map and its restrictions to each of the three affine planes $\mathbb{A}_i = \{(x_0 : x_1 : x_2) \mid x_i \neq 0\}$ covering $\mathbb{P}^2$ define 3 charts that cover S, since the image of a projective variety by a regular map is always Zariski closed. Otherwise, say $\mathcal{A} = \{P_1, ..., P_k\}$; we need to make the following assumption:

(∗) If $\mathcal{A} \neq \emptyset$, then for every $P \in \mathcal{A} := \{P_1, ..., P_k\}$ the Jacobian matrix of F at P has rank 2.

Note that (∗) guarantees that I does not define multiple points (i.e. the base points of F are simple). We also assume without loss of generality that:

(a) no P_j is in any of the lines $\{x_0 = 0\}$, $\{x_1 = 0\}$ and $\{x_2 = 0\}$,
(b) no pair $\{P_i, P_j\}$, $i \neq j$, is aligned with any of the coordinate points $(1:0:0)$, $(0:1:0)$ and $(0:0:1)$.

We observe that the real constraint lays in (∗), since conditions (a) and (b) are satisfied after a general change of coordinates. In the following remark we discuss how these hypotheses can be computationally checked.

Remark 2.

1. Note that condition (∗) implies that the point P_j is regular in the projective scheme defined by the ideal I. Then, the intersection multiplicity is 1 at every P_j. Now, if we consider the ideal J defined by the 3×3 minors of the Jacobian matrix of F, the following methods, among others, can be applied to test (∗):

   (i) Check whether the ideal $I + J$ is zero-dimensional.
   (ii) Check whether $\sqrt{I+J} = (x_0, x_1, x_2)\mathbb{K}[x_0, x_1, x_2]$ (i.e., the irrelevant ideal) or, equivalently, whether $I + J$ contains a power of any of the variables.
   (iii) By means of resultants and gcds, using the formulas in (Theorems 2 and 3 [29]) (see also [30]).

2. Checking (i) without explicitly determining $P_1, ..., P_k$ can be carried out by certifying that all the ideals generated by $\{F_0, ..., F_n, x_i\}$ ($i = 0, 1, 2$) either are zero-dimensional or contain a power of the irrelevant ideal.

3. Checking (ii) can be done by computing the ideal bases B_1 and B_2 of Algorithm 3PatchSurface and checking whether they have adequate shape (see Proposition 1). However, it may be more efficient to check that, for all $i = 0, 1, 2$, the gcd of all resultants of couples (F_0, F_j) with respect to x_i is square free.

Now we consider the three affine planes $\mathbb{A}_i$ defined above. According to (a) all P_j lie in the intersection of the three affine planes. In this situation, the strategy is as follows. We will work with the parametrization

$$f := F|_{\mathbb{A}_0} : \mathbb{A}_0 \dashrightarrow S$$

as defined in $\mathbb{A}_0$, and we will blowup $\mathbb{A}_1$ and $\mathbb{A}_2$ at the base points of F to get new affine planes $\widetilde{\mathbb{A}_1}$ and $\widetilde{\mathbb{A}_2}$ with projections $\mathrm{Bl}_1 : \widetilde{\mathbb{A}_1} \to \mathbb{A}_1$ and $\mathrm{Bl}_2 : \widetilde{\mathbb{A}_2} \to \mathbb{A}_2$. Now, we introduce the compositions

$$g := F|_{\mathbb{A}_1} \circ \mathrm{Bl}_1 : \widetilde{\mathbb{A}_1} \to \mathbb{A}_1 \to S \text{ and } h := F|_{\mathbb{A}_2} \circ \mathrm{Bl}_2 : \widetilde{\mathbb{A}_2} \to \mathbb{A}_2 \to S,$$

and we prove that the union of the images of f, g and h is the whole S (see details in Proposition 1). During this process, we also need to keep track of what happens with the infinity line of $\mathbb{A}_0$, namely $L_\infty = \{(x_0 : x_1 : x_2) \mid x_0 = 0\}$. As a consequence, we derive the following Algorithm 1.

Algorithm 1: 3PatchSurface

Require: A map $F = (F_0 : \cdots : F_n)$ defined by coprime homogeneous polynomials in $\mathbb{K}[x_0, x_1, x_2]$, where the nonzero polynomials have the same degree, parametrizing a Zariski dense subset of a projective surface $S \subset \mathbb{P}^n$, such that conditions $(*)$, (a) and (b) are satisfied.

Ensure: Two maps $G = (G_0 : \cdots : G_n)$ and $H = (H_0 : \cdots : H_n)$ defined by homogeneous polynomials in $\mathbb{K}[x_0, x_1, x_2]$ where the nonzero polynomials have the same degree (while in the same list), such that the union of the images of $F(1 : \_ : \_), G(\_ : 1 : \_), H(\_ : \_ : 1) : \mathbb{A}^2 \to S$ cover S.

1: **if** the radical of the homogeneous ideal $(F_0, ..., F_n)\mathbb{K}[x_0, x_1, x_2]$ is irrelevant (i.e., F defines a regular morphism) **then**
2:     Return $G = F, H = F$.
3: **end if**
4: For the ideals $I_i = (F_0, ..., F_n, x_i - 1)$ for $i = 1, 2$, compute reduced Gröbner bases $B_1 = \{x_2 - q_1(x_0), x_1 - 1, p_1(x_0)\}$ for I_1 and $B_2 = \{x_2 - 1, x_1 - q_2(x_0), p_2(x_0)\}$ for I_2, with lexicographical order $x_0 < x_1 < x_2$.
5: Set $k = \deg(p_i) > \deg(q_i)$ for whatever $i = 1, 2$.
6: Set $\mathbf{P}_1(x_0 : x_1) = p_1(\frac{x_0}{x_1})x_1^k$ and $\mathbf{Q}_1(x_0 : x_1) = q_1(\frac{x_0}{x_1})x_1^{k-1}$.
7: Put $\widetilde{G} = (\widetilde{G}_0 : \cdots : \widetilde{G}_n) = F\left(x_0 x_1^k : x_1^{k+1} : x_1^2 \mathbf{Q}_1(x_0 : x_1) + x_2 \mathbf{P}_1(x_0 : x_1)\right)$,
8: Set $\mathbf{P}_2(x_0 : x_2) = p_2(\frac{x_0}{x_2})x_2^k$ and $\mathbf{Q}_2(x_0 : x_2) = q_2(\frac{x_0}{x_2})x_2^{k-1}$.
9: Set $\widetilde{H} = (\widetilde{H}_0 : \cdots : \widetilde{H}_n) = F\left(x_0 x_2^k : x_2^2 \mathbf{Q}_2(x_0 : x_2) + x_1 \mathbf{P}_2(x_0 : x_2) : x_2^{k+1}\right)$
10: Set $\widehat{G} = \gcd(\widetilde{G}_0, ..., \widetilde{G}_n)$, $\widehat{H} = \gcd(\widetilde{H}_0, ..., \widetilde{H}_n)$.
11: Return $G = \widetilde{G}/\widehat{G}, H = \widetilde{H}/\widehat{H}$.

Remark 3. *In Proposition 1 we show that the integer k, introduced in the algorithm, is exactly the number of base points, that is the cardinality of $\mathcal{A}$ (see above), so we have not introduced equivocal notation.*

In the following, we see that the output of Algorithm 3PatchSurface is correct (see Theorem 4). We also recall that the required conditions for the algorithm can be checked computationally according to Remark 2. We start by proving that Step 4 works properly, assuming that conditions $(*)$, (a) and (b) are satisfied. This is probably a well-known result in a more general setting but, up to the authors' knowledge, there are no suitable references for the proof.

Proposition 1. *Let $F = (F_0 : \cdots : F_n)$, I_1 and I_2 be as in Algorithm 1. There exist $p_1, q_1 \in \mathbb{K}[x_2]$, $p_2, q_2 \in \mathbb{K}[x_1]$ such that $k = \deg(p_i) > \deg(q_i)$ for all $i = 1, 2$, and the reduced Gröbner basis B_1 and B_2 of I_1 and I_2 respectively, have the following shape:*

$$B_1 = \{x_2 - q_1(x_0), x_1 - 1, p_1(x_0)\} \quad \text{and} \quad B_2 = \{x_2 - 1, x_1 - q_2(x_0), p_2(x_0)\}.$$

Proof. Observe that both I_1 and I_2 define finite sets in $\mathbb{A}_1$ and $\mathbb{A}_2$, respectively. By Theorem 3, this implies that, since B_1 and B_2 are Gröbner bases, with respect to the lexicographical order $x_0 < x_1 < x_2$, there is a polynomial in each B_i just involving x_0. This is p_1 for I_1 and p_2 for I_2. Due to $(*)$, (a) and (b), each p_i defines k different parallel lines in $\mathbb{A}_i$, so its degree is k.

Applying again Theorem 3, there is another monic polynomial in $\mathbb{K}[x_0][x_1] \cap I_1$. Since the basis is reduced, and $x_1 - 1$ was originally among the generators of I_1, this polynomial in $\mathbb{K}[x_0][x_1]$ for I_1 is precisely $x_1 - 1$.

Now, let q_i be the interpolation polynomial whose graph goes through all the base points $P_j \in \mathbb{A}_i$ of F. It does exist because (b) holds (so there are no two different P_j in the same vertical line) and its degree is at most $k-1$. Then $x_1 - q_2(x_0)$ vanishes at every P_j so, by Hilbert's Nullstellensatz, it belongs to $\sqrt{I_2}$. Since $\deg(p_2) = k > \deg(q_2)$, $x_1 - q(x_0)$ cannot be reduced by $p(x_0)$, so, since B_1 is reduced, this is the monic polynomial in $\mathbb{K}[x_0][x_1] \cap I_2$.

We apply an analogous argument and reduction by $x_1 - 1$ to deduce that the monic polynomial in $\mathbb{K}[x_0, x_1][x_2]$ for I_1 must be $x_2 - q_2(x_0)$, and we know that $x_2 - 1$ is a reduced monic polynomial for I_2, so it is in B_2.

Since B_1 and B_2 are reduced Gröbner bases with respect to the lexicographical order $x_0 < x_1 < x_2$ and the ideals they generate define, precisely, the fundamental locus, they are the reduced Gröbner bases of I_1 and I_2. □

Before continuing, we state a Lemma.

Lemma 1. *In the conditions of Algorithm 1, neither $G(\_ : 1 : \_)$ nor $H(\_ : \_ : 1)$ have affine base points.*

Proof. By the properties of the rational map $F : \mathbb{P}^2 \dashrightarrow \mathbb{P}^n$, defined by F, and the fact that a blow up is bijective outside its blown up points, we know that any base point of $G(\_ : 1 : \_) = F \circ \mathrm{Bl}_1(\_ : 1 : \_)$ would be in one of the lines $\mathrm{Bl}_1(P_j)$. Now, we fix $P_j = (1 : \alpha_j : \beta_j)$ and we then prove that $G(\_ : 1 : \_)$ has no base points in the line $\{x_0 = \frac{1}{\alpha_j}\} \subset \mathbb{A}_1$.

Since all $F_i(x_0 : 1 : x_2)$ vanish at P_j, they have $\alpha_j x_0 - 1$ as a common factor (note that $\alpha_j \neq 0$ because (a) holds). Then, since $G_i = \widetilde{G}_i / \widehat{G}$, we have that $G_i(x_0 : 1 : x_2)$ is a divisor of

$$\overline{G}_i(x_0 : 1 : x_2) := \frac{\widetilde{G}_i(x_0 : 1 : x_2)}{x_0 - \frac{1}{\alpha_j}} = \frac{F_i(x_0 : 1 : q_1(x_0) + p_1(x_0)x_2)}{x_0 - \frac{1}{\alpha_j}} =$$

$$= \frac{F_i(x_0 : 1 : q_1(x_0) + p_1(x_0)x_2) - F_i(\frac{1}{\alpha_j} : 1 : q_1(\frac{1}{\alpha_j}) + p_1(\frac{1}{\alpha_j})\frac{\beta_j}{\alpha_j})}{x_0 - \frac{1}{\alpha_j}} =$$

$$= \frac{F_i(x_0 : 1 : q_1(x_0) + p_1(x_0)x_2) - F_i(\frac{1}{\alpha_j} : 1 : q_1(\frac{1}{\alpha_j}))}{x_0 - \frac{1}{\alpha_j}}.$$

This means that

$$\overline{G}_i\left(\frac{1}{\alpha_j} : 1 : x_2\right) =$$

$$= \frac{\partial F_i}{\partial x_0}\left(\frac{1}{\alpha_j} : 1 : q_1\left(\frac{1}{\alpha_j}\right)\right) + \frac{\partial F_i}{\partial x_2}\left(\frac{1}{\alpha_j} : 1 : q_1\left(\frac{1}{\alpha_j}\right)\right)\left(\frac{\partial q_1}{\partial x_0}\left(\frac{1}{\alpha_j}\right) + \frac{\partial p_1}{\partial x_0}\left(\frac{1}{\alpha_j}\right)x_2\right) =$$

$$= \begin{pmatrix} \frac{\partial F_i}{\partial x_0}(P_j) & \frac{\partial F_i}{\partial x_1}(P_j) & \frac{\partial F_i}{\partial x_2}(P_j) \end{pmatrix} \cdot \begin{pmatrix} 1 \\ 0 \\ \frac{\partial q_1}{\partial x_0}\left(\frac{1}{\alpha_j}\right) + \frac{\partial p_1}{\partial x_0}\left(\frac{1}{\alpha_j}\right)x_2 \end{pmatrix}.$$

On the other side, the vector $(1, \alpha_j, \beta_j)$ is also in the kernel of the Jacobian matrix of F at P_j, due to Euler's formula for homogeneous polynomials: $x_0 \frac{\partial F}{\partial x_0} + x_1 \frac{\partial F}{\partial x_1} + x_2 \frac{\partial F}{\partial x_2} = \deg(F)F$.

By $(*)$, the Jacobian matrix of F has rank 2 at P_j, so $(1, \alpha_j, \beta_j)$ generates its kernel. Then, $\left(1, 0, \frac{\partial q_1}{\partial x_0}\left(\frac{1}{\alpha_j}\right) + \frac{\partial p_1}{\partial x_0}\left(\frac{1}{\alpha_j}\right)x_2\right)$ is not in such kernel, since $\alpha_j \neq 0$. Therefore,

$$\overline{G}\left(\frac{1}{\alpha_j} : 1 : x_2\right) = \operatorname{Jac}(F) \cdot \begin{pmatrix} 1 \\ 0 \\ \frac{\partial q_1}{\partial x_0}\left(\frac{1}{\alpha_j}\right) + \frac{\partial p_1}{\partial x_0}\left(\frac{1}{\alpha_j}\right)x_2 \end{pmatrix} \neq 0$$

for all $x_2 \in \mathbb{C}$. Since all entries of $G\left(\frac{1}{\alpha_j} : 1 : x_2\right)$ are divisors of those of $\overline{G}\left(\frac{1}{\alpha_j} : 1 : x_2\right)$, then G is nonzero throughout the whole line $x_0 = \frac{1}{\alpha_j}$ in $\widetilde{\mathbb{A}_1}$.

Repeating the argument with H finishes the proof. □

The next result states the correctness of Algorithm 1:

Theorem 4. *The three parametrizations F, G and H output by Algorithm 1 satisfy that the union of the images of $F(1 : \_ : \_)$, $G(\_ : 1 : \_)$ and $H(\_ : \_ : 1)$ covers S completely.*

Proof. We devote these first lines to sketch the proof. Keeping in mind Theorem 2, when $(*), (a)$ and (b) hold, we construct the following diagram:

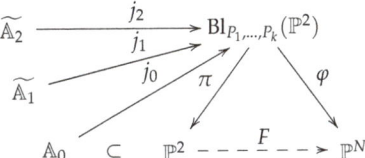

Here, $\operatorname{Bl}_{P_1,\ldots,P_k}(\mathbb{P}^2)$ is the blowup of the base points. To prove that the output of Algorithm 3PatchSurface works as expected, we need to show that

- $F(1 : \_ : \_) = \varphi \circ j_0$, $G(\_ : 1 : \_) = \varphi \circ j_1$ and $H(\_ : \_ : 1) = \varphi \circ j_2$.
- $j_0(\mathbb{A}_0) \cup j_1(\widetilde{\mathbb{A}_1}) \cup j_2(\widetilde{\mathbb{A}_2}) = \operatorname{Bl}_{P_1,\ldots,P_k}(\mathbb{P}^2)$.
- φ has no base points.

In this situation, we observe that, since φ has no base points, then it is a regular morphism, so $\varphi(\operatorname{Bl}_{P_1,\ldots,P_k}(\mathbb{P}^2))$ is an algebraic subset of $\mathbb{P}^n$. On the other side, F is dominant over S, so

$$S = \varphi(\operatorname{Bl}_{P_1,\ldots,P_k}(\mathbb{P}^2)) = F(\mathbb{A}_0) \cup G(\widetilde{\mathbb{A}_1}) \cup H(\widetilde{\mathbb{A}_2}),$$

and, hence, the theorem holds.

First of all, we define j_1 and j_2. The blow up of all the base points of F is, locally, the blow up of the ideal I_i in $\mathbb{A}_i$ with i being 1 or 2. Knowing the bases of I_1 and I_2, we have (see Section 4.2 [26]):

$$\operatorname{Bl}_{I_1}(\mathbb{A}_1) = \{((x_0 : 1 : x_2), (y_0 : y_1)) \mid p_1(x_0)y_0 = (x_2 - q_1(x_0))y_1\} \subset \mathbb{A}_1 \times \mathbb{P}^1,$$

$$\operatorname{Bl}_{I_2}(\mathbb{A}_2) = \{((x_0 : x_1 : 1), (y_0 : y_1)) \mid p_2(x_0)y_0 = (x_1 - q_2(x_0))y_1\} \subset \mathbb{A}_2 \times \mathbb{P}^1.$$

While the way to glue the two open subsets is not interesting for the purpose of this proof, it is easy to see that $\operatorname{Bl}_{I_1}(\mathbb{A}_1) \cup \operatorname{Bl}_{I_2}(\mathbb{A}_2)$ is the whole $\operatorname{Bl}_{\{P_1,\ldots,P_k\}}(\mathbb{P}^2)$ minus the point that is the strict transform of $(1 : 0 : 0)$. We now consider the inclusions

$$j_1 : \begin{array}{c} \widetilde{\mathbb{A}_1} \\ (x_0, x_2) \end{array} \begin{array}{c} \to \\ \mapsto \end{array} \begin{array}{c} \operatorname{Bl}_{I_1}(\mathbb{A}_1) \\ ((x_0 : 1 : q_1(x_0) + x_2 p_1(x_0)), (x_2 : 1)) \end{array},$$

$$j_2 : \begin{array}{c} \widetilde{\mathbb{A}_2} \\ (x_0, x_1) \end{array} \begin{array}{c} \to \\ \mapsto \end{array} \begin{array}{c} \operatorname{Bl}_{I_2}(\mathbb{A}_2) \\ ((x_0 : q_2(x_0) + x_1 p_2(x_0) : 1), (x_1 : 1)) \end{array}.$$

If $\pi_i : \mathrm{Bl}_{I_i}(\mathbb{A}_i) \to \mathbb{A}_i$ is the first projection in any of the cases, it is clear that $\mathrm{Bl}_i = \pi_i \circ j_i$ is defined by

$$\mathrm{Bl}_1(x_0, x_2) = (x_0, q_1(x_0) + x_2 p_1(x_0)) \text{ and } \mathrm{Bl}_2(x_0, x_1) = (x_0, q_2(x_0) + x_1 p_2(x_0)).$$

Therefore, we have that

$$G(x_0 : 1 : x_2) = F(\mathrm{Bl}_1(x_0, x_2)) \text{ and } H(x_0 : x_1 : 1) = F(\mathrm{Bl}_2(x_0, x_1)).$$

On the other hand, $\widetilde{\mathbb{A}_i}$ covers the whole $\mathbb{A}_i$ except the vertical lines through the base points. These affine lines are completely contained in $\mathbb{A}_0$, since the infinity point is $(0 : 0 : 1)$ for $\mathbb{A}_1$ and $(0 : 1 : 0)$ for $\mathbb{A}_2$. This means that, to show that $\widetilde{\mathbb{A}_1}$, $\widetilde{\mathbb{A}_2}$ and $\mathbb{A}_0 \backslash \{P_1, ..., P_k\}$ —through j_1, j_2 and the blowup j_0 of the base points— cover $\mathrm{Bl}_{\{P_1,...,P_k\}}(\mathbb{P}^2)$, we just need to prove that $\widetilde{\mathbb{A}_1}$ and $\widetilde{\mathbb{A}_2}$ cover the exceptional divisor. So we fix $P_j = (1 : \alpha_j : \beta_j)$ and we call $E_j \simeq \mathbb{P}^1$ the component of the exceptional divisor corresponding to P_j. Note that Bl_i covers a full neighborhood of P_j in $\mathbb{A}_i$ minus the vertical line. This corresponds to the line joining P_j with $(0 : 0 : 1)$ for the case $i = 1$ and the line joining P_j with $(0 : 1 : 0)$ for the case $i = 2$. By condition (b), these lines do not contain other base points. These two lines represent two different directions at P_j (i.e., two different points in E_j). This means that $j_1(\{x_0 = \frac{1}{\alpha_j}\})$ covers all E_j except one point and that $j_2(\{x_0 = \frac{1}{\beta_j}\})$ is an affine line passing through that point.

The only task remaining is proving that φ has no base points. Such base points would be in the exceptional divisor, which is covered by $\widetilde{\mathbb{A}_1}$ and $\widetilde{\mathbb{A}_2}$, but Lemma 1 states that there are no base points of φ in $j_1(\widetilde{\mathbb{A}_1}) \cup j_2(\widetilde{\mathbb{A}_2})$. □

4. Examples

This section is devoted to illustrating Algorithm 3PatchSurface by examples. In all of them, and in those of the next section, we have used Sage [31] for the calculations. We start with a toy example in which we explicitly show that the three parametrizations cover the whole projective surface.

Example 2. *Whitney's umbrella has implicit equation $y_0 y_1^2 - y_2^2 y_3 = 0$. The usual parametrization $(x_0 : x_1 : x_2) \mapsto (x_0^2 : x_1 x_2 : x_0 x_1 : x_2^2)$ does not satisfy condition (a), so we change coordinates to get a new parametrization*

$$F(x_0 : x_1 : x_2) = (x_2^2 - 2x_0 x_2 + x_0^2 : x_1^2 + x_1 x_2 - x_0 x_1 - x_0 x_2 :$$
$$- x_1 x_2 + x_0 x_1 + x_0 x_2 - x_0^2 : x_1^2 + 2x_1 x_2 + x_2^2).$$

The only base point of the new parametrization is $(1 : -1 : 1)$. We then compute the bases

$$B_1 = \{x_2 + 1, x_1 - 1, x_0 + 1\} \text{ and } B_2 = \{x_2 - 1, x_1 + 1, x_0 - 1\}.$$

So we get:

$$\widetilde{G}(x_0 : x_1 : x_2) = (x_0 + x_1) G(x_0 : x_1 : x_2) \text{ and } \widetilde{H}(x_0 : x_1 : x_2) = (x_2 - x_0) H(x_0 : x_1 : x_2)$$

where

$$G = (x_1^3 - 2x_1^2 x_2 + x_1^2 x_0 + x_1 x_2^2 - 2x_1 x_2 x_0 + x_2^2 x_0 : x_1^2 x_2 - x_1 x_2 x_0 :$$
$$x_1^3 - x_1^2 x_2 - x_1^2 x_0 + x_1 x_2 x_0 : x_1 x_2^2 + x_2^2 x_0)$$

and

$$H = (x_2^3 - x_2^2 x_0 : x_1^2 x_2 - x_1^2 x_0 + x_1 x_2^2 + x_0 x_1 x_2 : x_1 x_2^2 - x_0 x_1 x_2 + x_2^3 + x_0 x_2^2 : x_1^2 x_2 - x_0 x_1^2).$$

We now prove that the whole surface is covered by $F(1 : \_ : \_)$, $G(\_ : 1 : \_)$ and $H(\_ : \_ : 1)$. Let $A = (y_0 : y_1 : y_2 : y_3)$ be such that $y_0 y_1^2 - y_2^2 y_3 = 0$. Then:

1. if $y_0 y_2 (y_0 y_1 + y_0 y_2 - y_2^2) \neq 0$, then, taking $y_0 = 1$,

$$A = F\left(1 : \frac{y_2 + y_1 + y_2^2}{y_2 + y_1 - y_2^2} : \frac{-y_2 + y_1 - y_2^2}{y_2 + y_1 - y_2^2}\right)$$

Note here that $y_3 = \frac{y_1^2}{y_2^2}$, due to the equation of the umbrella and $y_0 = 1$.

2. if $y_0 y_1 + y_0 y_2 - y_2^2 = 0$ and $y_0 y_2 \neq 0$, taking $y_0 = 1$, we get $y_1 = y_2^2 - y_2$. This equality transforms the equation of the umbrella in $y_3 = (y_2 - 1)^2$. Then, $A = G(0 : 1 : 1 - 1/y_2)$.

3. if $y_2 = 0$ and $y_0 \neq 0$, then, taking $y_0 = 1$,

$$A = F\left(1 : 1 : \frac{-1 + \sqrt{y_3}}{1 + \sqrt{y_3}}\right).$$

Note here that every A in this case is gotten twice. Even $(1 : 0 : 0 : 1)$ is both $F(1 : 1 : 0)$ and $H(0 : -1 : 1)$.

4. if $y_0 = 0$, the infinity hyperplane section of Whitney's umbrella has two components: $y_2 = 0$ and $y_3 = 0$. Then:

$$A = (0 : y_1 : 0 : y_3) = F\left(1 : 0 : \frac{(y_3 + y_1)}{(y_3 - y_1)}\right)$$

when $y_3 - y_1 \neq 0$, and $A = (0 : 1 : 0 : 1) = G(0 : 1 : 1)$. Moreover,

$$A = (0 : y_1 : y_2 : 0) = G(-1 : y_1 + y_2 : y_1)$$

when $y_1 + y_2 \neq 0$. Finally, $A = (0 : -1 : 1 : 0) = H(1 : -1 : 1)$.

Observe that $F(1 : \_ : \_)$ covers the whole umbrella minus a couple of rational curves. Then, $G(\_ : 1 : \_)$ covers these curves minus just a point, that is covered by $H(\_ : \_ : 1)$.

The following example applies Algorithm 1 to a classic surface. While the computation time is not long, the output is too large, so we just sketch some computations.

Example 3. *Let us consider the Clebsch cubic, given by the equation*

$$z_0^3 + z_1^3 + z_2^3 + z_3^3 - (z_0 + z_1 + z_2 + z_3)^3 = 0.$$

A parametrization is defined by

$$(-x_0^2 x_1 + x_0^2 x_2 + x_0 x_1^2 - x_0 x_2^2 : x_0^3 - x_0^2 x_1 - x_0^2 x_2 + x_1 x_2^2 :$$
$$- x_0^3 + x_0^2 x_1 + x_0^2 x_2 - x_1^2 x_2 : -x_0 x_1^2 + x_0^2 x_1 - x_1 x_2^2).$$

This parametrization, however, does not satisfy conditions (a) and (b). After the coordinate change in $\mathbb{P}^2$ given by

$$(x_0 : x_1 : x_2) \mapsto (x_0 + 3x_1 + 2x_2 : x_0 + x_1 + 3x_2 : -x_0 - x_1 + x_2),$$

we get the parametrization $F = (F_0 : F_1 : F_2 : F_3)$, where

$F_0 = 8x_2^3 + 8x_2^2 x_1 + 8x_2^2 x_0 - 18x_2 x_1^2 - 12x_2 x_1 x_0 - 2x_2 x_0^2 - 18x_1^3 - 30x_1^2 x_0 - 14x_1 x_0^2 - 2x_0^3,$

$F_1 = -5x_2^3 - 17x_2^2 x_1 - 9x_2^2 x_0 + 19x_2 x_1^2 + 14x_2 x_1 x_0 + 3x_2 x_0^2 + 28x_1^3 + 30x_1^2 x_0 + 12x_1 x_0^2 + 2x_0^3,$

$F_2 = -x_2^3 + 15x_2^2 x_1 + 7x_2^2 x_0 - 13x_2 x_1^2 - 2x_2 x_1 x_0 + 3x_2 x_0^2 - 26x_1^3 - 24x_1^2 x_0 - 6x_1 x_0^2,$

$$F_3 = -9x_2^3 + 6x_2^2x_1 + 18x_2x_1^2 + 4x_2x_1x_0 - 2x_2x_0^2 + 5x_1^3 + 5x_1^2x_0 - x_1x_0^2 - x_0^3.$$

One can check that the base points are $P_1 = (5 : 2 : -3)$, $P_2 = (11 : -2 : -5)$, $P_3 = (7 : -2 : -1)$, $P_4 = (1 : 2 : 1)$, $P_{5,6} = (\pm 2 - \sqrt{5} : 2 : \sqrt{5})$. This agrees with the well known fact that a cubic smooth surface is a plane blown up at 6 general points.

Now we take $x_1 = 1$ or $x_2 = 1$ and get the two Gröbner bases $B_i = \{x_i - 1, x_{3-i} - q_i(x_0), p_i(x_0)\}$, where $i \in \{1, 2\}$ and:

$$p_1(x_0) = x_0^6 + 8x_0^5 + \frac{21}{4}x_0^4 - 61x_0^3 - \frac{1077}{16}x_0^2 + \frac{239}{4}x_0 - \frac{385}{64},$$

$$q_1(x_0) = -\frac{393}{8360}x_0^5 - \frac{18511}{50160}x_0^4 - \frac{12941}{50160}x_0^3 + \frac{13537}{5280}x_0^2 + \frac{1123141}{401280}x_0 - \frac{21649}{14592},$$

$$p_2(x_0) = x_0^6 + \frac{178}{15}x_0^5 + \frac{199}{5}x_0^4 + \frac{916}{25}x_0^3 - \frac{2387}{75}x_0^2 - \frac{3926}{75}x_0 - \frac{77}{15},$$

$$q_2(x_0) = -\frac{7075}{11264}x_0^5 - \frac{212195}{33792}x_0^4 - \frac{221885}{16896}x_0^3 + \frac{49613}{16896}x_0^2 + \frac{612691}{33792}x_0 + \frac{2987}{3072}.$$

Performing substitutions

$$\widetilde{G}(x_0 : 1 : x_2) = F(x_0 : 1 : q_1(x_0) + x_2 p_1(x_0))$$

and

$$\widetilde{H}(x_0 : x_1 : 1) = F(x_0 : q_2(x_0) + x_1 p_2(x_0) : 1)$$

then dividing by the gcd of all entries in each case produces two parametrizations of degree 15, $G(x_0 : 1 : x_2)$ and $H(x_0 : x_1 : 1)$, with about 45 coefficients per polynomial and about 70 bits per coefficient, that cover, together with $F(1 : x_1 : x_2)$, the whole cubic.

5. The Affine Case

In this section, we slightly change our point of view and we consider the problem of covering a rational affine surface by means of the images of several affine parametrizations. So, in the sequel we consider that we are given F and S as in Section 5, and we deal with the problem of covering $S \cap \mathbb{A}^n$, where $\mathbb{A}^n$ is the open subset of $\mathbb{P}^n$ defined by the first variable not vanishing. Equivalently, one may consider that we are indeed given an affine parametrization and the affine surface that it defines. Nevertheless, to be consistent with the notation used throughout the paper, we will use the first notational statement of the problem.

In this section, we prove that to cover a rational affine surface, only two patches are necessary (see Theorem 5 and Corollary 3). The basic idea is as follows. The given parametrization $F(1 : x_1 : x_2)$ covers a constructible subset. The complement of such subset is contained in the image of L_∞ (that is, $F(0 : x_1 : x_2)$) and the base points, which is a finite union of affine rational curves (see Corollary 2) and, maybe, an isolated point corresponding to a contracted L_∞. The parametrization $G(x_0 : 1 : x_2)$ of Algorithm 1, restricted to certain vertical lines, covers all such affine curves except at most one point. Since such curves also have a point at infinity, we want such point to be the image of the point at infinity of the parameter line. Based on these ideas, we derive Algorithm 2 that, when the original parametrization satisfies (∗), (a) and (b), covers affine surfaces using just two parametrizations.

Algorithm 1: 2PatchForAffine

Require: A list $F = (F_0 : \cdots : F_n)$ of coprime homogeneous polynomials of the same degree in $\mathbb{K}[x_0, x_1, x_2]$, parametrizing a Zariski dense subset of a projective surface $S \subset \mathbb{P}^n_{\mathbb{C}}$, such that conditions $(*)$, (a) and (b) are satisfied.

Ensure: A list $g' = (g'_0 : \cdots : g'_n)$ of rational functions of two variables such that $F(1 : \_ : \_)$ and $g'(\_,\_)$ cover $S \cap \mathbb{A}^n$.

1: Compute $B_1 = \{x_2 - q_1(x_0), x_1 - 1, p_1(x_0)\}$ and $G = (G_0 : \cdots : G_n)$ as in Algorithm 1.
2: **for** α root of $p_1(x_0)$ or $\alpha = 0$ **do**
3:    **if** $\deg G_0(\alpha : 1 : x_2) < \max\{\deg G_i(\alpha : 1 : x_2) \mid i = 1, ..., n\}$ or $G_0(\alpha : 1 : x_2)$ is constant **then**
4:       Include α in set A and $\beta_\alpha := \infty$.
5:    **else**
6:       Include α in set B. Choose β_α among the roots of $G_0(\alpha : 1 : x_2)$.
7:    **end if**
8: **end for**
9: Let $s(x_0)$ be a polynomial vanishing at all the $\alpha \in A$ and not vanishing at any of the $\alpha \in B$. See Remark 4 below, for suggestions on how to find one.
10: Choose $r(x_0)$, a polynomial such that $r(\alpha) = \beta_\alpha s(\alpha)$ for all $\alpha \in B$ and $r(\alpha) \neq 0$ for all $\alpha \in A$.
11: Find Bezout coefficients u and v such that $\gcd(r, s) = u \cdot r + v \cdot s$.
12: **return** $g'(x_0, x_2) = G\left(x_0 : 1 : \dfrac{r(x_0)x_2 + v(x_0)}{s(x_0)x_2 - u(x_0)}\right)$.

Remark 4. *Let us comment some computational aspects of Algorithm 2.*

1. *For $s(x_0)$, one may proceed as follows: collect the coefficients of $G_0(x_0 : 1 : x_2)$ for x_2 except the one for x_2^0, and compute the gcd, $d(x_0)$, of these coefficients; then $s(x_0) = \gcd(d(x_0), x_0 p_1(x_0))$.*
2. *Note that Algorithm 1 does not extend the field that is used to define F, However, Algorithm 2 needs to consider possibly algebraic coordinates for the points that need to be sent to the infinity in the parameter plane $\mathbb{A}_1$. Example 4 shows that the application of the algorithm forces the usage of algebraic coefficients.*

In order to state the correctness of Algorithm 2, we start with a technical lemma.

Lemma 2. *In the setting of Algorithm 2, if no component of $G(\alpha : 1 : x_2)$ is constant, it holds that*

1. *If $\alpha \in A$, then $\lim_{x_2 \to \infty} G(\alpha : 1 : x_2)$ is a point at infinity.*
2. *If $\alpha \in B$, then $G(\alpha : 1 : \beta_\alpha)$ is a point at infinity.*

Proof. Let α be an element of A. If $G_0(\alpha : 1 : x_0)$ is identically zero, then the result is obvious. Otherwise, the map $(\frac{G_1}{G_0}(\alpha : 1 : x_2), ..., \frac{G_n}{G_0}(\alpha : 1 : x_2))$, from the affine plane to the surface, has degree strictly higher in the numerator of at least one of its components, because $\deg(G_0(\alpha : 1 : x_2)) < \max\{\deg(G_i(\alpha : 1 : x_2)) \mid i = 1, ..., n\}$. So the limit, when x_2 tends to ∞, is at infinity. Note that the full $(n+1)$–tuple $G(\alpha : 1 : x_2)$ is not constant, so the case of constant first entry $G_0(\alpha : 1 : x_2)$ satisfies the inequality too.

Now, let α be an element of B. Then, $G_0(\alpha : 1 : \beta_\alpha) = 0$, so the statement holds. □

Let us, now, prove that Algorithm 2 works as expected.

Theorem 5. *Let F, S and g' be as in Algorithm 2. Then $F(1 : x_1 : x_2)$ and $g'(x_0, x_2)$ cover the whole $S \cap \{y_0 \neq 0\}$.*

Proof. Since the input of Algorithms 1 and 2 are the same, by Theorem 4, one deduces that $F(1 : x_1 : x_2)$, $G(x_0 : 1 : x_2)$ and $H(x_0 : x_1 : 1)$ cover the projective surface S.

Taking into account how G and H in Algorithm 1 are defined, any point in $\widetilde{\mathbb{A}_1}$ and $\widetilde{\mathbb{A}_2}$, not in L_∞ or in $\{p_i(x_0) = 0\}$, is sent by $G(x_0 : 1 : x_2)$ or $H(x_0 : x_1 : 1)$ into the image by F of a point in $\mathbb{A}_0$. Therefore, we need to check that g' covers any affine point in $G(\{x_0 = 0\} \cup \{p_1(x_0) = 0\}) \cup H(\{x_0 = 0\} \cup \{p_2(x_0) = 0\})$.

Any component C of $G(\{x_0 = 0\} \cup \{p_1(x_0) = 0\}) \cup H(\{x_0 = 0\} \cup \{p_2(x_0) = 0\})$ is either a point or a rational curve covered by $G(\alpha : 1 : x_2) \cup H(\alpha' : x_1 : 1)$, where either $\alpha = \alpha' = 0$ or $(1 : \frac{1}{\alpha} : \frac{1}{\alpha'})$ is a base point of F. Moreover, such component is the Zariski closure of the image of the restriction $G|_{\{x_0 = \alpha, x_1 \neq 0\}}$, which coincides with the Zariski closure of $g'|_{\{x_0 = \alpha\}}$.

If C is just a point, then it is covered by $g'|_{\{x_0=\alpha\}}$. Otherwise, it is well known that any morphism defined in an open subset of a projective smooth curve can be extended regularly to the whole curve and that the image of a projective curve by a regular morphism is a Zariski closed subset. Then, we can extend $g'|_{\{x_0=\alpha\}}$ to the Zariski closure of the line where it is defined and we would cover completely C. This means that $g'|_{\{x_0=\alpha\}}$ covers all C minus, at most, just a point (the image of the infinity point of the affine line). However, this point is $G(\alpha : 1 : \beta_\alpha)$, which is at infinity by Lemma 2. Therefore, any point in $C \cap \{y_0 \neq 0\}$ is in $g'(\{x_0 = \alpha\}) \subset g'(\mathbb{A}^2)$. □

Example 4. *Consider the following projective transformation of the Veronese morphism:*

$$F : \quad \mathbb{P}^2 \quad \to \quad \mathbb{P}^5$$
$$(x_0 : x_1 : x_2) \mapsto (x_0^2 + x_1^2 + x_2^2 : x_0 x_1 : x_0 x_2 : x_1^2 : x_1 x_2 : x_2^2)$$

Since there are no base points, Algorithm 2 generates $G = F$, and then, for just $\alpha = 0$, it computes β_0 such that $\beta_0^2 + 1 = 0$. This means that β_0 must be imaginary, so it is not rational. The output

$$g'(x_0, x_2) = G\left(x_0 : 1 : \frac{ix_2 + 1}{x_2}\right)$$

has imaginary coefficients.

We observe that any choice of two rationally defined affine planes $\mathbb{A}_0$ and $\mathbb{A}_1$ of the projective plane will leave a point P in the projective plane over $\mathbb{Q}$ out of the union, and then $F(P)$, which is not at infinity, will not be covered. Observe, however, that the surface is isomorphic to the projective plane, so one can compose the two maps appearing in Example 1 with the Veronese morphism to cover, not just the affine part, but the whole projective surface without extending the field

Example 5. *Let us again consider the cubic of Example 3. We recall that*

$$p_1(x_0) = x_0^6 + 8x_0^5 + \frac{21}{4}x_0^4 - 61x_0^3 - \frac{1077}{16}x_0^2 + \frac{239}{4}x_0 - \frac{385}{64}.$$

One can factor G_0, as obtained in Example 3, and one of the factors is

$$200640 x_2 x_0^4 + 1203840 x_2 x_0^3 - 1304160 x_2 x_0^2 - 9329760 x_2 x_0 +$$
$$4827900 x_2 - 9432 x_0^3 - 55180 x_0^2 + 56238 x_0 + 388135.$$

Here, one can easily get x_2 as a rational function on x_0, so we have a point going to infinity at each vertical line:

$$x_2 = \beta_{x_0} = \frac{9432 x_0^3 + 55180 x_0^2 - 56238 x_0 - 388135}{200640 x_0^4 + 1203840 x_0^3 - 1304160 x_0^2 - 9329760 x_0 + 4827900}. \quad (1)$$

The set A is given by the common roots of p_1 and the denominator in (1), so

$$s(x_0) = \gcd(p_1, \text{denominator}(\beta_{x_0})) = 16 x_0^4 + 96 x_0^3 - 104 x_0^2 - 744 x_0 + 385.$$

We need a polynomial $r(x_0)$, coprime with $s(x_0)$, whose values at the roots of $\frac{x_0 p_1(x_0)}{s(x_0)}$ equal $\beta_{x_0} s(x_0)$. Such roots are 0 and $1 \pm \frac{\sqrt{5}}{2}$, and the interpolating polynomial

$$r(x_0) = \frac{18511}{3135} x_0^2 - \frac{898}{209} x_0 - \frac{7057}{228}$$

works. We now need a Bezout identity $ur + vs = 1$, so we get

$$u(x_0) = -\frac{26198287461}{21693258568709}x_0^3 - \frac{1732666485971}{130159551412254}x_0^2 - \frac{487041584557}{12396147753548}x_0 - \frac{632041996387}{74376886521288}$$

and

$$v(x_0) = \frac{1510767910251}{3389825077278640}x_0 + \frac{357074564524303}{1865372313953 90304}.$$

Then, we have that the images of $F(1 : x_1 : x_2)$ and $G\left(x_0 : 1 : \frac{r(x_0)x_2 + v(x_0)}{s(x_0)x_2 - u(x_0)}\right)$ cover the whole affine cubic.

In [24], it is proved that there exist affine surfaces that cannot be covered by means of a unique map from the affine plane. In fact, the surface in Example 5 is proved to be one of them. Now, the following corollary of Theorem 5 shows that, under hypotheses (*), (a), (b), one can always cover the affine surface with two affine parametrization images.

Corollary 3. *Let S be an affine surface such that there exists a parametrization $f : \mathbb{A}^2 \dashrightarrow S$ with a projectivization F satisfying (*), (a) and (b). Then S can be covered with just two parametrizations.*

In order to prove that two affine patches are enough we have had to impose, to the projectivization of the input affine parametrization, hypotheses (*), (a) and (b). If we do not impose (*), we cannot ensure this general result. However, it is interesting to observe that there are affine surfaces not satisfying (*) that can be covered by only one map. To create an example, it is enough to send the exceptional divisor to the infinity hyperplane together with the image of L_∞.

Example 6. *Consider the transformation of the plane $t(x_0 : x_1 : x_2) = (x_0 x_1 x_2 : x_1^3 : x_0 x_2^2)$ and let F be its composition with the degree-3 Veronese morphism $v_3(x_0 : x_1 : x_2) = (x_0 x_1 x_2 : x_0^3 : \cdots : x_2^3)$. We can observe that $\mathbb{P}^2 - t(\mathbb{A}_0) = \{x_0 x_1 x_2 = 0\} = v_3^{-1}(\{y_0 = 0\})$. On the other hand, the fundamental locus is defined by the ideal $I = (x_0 x_1 x_2, x_1^3, x_0 x_2^2)$, which is singular, so F does not satisfy (*), but $F(1, x_1, x_2)$ covers the whole affine surface.*

6. Conclusions

In the introduction we have commented the importance for applications of having surjective parametrizations of curves and surfaces. For the curve case, the problem can be solved by means of a single parametrization. For the surface case, the situation is much more complicated and, in general, one may need more that one parametrization to cover it. So the problem is reformulated by, on one hand, asking for the minimization of the number of parametrizations in the cover and, on the other, by requiring the actual computation of covers. There were previous results for some particular types of surfaces, specially of those whose construction is directly related to curves as ruled surfaces, revolution surfaces, etc.

In this paper we present theoretical results on the number of parametrizations that one may need to cover a surface, and we enlarge the class of surfaces where this approach is valid. More precisely, we present two algorithms, one for the projective case and the other for the affine case. In the affine case, we are able to cover any rational affine surface satisfying certain mild hypotheses on the base locus of the input parametrization, in a way that is optimal in the number of cover elements, namely, two. For the projective case, the answer provides three cover parametrizations. Two open problems are the extension of the results to the case where no condition of the base locus is imposed, and the optimality on the number of cover parametrizations in the projective case.

Author Contributions: All the stages of work have been participated by all the authors of the article. All authors have read and agreed to the published version of the manuscript.

Funding: The authors are partially supported by FEDER/Ministerio de Ciencia, Innovación y Universidades-Agencia Estatal de Investigación/MTM2017-88796-P (Symbolic Computation: new challenges in Algebra and Geometry together with its applications). J. Caravantes, J.R. Sendra and C. Villarino belong to the Research Group ASYNACS (Ref. CT-CE2019/683). D. Sevilla is a member of the research group GADAC and is partially supported by Junta de Extremadura and FEDER funds (group FQM024).

Institutional Review Board Statement: Not applicable.

Informed Consent Statement: Not applicable.

Data Availability Statement: Not applicable.

Conflicts of Interest: The authors declare no conflict of interest.

References

1. Farin, G.; Hoschek, J.; Kim, M.S. (Eds.) *Handbook of Computer Aided Geometric Design*; North-Holland: Amsterdam, The Netherlands, 2002; p. xxviii+820.
2. Hoschek, J.; Lasser, D. *Fundamentals of Computer Aided Geometric Design*; A.K. Peters: Wellesley, MA, USA; Berlin, Germany, 1993.
3. Agoston, M. *Computer Graphics and Geometric Modelling*; Computer Graphics and Geometric Modeling; Springer: Berlin/Heidelberg, Germany, 2005.
4. Marsh, D. *Applied Geometry for Computer Graphics and CAD*; Springer Undergraduate Mathematics Series; Springer: London, UK, 2005.
5. Sendra, J.R.; Villarino, C.; Sevilla, D. Missing sets in rational parametrizations of surfaces of revolution. *Comput. Aided Des.* **2015**, *66*, 55–61. [CrossRef]
6. Sendra, J.R.; Sevilla, D.; Villarino, C. Covering of surfaces parametrized without projective base points. In Proceedings of the International Symposium on Symbolic and Algebraic Computation, ISSAC '14, Kobe, Japan, 23–25 July 2014; Nabeshima, K., Nagasaka, K., Winkler, F., Szántó, Á., Eds.; ACM: New York, NY, USA, 2014; pp. 375–380. [CrossRef]
7. Sendra, J.R.; Sevilla, D.; Villarino, C. Covering rational ruled surfaces. *Math. Comput.* **2017**, *86*, 2861–2875. [CrossRef]
8. Sendra, J.R.; Winkler, F. Parametrization of algebraic curves over optimal field extensions. *J. Symb. Comput.* **1997**, *23*, 191–207.
9. Recio, T.; Sendra, J.R.; Tabera, L.F.; Villarino, C. Generalizing circles over algebraic extensions. *Math. Comp.* **2010**, *79*, 1067–1089. [CrossRef]
10. Andradas, C.; Recio, T.; Sendra, J.R.; Tabera, L.F. On the simplification of the coefficients of a parametrization. *J. Symb. Comput.* **2009**, *44*, 192–210. [CrossRef]
11. Sendra, J.R.; Winkler, F.; Pérez-Díaz, S. *Rational Algebraic Curves. A Computer Algebra Approach*; Algorithms and Computation in Mathematics; Springer: Berlin, Germany, 2008; Volume 22, p. x+267.
12. Hartshorne, R. *Algebraic Geometry (Graduate Texts in Mathematics, No. 52)*; Springer: New York, NY, USA; Heidelberg, Germany, 1977; p. xvi+496.
13. Alonso, C.; Gutiérrez, J.; Recio, T. Reconsidering algorithms for real parametric curves. *Appl. Algebra Engrg. Comm. Comput.* **1995**, *6*, 345–352. [CrossRef]
14. Pérez-Díaz, S. On the problem of proper reparametrization for rational curves and surfaces. *Comput. Aided Geom. Des.* **2006**, *23*, 307–323. [CrossRef]
15. Sederberg, T.W. Improperly parametrized rational curves. *Comput. Aided Geom. Des.* **1986**, *3*, 67–75. [CrossRef]
16. Pérez-Díaz, S. A partial solution to the problem of proper reparametrization for rational surfaces. *Comput. Aided Geom. Des.* **2013**, *30*, 743–759. [CrossRef]
17. Schicho, J. Rational parametrization of surfaces. *J. Symb. Comput.* **1998**, *26*, 1–29. [CrossRef]
18. Andradas, C.; Recio, T. Plotting missing points and branches of real parametric curves. *Appl. Algebra Engrg. Comm. Comput.* **2007**, *18*, 107–126. [CrossRef]
19. Rubio, R.; Serradilla, J.; Vélez, M.P. A note on implicitization and normal parametrization of rational curves. In Proceedings of the 2006 International Symposium on Symbolic and Algebraic Computation, Genoa, Italy, 9–12 July 2006; ACM Press: New York, NY, USA, 2006; pp. 306–309.
20. Sendra, J.R. Normal Parametrizations of Algebraic Plane Curves. *J. Symb. Comput.* **2002**, *33*, 863–885. [CrossRef]
21. Bajaj, C.L.; Royappa, A.V. Finite representations of real parametric curves and surfaces. *Internat. J. Comput. Geom. Appl.* **1995**, *5*, 313–326. [CrossRef]
22. Gao, X.S.; Chou, S.C. On the normal parameterization of curves and surfaces. *Internat. J. Comput. Geom. Appl.* **1991**, *1*, 125–136. [CrossRef]
23. Pérez-Díaz, S.; Sendra, J.R.; Villarino, C. A first approach towards normal parametrizations of algebraic surfaces. *Internat. J. Algebra Comput.* **2010**, *20*, 977–990. [CrossRef]

24. Caravantes, J.; Sendra, J.R.; Sevilla, D.; Villarino, C. On the existence of birational surjective parametrizations of affine surfaces. *J. Algebra* **2018**, *501*, 206–214. [CrossRef]
25. Bodnár, G.; Hauser, H.; Schicho, J.; Villamayor U., O. Plain varieties. *Bull. Lond. Math. Soc.* **2008**, *40*, 965–971. Available online: https://academic.oup.com/blms/article-pdf/40/6/965/772642/bdn078.pdf (accessed on 8 February 2021). [CrossRef]
26. Shafarevich, I.R.T. *Basic Algebraic Geometry 1: Varieties in Projective Space*, 2nd ed.; Translated from the 1988 Russian edition and with notes by Miles Reid; Springer: Berlin, Germany, 1994; p. xx+303.
27. Beauville, A. *Complex Algebraic Surfaces*; London Mathematical Society Lecture Note Series; Barlow, R., Shepherd-Barron, N.I., Reid, M., transed.; Cambridge University Press: Cambridge, UK, 1983; Volume 68, p. iv+132.
28. Cox, D.A.; Little, J.; O'Shea, D. *Ideals, Varieties, and Algorithms: An Introduction to Computational Algebraic Geometry and Commutative Algebra*, 4th ed.; Undergraduate Texts in Mathematics; Springer: Cham, Switzerland, 2015; p. xvi+646. [CrossRef]
29. Cox, D.A.; Pérez-Díaz, S.; Sendra, J.R. On the base point locus of surface parametrizations: Formulas and consequences. *arXiv* **2020**, arXiv:2008.08009.
30. Pérez-Díaz, S.; Sendra, J.R. Computing Birational Polynomial Surface Parametrizations without Base Points. *Mathematics* **2020**, *8*, 2224. [CrossRef]
31. The Sage Developers. SageMath, the Sage Mathematics Software System (Version 9.0). 2020. Available online: http://www.sagemath.org (accessed on 8 February 2021).

Article

The μ-Basis of Improper Rational Parametric Surface and Its Application

Sonia Pérez-Díaz [1,*,†] **and Li-Yong Shen** [2,3,†]

1. Dpto de Física y Matemáticas, Universidad de Alcalá, E-28871 Madrid, Spain
2. School of Mathematical Sciences, University of Chinese Academy of Sciences, Beijing 100049, China; lyshen@ucas.ac.cn
3. Key Laboratory of Big Data Mining and Knowledge Management, CAS, Beijing 100190, China
* Correspondence: sonia.perez@uah.es
† These authors contributed equally to this work.

Abstract: The μ-basis is a newly developed algebraic tool in curve and surface representations and it is used to analyze some essential geometric properties of curves and surfaces. However, the theoretical frame of μ-bases is still developing, especially of surfaces. We study the μ-basis of a rational surface $\mathcal{V}$ defined parametrically by $\mathcal{P}(\bar{t})$, $\bar{t} = (t_1, t_2)$ not being necessarily proper (or invertible). For applications using the μ-basis, an inversion formula for a given proper parametrization $\mathcal{P}(\bar{t})$ is obtained. In addition, the degree of the rational map $\phi_\mathcal{P}$ associated with any $\mathcal{P}(\bar{t})$ is computed. If $\mathcal{P}(\bar{t})$ is improper, we give some partial results in finding a proper reparametrization of $\mathcal{V}$. Finally, the implicitization formula is derived from $\mathcal{P}$ (not being necessarily proper). The discussions only need to compute the greatest common divisors and univariate resultants of polynomials constructed from the μ-basis. Examples are given to illustrate the computational processes of the presented results.

Keywords: μ-basis; rational surfaces; inversion; improper; reparametrization; implicitization; resultant

Citation: Pérez-Díaz, S.; Shen, L.-Y. The μ-Basis of Improper Rational Parametric Surface and Its Application. *Mathematics* **2021**, *9*, 640. https://doi.org/10.3390/math9060640

Academic Editor: Gabriel Eduard Vîlcu

Received: 12 February 2021
Accepted: 12 March 2021
Published: 17 March 2021

Publisher's Note: MDPI stays neutral with regard to jurisdictional claims in published maps and institutional affiliations.

Copyright: © 2021 by the authors. Licensee MDPI, Basel, Switzerland. This article is an open access article distributed under the terms and conditions of the Creative Commons Attribution (CC BY) license (https://creativecommons.org/licenses/by/4.0/).

1. Introduction

The study of representations of rational curves and surfaces is a fundamental task in computer aided geometric design (CAGD) and computer algebra. There exist two typical problems in the study of representations.

- Implicitization: for a rational parametric curve or surface, implicitization is to find an algebraic expression of the curve or surface.
- Proper Reparametrization: for an improper rational parametric curve or surface, proper reparametrization is to find a proper parametric expression of the curve or surface.

The parametric expression of a curve or surface is widely used in geometric modeling, such as NURBS representations. The algebraic equation, which is also called implicit equation, is another important representation, and this is much better than the parametric expression in determining whether or not a point is on the curve or surface. Hence the implicitization problem is classical in CAGD and there are implicitization algorithms for rational curves and surfaces proposed over the past several decades [1–14]. Among all of these techniques, the Gröbner bases [2] is well-known, since it is theoretically complete. However, this method has exponential computational complexity and, thus, it is inefficient. This is the reason that people can not apply the Gröbner basis method for practical implicitization in application. Alternatively, in computational application, people prefer to find the implicit equations from certain implicit matrices. The implicit matrices can be constructed as resultant matrices or matrices of moving curves/surfaces. The implicit matrix of the curve is much simpler than that of the surface, since the curve only introduces one variable. Actually, the construction of bivariate resultant is not uniform and it is still a developing technique in computer algebra [15–21]. The implicit matrix of moving curves/surfaces was introduced in [10] and developed by more researchers [5,11,22–25]. The μ-basis of a curve or surface was later defined by moving lines or moving planes with certain properties [5,23].

The implicit equation of the surface is generally included as a factor in its constructed resultants, but a constructed resultant may have extraneous factors that are not easy to identify and remove. For implicit matrices, some works attempt to construct the matrix whose determinant is exactly the implicit equation [10,24,26,27], but the ways to construct such implicit matrices are not complete for general surfaces. In the implicit matrix, there is more information than the implicit equation, such as singularities with multiplicity counting. Accordingly, people sometimes construct the implicit matrix by simple way and then the determinant of this matrix may have extraneous factors other than the implicit equation [22]. For the matrix constructed from the μ-basis of arbitrary three linearly independent syzygies, the extraneous factors are identified completely based on the analysis of base points or to parameters at infinity of tensor product surfaces [11,25], but some computations need the Gröbner bases of a zero dimensional algebraic variety.

When considering the proper reparametrization problem, an essential question is to decide whether a rational parametrization is proper. If a given rational parametrization is not proper, also called improper, a generic point lying on the variety corresponds to more than one parameter. On the other side, if a rational parametrization is improper, we ask whether it can be reparameterized, such that we can get a new proper parametrization. For algebraic curves, is well-known that the existence of a proper reparametrization for a given improper rational parametrization is certified by Lüroth's Theorem [28]. One can have a look, for instance, at a previous bibliography, as, for instance, [29–31], where some efficient methods are proposed to find a proper reparametrization for an improper parametrization of an algebraic curve. For a given algebraic surface, Castelnuovo's Theorem states that unirationality and rationality are equivalent over algebraically closed fields, but only some partial algorithmic methods approaching the problem are known (see [30,32]).

The μ-basis was first used in [17]. Here, the authors provided a representation for the implicit equation of a given curve defined parametrically. The μ-basis developing as a new algebraic tool can be used to obtain the parametric equation of a rational curve or a rational surface, in order to compute the implicit equation defining these varieties, and to study singularities and intersections [33]. There are several methods to compute the μ-basis for rational curves by computing two moving lines that satisfy the required properties [17], based on Gröbner basis [34] or based on vector elimination [23]. The μ-basis has also been generalized to rational surfaces [5], although the case of rational surfaces is different; for instance, the degrees of the μ-basis elements can be different. An algorithm to compute a μ-basis of a rational surface is designed that is based on polynomial matrix factorization [35]. Another possible way is to compute a basis of the syzygy module of the surface with the application of Quillen–Suslin Theorem [36,37]. In order to avoid the extraneous factor in implicitization, people tried to find the strong μ-bases of surfaces that have the very similar properties of the μ-bases of curves. However, the surfaces with strong μ-bases are relatively rare [25,38].

The μ-basis has shown different advantages by assuming that the given parametrization is always proper. A recent result attempts to find inverse formula, proper reparametrization, and algebraic equation for an improper parametrization of an algebraic curve by using μ-basis [39]. In this paper, we pay attention to the μ-basis of an improper parametrization of an algebraic surface and then apply the μ-basis in the problems of proper reparametrization and implicitization. There are intrinsical differences in the discussions between the surface and the curve; hence, some results of the curve can not be extended to the surface straightforward. After we give the definition of μ-basis of a rational parametric surface defined parametrically by $\mathcal{P}(\bar{t})$, we find the inversion formula (if $\mathcal{P}$ is proper) and the degree of the rational map that is induced by $\mathcal{P}$ while using the μ-basis. Although the proper reparametrization problem is still opening, we address the problem of proper reparametrization partially based on the latest results and the properties of the μ-basis. As an important application, we derive the implicitization formula from the μ-basis from a given parametrization not being necessarily proper. Starting from the μ-basis, the computations only involve greatest common divisors (gcds) and univariate resultants of some polynomials constructed from the μ-basis. While the surface implicitization form μ-basis involved the computation of Gröbner bases in [5].

We have structured the article, as indicated below. In Section 2, we present some important definitions and properties for the μ-basis of rational surfaces. In Section 3, given a rational parametrization $\mathcal{P}$, we study the inversion computation (if $\mathcal{P}$ is proper) and the degree of the induced rational map. In Section 4, we focus on the proper reparametrization problem while using μ-basis. In Section 5, we come to the implicitization problem from a given rational parametrization not necessarily proper using μ-basis. We finish the paper with Section 6, where we present a brief summary of our work.

2. μ-Basis for Rational Surfaces: Definition and Previous Results

Let R denote the polynomial ring $\mathbb{K}[t_1, t_2]$ over an algebraically closed field $\mathbb{K}$ of characteristic zero and R^m denote the set of m-dimensional row vectors with entries in the polynomial ring R. A submodule M of R^m is a subset of R^m, for which this condition holds: for any $\mathbf{f}_1, \mathbf{f}_2 \in M$ and $h_1, h_2 \in R$, we have $h_1\mathbf{f}_1 + h_2\mathbf{f}_2 \in M$. A set of elements $\mathbf{f}_i \in M$, for $i = 1, \ldots, k$, is called a generating set of M if for any $\mathbf{m} \in M$, there exist $h_i \in R$, for $i = 1, \ldots, k$ satisfying that

$$\mathbf{m} = h_1 \mathbf{f}_1 + \ldots + h_k \mathbf{f}_k.$$

The Hilbert Basis Theorem states that every submodule $M \subset R^m$ has a finite generating set. If, for any $m \in M$, the above expression is unique, then $\{\mathbf{f}_1, \ldots, \mathbf{f}_k\}$ is called a basis of the module M. If a module has a basis, then it is called a free module. For any $(\mathbf{f}_1, \ldots, \mathbf{f}_k) \in R^k$, the set

$$\mathrm{syz}(\mathbf{f}_1, \ldots, \mathbf{f}_k) := \{(h_1, \ldots, h_k) \in R^k | h_1 \mathbf{f}_1 + \ldots + h_k \mathbf{f}_k = 0\}$$

is a module over R, called a syzygy module [40]. An important result regarding syzygy modules is that if $a, b, c, d \in \mathbb{R}[t_1, t_2]$ are four relatively prime polynomials then, the syzygy module $\mathrm{syz}(a, b, c, d)$ is a free module of rank 3 (see [5]).

Let $\mathcal{V}_a$ be a rational affine irreducible surface, and let

$$\mathcal{P}_a(\bar{t}) = \left(\frac{\wp_1(\bar{t})}{\wp_4(\bar{t})}, \frac{\wp_2(\bar{t})}{\wp_4(\bar{t})}, \frac{\wp_3(\bar{t})}{\wp_4(\bar{t})} \right) \in \mathbb{K}(\bar{t})^3, \ \bar{t} = (t_1, t_2)$$

be a rational affine parametrization of $\mathcal{V}_a$, where $\gcd(\wp_1, \wp_2, \wp_3, \wp_4) = 1$. Sometimes, we write the parametrization in the homogenous coordinate form $\mathcal{P}(\bar{t}) = (\wp_1(\bar{t}) : \wp_2(\bar{t}) : \wp_3(\bar{t}) : \wp_4(\bar{t}))$ and, in this case, we denote the surface in the projective space as $\mathcal{V}$.

A moving surface of degree l is a family of algebraic surfaces with parameter pairs (t_1, t_2)

$$S(\bar{x}, \bar{t}) = \sum_{i=1}^{\sigma} f_i(\bar{x}) b_i(\bar{t})$$

, where $f_i(\bar{x}), i = 1, \ldots, \sigma$ are polynomials of degree l, and $b_i(\bar{t}) \in \mathbb{R}[\bar{t}], i = 1, \ldots, \sigma$ (called blending functions) are linearly independent. We say that a moving surface follows the rational surface $\mathcal{P}$ if

$$\wp_4(\bar{t})^l S(\mathcal{P}_a(\bar{t}), \bar{t}) = 0.$$

We observe that the implicit equation of a given rational surface $\mathcal{V}$ is a moving surface of $\mathcal{P}$. A moving plane is a moving surface of degree one. We denote the next moving plane

$$A(\bar{t})x_1 + B(\bar{t})x_2 + C(\bar{t})x_3 + D(\bar{t})x_4$$

by $\mathbf{L}(\bar{t}) := (A(\bar{t}), B(\bar{t}), C(\bar{t}), D(\bar{t})) \in \mathbb{R}[\bar{t}]^4$. In the following, we denote, by $\mathbf{L}_{\bar{t}}$, the set of the moving planes that follow the rational surface that is parametrized by $\mathcal{P}$. Thus, $\mathbf{L}_{\bar{t}}$

is exactly the syzygy module syz($\wp_1, \wp_2, \wp_3, \wp_4$). Now, we define the μ-basis of the rational surface $\mathcal{P}$.

Definition 1. *Let $\mathbf{p}, \mathbf{q}, \mathbf{r} \in \mathbf{L}_{\bar{t}}$ be three moving planes satisfying that $[\mathbf{p}, \mathbf{q}, \mathbf{r}] = k\mathcal{P}(\bar{t})$, where k is a nonzero constant. Subsequently, it is said that $\mathbf{p}, \mathbf{q}, \mathbf{r}$ form a μ-basis of the rational surface $\mathcal{P}$. In the following, $[\mathbf{p}, \mathbf{q}, \mathbf{r}]$ denotes the outer product of $\mathbf{p}, \mathbf{q}, \mathbf{r}$.*

Geometrically, the above definition means that the point of rational surface $\mathcal{P}$ can be represented as the intersection of three moving planes $\mathbf{p}, \mathbf{q}, \mathbf{r}$. This definition is generalized from the moving lines of a rational curves [23]. Notice that the result in the curve case was proposed twenty years ago, but the surface case has been a mystery for a long time. The μ-bases surfaces that have the very similar properties of the μ-bases of curves is called strong μ-basis, but the strong μ-bases are relatively rare [25]. Therefore, we have to study the μ-basis of the surface from initial definition and, here, we review some basic theorems in [5].

Theorem 1. *For any rational surface $\mathcal{P}$, there always exist three moving planes $\mathbf{p}, \mathbf{q}, \mathbf{r}$, such that $[\mathbf{p}, \mathbf{q}, \mathbf{r}] = k \cdot \mathcal{P}(\bar{t})$ holds. In fact, any basis $\mathbf{p}, \mathbf{q}, \mathbf{r}$ of syz($\wp_1(\bar{t}), \wp_2(\bar{t}), \wp_3(\bar{t}), \wp_4(\bar{t})$) satisfies the above equality.*

Theorem 2. *Let $\mathbf{p}, \mathbf{q}, \mathbf{r}$ be a μ-basis of the rational surface $\mathcal{P}$. Thus, $\mathbf{p}, \mathbf{q}, \mathbf{r}$ provide a basis for the module $\mathbf{L}_{\bar{t}}$ (hence, $\mathbf{L}_{\bar{t}}$ is a free module). That is, for any $\mathbf{l}(\bar{t}) \in \mathbf{L}_{\bar{t}}$, there exist some polynomials $h_i(\bar{t}), i = 1, 2, 3$, satisfying that*

$$\mathbf{l}(\bar{t}) = h_1 \mathbf{p} + h_2 \mathbf{q} + h_3 \mathbf{r}.$$

In addition, the above expression is unique.

An immediate consequence of the above theorems is that if $\mathbf{p}, \mathbf{q}, \mathbf{r}$ form a μ-basis if and only if $\mathbf{p}, \mathbf{q}, \mathbf{r}$ are a basis of syz($\wp_1(\bar{t}), \wp_2(\bar{t}), \wp_3(\bar{t}), \wp_4(\bar{t})$)

3. Inversion and Degree Using μ-Basis

Let $\mathbb{K}$ be an algebraically closed field of characteristic zero. We denote, by $f(x_1, x_2, x_3) \in \mathbb{K}[x_1, x_2, x_3]$, the defining polynomial of a rational affine irreducible surface $\mathcal{V}_a$ defined by the rational affine parametrization

$$\mathcal{P}_a(\bar{t}) = \left(\frac{\wp_1(\bar{t})}{\wp_4(\bar{t})}, \frac{\wp_2(\bar{t})}{\wp_4(\bar{t})}, \frac{\wp_3(\bar{t})}{\wp_4(\bar{t})} \right) \in \mathbb{K}(\bar{t})^3, \quad \bar{t} = (t_1, t_2).$$

The homogeneous implicit polynomial defining the corresponding the projective rational surface $\mathcal{V}$ will be denoted as $F(x_1, x_2, x_3, x_4) \in \mathbb{K}[x_1, x_2, x_3, x_4]$, where $F(x_1, x_2, x_3, x_4) = x_4^{\deg(f)} f(x_1/x_4, x_2/x_4, x_3/x_4)$, and the parametrization in the homogenous coordinate form is given as $\mathcal{P}(\bar{t}) = (\wp_1(\bar{t}) : \wp_2(\bar{t}) : \wp_3(\bar{t}) : \wp_4(\bar{t}))$.

Besides implicitization, other applications of μ-basis include, as in the case of algebraic curves, point inversion and, in general, the computation of the fiber. The point inversion problem can be stated, as follows: given a point Q on the space, decide whether the point is on a rational surface $\mathcal{V}$ defined parametrically by $\mathcal{P}(\bar{t})$ or not, and, in the affirmative case, compute the corresponding parameter values t_1, t_2, such that $\mathcal{P}(t_1, t_2) = Q$. In this section, we recall some efficient algorithms that allow for computing the point inversion and, in general, the computation of the degree of the rational map that is induced by $\mathcal{P}$. For this purpose, we will use μ-basis.

In order to deal with these problems, we first recall that, associated with the parametrization $\mathcal{P}_a(\bar{t})$, we consider the induced rational map $\phi_\mathcal{P} : \mathbb{K} \longrightarrow \mathcal{V}_a \subset \mathbb{K}^3; \bar{t} \longmapsto \mathcal{P}_a(\bar{t})$. We denote, by $\deg(\phi_\mathcal{P})$, the degree of the induced rational map $\phi_\mathcal{P}$ (see [41] p. 143, and [42] p. 80). Observe that the birationality of $\phi_\mathcal{P}$, which is the properness of the input parametrization, is characterized by $\deg(\phi_\mathcal{P}) = 1$ (see [41,42]). We additionally remind that $\deg(\phi_\mathcal{P})$ determines the cardinality of the fiber of a generic element (see Theorem 7, p. 76 in [41]). The degree measures the number of times the parametrization traces the curve when the parameter takes values in $\mathbb{K}^2$. Finally, let $\mathcal{F}_\mathcal{P}(Q)$ be the fiber of a point $Q \in \mathcal{V}_a$; that is $\mathcal{F}_\mathcal{P}(Q) = \mathcal{P}_a^{-1}(Q) = \{\bar{t} \in \mathbb{K}^2 \,|\, \mathcal{P}_a(\bar{t}) = Q\}$.

In the following, given the projective parametrization $\mathcal{P}(\bar{t})$ of a surface $\mathcal{V}$ and

$$\mathbf{p}(\bar{t}) = (p_1, p_2, p_3, p_4),\ \mathbf{q}(\bar{t}) = (q_1, q_2, q_3, q_4),\ \mathbf{r}(\bar{t}) = (r_1, r_2, r_3, r_4)$$

a μ-basis for $\mathcal{P}(\bar{t})$, we consider a generic point $Q = (x_1, x_2, x_3, x_4)$ on the surface, and the polynomials

$$p^\mathcal{P}(\bar{t}, \bar{x}) = \mathbf{p}(\bar{t}) \cdot Q,\ q^\mathcal{P}(\bar{t}, \bar{x}) = \mathbf{q}(\bar{t}) \cdot Q,\ r^\mathcal{P}(\bar{t}, \bar{x}) = \mathbf{r}(\bar{t}) \cdot Q,\ \bar{x} = (x_1, x_2, x_3, x_4).$$

Remind that $p^\mathcal{P}(\bar{t}, \mathcal{P}(\bar{t})) = q^\mathcal{P}(\bar{t}, \mathcal{P}(\bar{t})) = r^\mathcal{P}(\bar{t}, \mathcal{P}(\bar{t})) = 0$ ($\mathbf{p}, \mathbf{q}, \mathbf{r}$ is a μ-basis for $\mathcal{P}(\bar{t})$). We denote, by V_1, V_2, V_3, the auxiliary curves over $\mathbb{K}(\mathcal{V})$, defined, respectively, by the polynomials $p^\mathcal{P}(\bar{t}, \bar{x}), q^\mathcal{P}(\bar{t}, \bar{x}), r^\mathcal{P}(\bar{t}, \bar{x}) \in \mathbb{K}(\mathcal{V})[\bar{t}]$, where $\mathbb{K}(\mathcal{V})$ is the field of rational functions of the given surface.

Finally, let

$$S^\mathcal{P}(t_1, \bar{x}) = \mathrm{pp}_{\bar{x}}(\mathrm{Cont}_Z(\mathrm{Res}_{t_2}(p^\mathcal{P}, q^\mathcal{P} + Zr^\mathcal{P}))) \in \mathbb{K}(\mathcal{V})[t_1],$$

where $\mathrm{Cont}_x(h)$ returns the content of a polynomial h w.r.t. some variable x, $\mathrm{pp}_x(h)$ returns the primitive part of a polynomial h with respect to a variable x and $\mathrm{Res}_x(h_1, h_2)$ returns the resultant of two polynomials h_1 and h_2 w.r.t. some variable x. Similarly, one also considers the polynomial

$$T^\mathcal{P}(t_2, \bar{x}) = \mathrm{pp}_{\bar{x}}(\mathrm{Cont}_Z(\mathrm{Res}_{t_1}(p^\mathcal{P}, q^\mathcal{P} + Zr^\mathcal{P}))) \in \mathbb{K}(\mathcal{V})[t_2].$$

The computation of $S^\mathcal{P}, T^\mathcal{P}$ can be done in two different ways. First, we consider that the implicit equation defining the input surface is known. In this case, we carry out the arithmetic over $\mathbb{K}(\mathcal{V})$ while using this implicit equation. We observe that, since $I(\mathcal{V}) = \langle F \rangle$ ($I(\mathcal{V})$ represents the ideal of $\mathcal{V}$), the basic arithmetic on $\mathbb{K}[\mathcal{V}]$ can be carried out by computing polynomial remainders. Thus, we conclude that the quotient field $\mathbb{K}(\mathcal{V})$ is computable. Furthermore, we note that we calculate the resultants of polynomials in $\mathbb{K}(\mathcal{V})[\bar{t}]$, which is a unique factorization domain, and we compute gcds of univariate polynomials over $\mathbb{K}(\mathcal{V})$ and, thus, in an Euclidean domain. In the second way, we avoid the requirement on the implicit equation. More precisely, the elements are represented (not uniquely) as function of polynomials in the variables x_1, x_2, x_3, x_4. We check the zero equality while using the input rational parametrization. This way could be too time consuming. In order to avoid this problem, one may test zero–equality by substituting a random point on the surface. The result of this test is correct with probability almost one. Additionally, one may also test the correctness of the computation of the inverse by checking it on a randomly chosen point on the given surface. In this way, we avoid the computation of the implicit polynomial.

In the following theorem, we provide the technique for computing the components of the inverse of a given rational proper parametrization $\mathcal{P}(\bar{t})$. Additionally, we characterize the properness of $\mathcal{P}(\bar{t})$. We illustrate this result in Example 1.

Theorem 3. *The rational parametrization $\mathcal{P}(\bar{t})$ is proper if and only if for a generic point $Q = (x_1, x_2, x_3, x_4)$ on the surface, it holds that $\deg_{t_1}(S^\mathcal{P}) = 1$. In this case, the t_1-coordinate of the inverse of $\mathcal{P}(\bar{t})$ is given by solving $S^\mathcal{P}(t_1, \bar{x}) = 0$ w.r.t the variable t_1.*

Proof. Using the results shown in [43] (see Proposition 1), we deduce that the non-constant t_1-coordinates of the intersections points in V_i, $i = 1, 2, 3$ are given by the roots of the polynomial $S^\mathcal{P}(t_1, \bar{x})$. Thus, we only have to prove that $\mathcal{P}(\bar{t})$ is proper if and only there exists one unique point $A = (A_1, A_2) \in (V_1 \cap V_2 \cap V_3) \cap (K \setminus \mathbb{K})^2$ (K denotes the algebraic closure of the field $\mathbb{K}(\mathcal{V})$; i.e $K = \overline{\mathbb{K}(\mathcal{V})}$). Indeed: first, let $M = (M_1(\overline{x}), M_2(\overline{x}))$ be the inverse of the rational proper parametrization $\mathcal{P}(\overline{t})$. Subsequently, $M(\mathcal{P}) = \overline{t}$ and, thus, $p^{\mathcal{P}}(M(\mathcal{P}), \mathcal{P}) = q^{\mathcal{P}}(M(\mathcal{P}), \mathcal{P}) = r^{\mathcal{P}}(M(\mathcal{P}), \mathcal{P}) = 0$, which implies that

$$M \in V_1 \cap V_2 \cap V_3 \cap K^2.$$

In addition, since M is the inverse of $\mathcal{P}(\overline{t})$, one has that $M \in (K \setminus \mathbb{K})^2$. Hence, $M \in (V_1 \cap V_2 \cap V_3) \cap (K \setminus \mathbb{K})^2$. Now, let us see that M is unique. Let $M^* \in (V_1 \cap V_2 \cap V_3) \cap (K \setminus \mathbb{K})^2$. The equalities $p^{\mathcal{P}}(M^*, \overline{x}) = q^{\mathcal{P}}(M^*, \overline{x}) = r^{\mathcal{P}}(M^*, \overline{x}) = 0$ imply that

$$p^{\mathcal{P}}(R(\overline{t}), \mathcal{P}) = q^{\mathcal{P}}(R(\overline{t}), \mathcal{P}) = r^{\mathcal{P}}(R(\overline{t}), \mathcal{P}) = 0, \quad R(\overline{t}) = M^*(\mathcal{P}).$$

Afterwards, by the properties of resultants and by Lemma 1, we get that $\mathcal{P}(\overline{t}) = k\mathcal{P}(R(\overline{t}))$ and since $\mathcal{P}$ is proper we deduce that $R(\overline{t}) = M^*(\mathcal{P}(\overline{t})) = \overline{t}$. Thus, left composing by $\mathcal{P}^{-1}$, we get that $M^* = \mathcal{P}^{-1} = M$.

Reciprocally, because there exists a unique point in $(V_1 \cap V_2 \cap V_3) \cap (K \setminus \mathbb{K})^2$, A is fixed under the action of the Galois group and, thus, $A \in \mathbb{K}(\mathcal{V})^2$. Reasoning similarly, as we did for the uniqueness in the above implication, one gets that $A \circ \mathcal{P} = \overline{t}$ and, then, we conclude that A is the inverse of $\mathcal{P}$. □

Remark 1. *Theorem 3 can be stated similarly for $T^{\mathcal{P}}(t_2, \overline{x})$. More precisely, $\mathcal{P}(\overline{t})$ is proper if and only if, for a generic point $Q = (x_1, x_2, x_3, x_4)$ on the surface, it holds that $\deg_{t_2}(T^{\mathcal{P}}) = 1$. In this case, the t_2-coordinate of the inverse of $\mathcal{P}(\overline{t})$ is given by solving $T^{\mathcal{P}}(t_2, \overline{x}) = 0$ w.r.t the variable t_2.*

We also note that we may work over the affine space (i.e., $x_4 = 1$) and the obtained results are the same, but over the affine space. If $x_4 = 1$, then the computations are more efficient.

Example 1. *Let $\mathcal{V}$ be the rational surface that is defined by the parametrization*

$$\mathcal{P}(\overline{t}) = (t_2 t_1 : t_2 + t_1 : t_2 - t_1 : t_2^2 + t_1^2 + 2).$$

First, we compute the polynomials

$$p^{\mathcal{P}}(\overline{t}, \overline{x}) = \mathbf{p}(\overline{t}) \cdot \overline{x} = -2x_1 + t_1 x_2 + t_1 x_3,$$
$$q^{\mathcal{P}}(\overline{t}, \overline{x}) = \mathbf{q}(\overline{t}) \cdot \overline{x} = -2t_1 x_1 + (t_1^2 + 1 + t_2 t_1)x_2 - x_3 - t_1 x_4,$$
$$r^{\mathcal{P}}(\overline{t}, \overline{x}) = \mathbf{r}(\overline{t}) \cdot \overline{x} = (-2t_2 + 2t_1)x_1 + (t_2^2 - t_1^2)x_2 + 2x_3 + (-t_2 + t_1)x_4,$$

where the μ-basis is given as

$$\mathbf{p}(\overline{t}) = (-2, t_1, t_1, 0),$$
$$\mathbf{q}(\overline{t}) = (-2t_1, t_1^2 + 1 + t_2 t_1, -1, -t_1),$$
$$\mathbf{r}(\overline{t}) = (-2t_2 + 2t_1, t_2^2 - t_1^2, 2, -t_2 + t_1).$$

Now, we determine $S^{\mathcal{P}}(t_1, \overline{x})$ and $T^{\mathcal{P}}(t_2, \overline{x})$. We obtain

$$S^{\mathcal{P}}(t_1, \overline{x}) = -2x_1 + t_1 x_2 + t_1 x_3,$$

$$T^{\mathcal{P}}(t_2, \overline{x}) = -2x_1 x_4 x_2 - 2x_1 x_4 x_3 + 2x_1 x_2^2 t_2 + 2x_1 x_2 t_2 x_3 - 4x_1^2 x_3 - x_2 x_3^2 - x_3^3 + x_2^3 + x_3 x_2^2.$$

Because $\deg_{t_1}(S^{\mathcal{P}}) = 1$, we conclude that $\mathcal{P}$ is proper and the first coordinate of the inverse is given as

$$I_1 = \frac{2x_1}{x_2 + x_3}.$$

Reasoning similarly with T^P, *we obtain the second coordinate of the inverse, which is given as*

$$I_2 = \frac{2x_1x_4x_2 + 2x_1x_4x_3 + x_3^3 - x_2^3 + 4x_1^2x_3 + x_2x_3^2 - x_3x_2^2}{2x_1x_2(x_2+x_3)}.$$

Based upon the above theorem, we may compute $\mathcal{F}_\mathcal{P}(Q)$ for a generic point Q and, thus, to obtain the degree of the rational map induced by $\deg(\phi_\mathcal{P})$. For this purpose, we consider $Q = \mathcal{P}_a(\bar{s})$, where $\bar{s} = (s_1, s_2)$ are new variables, and the polynomials

$$S^\mathcal{P}(t_1, \bar{s}) = \mathrm{pp}_{\bar{s}}(\mathrm{Cont}_Z(\mathrm{Res}_{t_2}(p^\mathcal{P}, q^\mathcal{P} + Zr^\mathcal{P}))) \in \mathbb{K}[t_1, \bar{s}],$$

$$T^\mathcal{P}(t_2, \bar{s}) = \mathrm{pp}_{\bar{s}}(\mathrm{Cont}_Z(\mathrm{Res}_{t_1}(p^\mathcal{P}, q^\mathcal{P} + Zr^\mathcal{P}))) \in \mathbb{K}[t_2, \bar{s}],$$

where

$$p^\mathcal{P}(\bar{t}, \bar{s}) = \mathbf{p}(\bar{t}) \cdot \mathcal{P}(\bar{s}), \quad q^\mathcal{P}(\bar{t}, \bar{s}) = \mathbf{q}(\bar{t}) \cdot \mathcal{P}(\bar{s}), \quad r^\mathcal{P}(\bar{t}, \bar{s}) = \mathbf{r}(\bar{t}) \cdot \mathcal{P}(\bar{s}).$$

Remind that $p^\mathcal{P}(\bar{t}, \bar{t}) = q^\mathcal{P}(\bar{t}, \bar{t}) = r^\mathcal{P}(\bar{t}, \bar{t}) = 0$ ($\mathbf{p}, \mathbf{q}, \mathbf{r}$ is a μ-basis for $\mathcal{P}(\bar{t})$). We denote, by V_1, V_2, V_3, the auxiliary curves over $\mathbb{K}(\bar{s})$ defined, respectively, by the polynomials $p^\mathcal{P}(\bar{t}, \bar{s}), q^\mathcal{P}(\bar{t}, \bar{s}), r^\mathcal{P}(\bar{t}, \bar{s}) \in \mathbb{K}[\bar{t}, \bar{s}]$. Thus, one obtains the following proposition.

Theorem 4. *For a generic point $Q = \mathcal{P}_a(\bar{s})$, it holds that*

$$\deg(\phi_\mathcal{P}) = \mathrm{Card}(\mathcal{F}_\mathcal{P}(Q)) = \deg_{t_1}(S^\mathcal{P}(t_1, \bar{s})) = \deg_{t_2}(T^\mathcal{P}(t_2, \bar{s})).$$

Proof. First, we use Proposition 1 in [43], and we deduce that the non-constant t_1-coordinates of the intersections points in V_i, $i = 1, 2, 3$ are given by the roots of the polynomial $S^\mathcal{P}(t_1, \bar{s})$. Thus, we only have to prove that $M \in \mathcal{F}_\mathcal{P}(Q)$ if and only if $M \in V_1 \cap V_2 \cap V_3 \cap (K \setminus \mathbb{K})^2$, where $K = \overline{\mathbb{K}(\bar{s})}$ is the algebraic closure of the field. Indeed, if $M \in V_1 \cap V_2 \cap V_3 \cap (K \setminus \mathbb{K})^2$ thus $p^\mathcal{P}(M, \bar{s}) = q^\mathcal{P}(M, \bar{s}) = r^\mathcal{P}(M, \bar{s}) = 0$, which implies that

$$\mathbf{p}(M) \cdot \mathcal{P}(\bar{s}) = \mathbf{q}(M) \cdot \mathcal{P}(\bar{s}) = \mathbf{r}(M) \cdot \mathcal{P}(\bar{s}) = 0.$$

Because

$$\mathbf{p}(M) \cdot \mathcal{P}(M) = \mathbf{q}(M) \cdot \mathcal{P}(M) = \mathbf{r}(M) \cdot \mathcal{P}(M) = 0$$

, we get that $\mathcal{P}(M) = k\mathcal{P}(\bar{s})$ with $k \neq 0$ (since $M \notin \mathbb{K}^2$), which implies that $\mathcal{P}_a(M) = \mathcal{P}_a(\bar{s})$. Hence, $M \in \mathcal{F}_\mathcal{P}(Q)$.

Reciprocally, let $M \in \mathcal{F}_\mathcal{P}(Q)$. Subsequently, $\mathcal{P}_a(M) = \mathcal{P}_a(\bar{s})$ which implies that $\mathcal{P}(M) = k\mathcal{P}(\bar{s})$ with $k \neq 0$. Because

$$\mathbf{p}(M) \cdot \mathcal{P}(M) = \mathbf{q}(M) \cdot \mathcal{P}(M) = \mathbf{r}(M) \cdot \mathcal{P}(M) = 0,$$

we get that

$$\mathbf{p}(M) \cdot \mathcal{P}(\bar{s}) = \mathbf{q}(M) \cdot \mathcal{P}(\bar{s}) = \mathbf{r}(M) \cdot \mathcal{P}(\bar{s}) = 0.$$

Thus, $p^\mathcal{P}(M, \bar{s}) = q^\mathcal{P}(M, \bar{s}) = r^\mathcal{P}(M, \bar{s}) = 0$ and, hence, $M \in V_1 \cap V_2 \cap V_3$. Furthermore, since $M \in \mathcal{F}_\mathcal{P}(Q)$, we also get that $M \in (K \setminus \mathbb{K})^2$. □

Clearly, Theorem 4 can be also stated for a generic point $Q = (x_1, x_2, x_3, x_4)$ on the surface, and the polynomials

$$p^\mathcal{P}(\bar{t}, \bar{x}) = \mathbf{p}(\bar{t}) \cdot Q, \quad q^\mathcal{P}(\bar{t}, \bar{x}) = \mathbf{q}(\bar{t}) \cdot Q, \quad r^\mathcal{P}(\bar{t}, \bar{x}) = \mathbf{r}(\bar{t}) \cdot Q, \quad \bar{x} = (x_1, x_2, x_3, x_4).$$

(remind that $p^\mathcal{P}(\bar{t}, \mathcal{P}(\bar{t})) = q^\mathcal{P}(\bar{t}, \mathcal{P}(\bar{t})) = r^\mathcal{P}(\bar{t}, \mathcal{P}(\bar{t})) = 0$ ($[\mathbf{p}, \mathbf{q}, \mathbf{r}] = k\mathcal{P}(\bar{t})$). For this purpose, one considers the polynomials

$$S^\mathcal{P}(t_1, \bar{x}) = \mathrm{pp}_{\bar{x}}(\mathrm{Cont}_Z(\mathrm{Res}_{t_2}(p^\mathcal{P}, q^\mathcal{P} + Zr^\mathcal{P}))) \in \mathbb{K}(\mathcal{V})[t_1],$$

and
$$T^{\mathcal{P}}(t_2, \bar{x}) = \text{pp}_{\bar{x}}(\text{Cont}_Z(\text{Res}_{t_1}(p^{\mathcal{P}}, q^{\mathcal{P}} + Zr^{\mathcal{P}}))) \in \mathbb{K}(\mathcal{V})[t_2].$$

where computation can be done, as we described in the paragraph before Theorem 3, i.e., over the field of rational functions $\mathbb{K}(\mathcal{V})$. Thus, one has the following corollary.

Corollary 1. *For a generic point $Q = (x_1, x_2, x_3, x_4) \in \mathcal{V}$, it holds that*
$$\deg(\phi_{\mathcal{P}}) = \text{Card}(\mathcal{F}_{\mathcal{P}}(Q)) = \deg_{t_1}(S^{\mathcal{P}}(t_1, \bar{x})) = \deg_{t_2}(T^{\mathcal{P}}(t_2, \bar{x})).$$

Remark 2. *From the proof of Theorem 4, we deduce that $\mathcal{F}_{\mathcal{P}}(Q) = V_1 \cap V_2 \cap V_3 \cap (K \setminus \mathbb{K})^2$.*

Example 2. *Let $\mathcal{V}$ be the rational surface that is defined by the parametrization*
$$\mathcal{P}(\bar{t}) = (t_2^2 t_1^2 - t_1^4 : -t_2 + t_2^3 + t_2 t_1^4 : -t_1^2 + t_2 t_1^2 + t_2^2 t_1^2 - t_1^4 : -t_1^2 + t_2^2 t_1^2 + t_1^4).$$

We determine the polynomials $p^{\mathcal{P}}(\bar{t}, \bar{s}) = \mathbf{p}(\bar{t}) \cdot \mathcal{P}(\bar{s}), q^{\mathcal{P}}(\bar{t}, \bar{s}) = \mathbf{q}(\bar{t}) \cdot \mathcal{P}(\bar{s}), r^{\mathcal{P}}(\bar{t}, \bar{s}) = \mathbf{r}(\bar{t}) \cdot \mathcal{P}(\bar{s})$, where the μ-basis is given by

$$\begin{aligned}
\mathbf{p}(\bar{t}) &= (t_2 t_1^2 + 2t_1^2 + 4t_2^4 + 6t_2^3 - 4t_2 - 4, -2t_1^2, \\
&\quad -2t_2 t_1^2 - t_1^2 - 4t_2^4 - 4t_2^3 + 4t_2 + 2, t_2 t_1^2 + t_1^2 + 2t_2^3 - 2) \\
\mathbf{q}(\bar{t}) &= (-2t_2 t_1^2 - 3t_1^2 + t_2 + 1, 0, 2t_2 t_1^2 + 2t_1^2 - t_2 - 1, -t_1^2 + 1) \\
\mathbf{r}(\bar{t}) &= (-2t_2^2 - t_2 + 2, 0, 2t_2^2 - 1, -t_2 + 1).
\end{aligned}$$

Now, we determine $S^{\mathcal{P}}(t_1, \bar{s})$ (similarly, if we compute $T^{\mathcal{P}}(t_2, \bar{s})$), and we obtain
$$S^{\mathcal{P}}(t_1, \bar{s}) = \text{pp}_{\bar{s}}(\text{Cont}_Z(\text{Res}_{t_2}(p^{\mathcal{P}}, q^{\mathcal{P}} + Zr^{\mathcal{P}}))) = t_1^2 - s_1^2 \in \mathbb{K}[t_1, \bar{s}].$$

Therefore, applying Theorem 4, we conclude that $\mathcal{P}$ is not proper and in fact $\deg(\phi_{\mathcal{P}}) = 2$. Furthermore,
$$\mathcal{F}_{\mathcal{P}}(\mathcal{P}_a(\bar{s})) = \{(s_1, s_2), (-s_1, s_2)\}.$$

4. On the Problem of the Reparametrization Using μ-Basis

In this section, we consider the problem of computing a rational proper reparametrization of a given algebraic surface defined by an improper parametrization. That is, given an algebraically closed field $\mathbb{K}$, and $\mathcal{P}(\bar{t})$, $\bar{t} = (t_1, t_2)$, a rational parametrization of a surface $\mathcal{V}$ over $\mathbb{K}$, we want to compute a proper parametrization of $\mathcal{V}$, $\mathcal{Q}(\bar{t})$, and $R(\bar{t}) \in (\mathbb{K}(\bar{t}) \setminus \mathbb{K})^2$, such that
$$\mathcal{P}(\bar{t}) = \mathcal{Q}(R(\bar{t})).$$

Notice that we consider $\mathcal{Q}(R(\bar{t}))$, with $R(\bar{t}) = (r_1(\bar{t})/r(\bar{t}), r_2(\bar{t})/r(\bar{t})) \in (\mathbb{K}(\bar{t}) \setminus \mathbb{K})^2$, in homogenous form, i.e., $\mathcal{P}(\bar{t}) = \mathcal{Q}(R(\bar{t}))$ means that $\mathcal{P}(\bar{t}) = \mathcal{Q}\left(\frac{r_1(\bar{t})}{r(\bar{t})}, \frac{r_2(\bar{t})}{r(\bar{t})}\right) r(\bar{t})^{\deg(\mathcal{Q})}$, which is a polynomial vector in homogenous form.

In this section, although we do not provide a solution to the general reparametrization problem, we show how the μ-basis can be used to provide some information concerning $R(\bar{t}) \in (\mathbb{K}(\bar{t}) \setminus \mathbb{K})^2$. We address the problem partially and the idea presented is based in the results in [32], but we trust that we could develop deeply these new approaches in future works, and more results concerning this topic allow us to get more advances.

The approach that is presented in this section is based on the computation of polynomial gcds and univariate resultants. These techniques always work and the time performance is very effective. The algorithm presented follows directly from the algorithm that was developed in [39], which solves the problem for the case of curves. Accordingly, we first outline this approach and illustrate it with an example.

Example 3. *Let $\mathcal{C}$ be the rational curve that is defined by the parametrization*

$$\mathcal{P}(t) = (-t^4 + t^3 + 2t^2 + 2t + 1 : (t+1)(3t^2 + 2t + 2) : 2t^4 + 3t^3 + 5t^2 + 4t + 2).$$

In Step 2 of the algorithm, we determine the polynomials $p^\mathcal{P}(t,s)$ and $q^\mathcal{P}(t,s)$, where the μ-basis is
$$\mathbf{p}(t) = \left(\; 40t^2 + 40t + 40, -30t^2 - 35t - 35, 20t^2 + 15t + 15 \; \right)^T$$
$$\mathbf{q}(t) = \left(\; 30t^2 + 20t + 20, -25t^2 - 20t - 20, 15t^2 + 10t + 10 \; \right)^T.$$

Now, we compute $G^\mathcal{P}(t,s)$,

$$G^\mathcal{P}(t,s) = C_0(t) + C_1(t)s + C_2(t)s^2,$$

where $C_0(t) = -10t^2$, $C_1(t) = -10t^2$, and $C_2(t) = 10 + 10t$. Because $m := \deg_t(G^\mathcal{P}) > 1$, we go to Step 5 of Algorithm 1, and we consider

$$R(t) = \frac{C_2(t)}{C_0(t)} = \frac{-1-t}{t^2}.$$

Note that $\gcd(C_0, C_2) = 1$. Now, we determine the polynomials

$$L_1(s, x_1) = \operatorname{Res}_t(x_1 \wp_3(t) - \wp_1(t), sC_0(t) - C_2(t)) = (x_3 + sx_3 + 2x_1 - s^2 x_3 - 3sx_1 + 2s^2 x_1)^2,$$

$$L_2(s, x_2) = \operatorname{Res}_t(x_2 \wp_3(t) - \wp_2(t), sC_0(t) - C_2(t)) = (3sx_3 + 2x_2 - 3sx_2 + 2s^2 x_2 - 2s^2 x_3)^2.$$

Finally, the algorithm outputs the proper parametrization $\mathcal{Q}(t)$, and the rational function $R(t)$ (see Step 7)

$$\mathcal{Q}(t) = \left(t^2 - 1 - t : t(-3 + 2t) : 2 - 3t + 2t^2 \right), \qquad R(t) = \frac{-1-t}{t^2}.$$

Algorithm 1 Proper Reparametrization for Curves using μ-Basis

Input a rational parametrization $\mathcal{P}(t) = (\wp_1(t) : \wp_2(t) : \wp_3(t))$, of a plane algebraic curve $\mathcal{C}$.
Output a rational proper parametrization $\mathcal{Q}(t)$ of $\mathcal{C}$, and a rational function $R(t)$ such that $\mathcal{P}(t) = \mathcal{Q}(R(t))$.
Steps
1. Compute a μ-basis of $\mathcal{P}$. Let $\mathbf{p}(t), \mathbf{q}(t)$ be this μ-basis.
2. Compute $p^\mathcal{P}(t,s) = \mathbf{p}(t) \cdot \mathcal{P}(s)$, $q^\mathcal{P}(t,s) = \mathbf{q}(t) \cdot \mathcal{P}(s)$.
3. Compute $G^\mathcal{P}(t,s) = \gcd(p^\mathcal{P}(t,s), q^\mathcal{P}(t,s)) = C_m(t)s^m + \cdots + C_0(t)$. Let $m := \deg_t(G^\mathcal{P}(t,s))$.
4. If $m = 1$, return $\mathcal{Q}(t) = \mathcal{P}(t)$, and $R(t) = t$. Otherwise go to Step 5.
5. Consider $R(t) = \frac{C_i(t)}{C_j(t)} \in \mathbb{K}(t)$, such that $C_j(t), C_i(t)$ are not associated polynomials (i.e., $C_j(t) \neq kC_i(t), k \in \mathbb{K}$).
6. For $i = 1, 2$, compute

$$L_i(s, x_i) = \operatorname{Res}_t(x_i \wp_3(t) - \wp_i(t), sC_j(t) - C_i(t)) = (q_{i2}(s)x_i - q_{i1}(s))^{\deg(R)}.$$

7. Return $\mathcal{Q}_a(t) = (q_{11}(t)/q_{12}(t), q_{21}(t)/q_{22}(t))$ or the equivalent projective parametrization $\mathcal{Q}(t)$, and $R(t) = C_i(t)/C_j(t)$.

The main idea of the result that we develop in this paper consists in computing a reparametrization of two auxiliary parametrizations (defining two space curves), $\mathcal{P}_1$ and $\mathcal{P}_2$, directly defined from a given rational parametrization of the surface $\mathcal{P}$ (see Definition 2). Moreover, using that the degree of a rational map is multiplicative under composition, we get some results that relate the degree of the rational map that is induced by $\mathcal{P}$ with the degree of $\mathcal{Q}$, and the degree of the rational maps induced by $\mathcal{P}_1$ and $\mathcal{P}_2$. In addition, we also prove the relation with the degree w.r.t. the variables t_1, t_2 of $R(\bar{t}) \in \mathbb{K}(\bar{t})^2$.

To start with, we first provide the following lemma that analyzes the behavior of the μ-basis under reparametrizations.

Lemma 1. *Let $\tilde{\mathbf{p}}(\bar{t})$, $\tilde{\mathbf{q}}(\bar{t})$ and $\tilde{\mathbf{r}}(\bar{t})$ be a μ-basis for a parametrization $\mathcal{Q}(\bar{t})$ of a surface $\mathcal{V}$. Let $R(\bar{t}) \in (\mathbb{K}(\bar{t}) \setminus \mathbb{K})^2$. Subsequently, $\mathbf{p}(\bar{t}) = \tilde{\mathbf{p}}(R(\bar{t}))$, $\mathbf{q}(\bar{t}) = \tilde{\mathbf{q}}(R(\bar{t}))$, $\mathbf{r}(\bar{t}) = \tilde{\mathbf{r}}(R(\bar{t}))$ is a μ-basis for the reparametrization $\mathcal{P}(\bar{t}) = \mathcal{Q}(R(\bar{t}))$.*

Proof. Taking into account that $\tilde{\mathbf{p}}(\bar{t})$, $\tilde{\mathbf{q}}(\bar{t})$ and $\tilde{\mathbf{r}}(\bar{t})$ is a μ-basis for $\mathcal{Q}(\bar{t})$, from Theorem 1, it follows that $[\tilde{\mathbf{p}}, \tilde{\mathbf{q}}, \tilde{\mathbf{r}}] = k\,\mathcal{Q}(\bar{t})$ for some non-zero constant k. Therefore, we easily get that $[\mathbf{p}, \mathbf{q}, \mathbf{r}] = k\,\mathcal{P}(\bar{t})$ for some non-zero constant k. Hence, from Theorem 1, we conclude that $\mathbf{p}(\bar{t})$, $\mathbf{q}(\bar{t})$, $\mathbf{r}(\bar{t})$ is a μ-basis for $\mathcal{P}(\bar{t})$. □

In the next proposition, we assume that we know $\mathbf{p}$, $\mathbf{q}$ and $\mathbf{r}$ and $R(\bar{t}) = (r_1(\bar{t}), r_2(\bar{t})) \in (\mathbb{K}(\bar{t}) \setminus \mathbb{K})^2$, and we present a method for computing $\tilde{\mathbf{p}}$, $\tilde{\mathbf{q}}$ and $\tilde{\mathbf{r}}$ from $\mathbf{p}$, $\mathbf{q}$ and $\mathbf{r}$, respectively. We state Proposition 1 for $\mathbf{p} = (p_1, p_2, p_3, p_4)$ and $\tilde{\mathbf{p}} = (\tilde{p}_1, \tilde{p}_2, \tilde{p}_3, \tilde{p}_4)$. One reasons, similarly, to obtain $\tilde{\mathbf{q}}$ from $\mathbf{q}$ and $\tilde{\mathbf{r}}$ from $\mathbf{r}$.

We assume w.l.o.g that $p_4 \neq 0$m which implies that $\tilde{p}_4 \neq 0$ (otherwise, we consider another non-zero component of $\mathbf{p}$). Thus, we may write

$$(p_1/p_4, p_2/p_4, p_3/p_4) = (\tilde{p}_1/\tilde{p}_4, \tilde{p}_2/\tilde{p}_4, \tilde{p}_3/\tilde{p}_4)(R(\bar{t})).$$

Let us assume that $\gcd(p_1, p_4) = \gcd(\tilde{p}_1, \tilde{p}_4) = 1$ (otherwise, we simplify these rational functions). In addition, we note that, if $p_1 = 0$, then we easily get that $\tilde{p}_1 = 0$. For the case of $p_1 \neq 0$, we consider $(r_1, r_2, p_1/p_4)$ that can be seen as an affine parametrization of the surface defined by the irreducible polynomial $\tilde{p}_1(x_1, x_2) - x_3 \tilde{p}_4(x_1, x_2) \in \mathbb{K}[x_1, x_2, x_3]$ (note that $\gcd(\tilde{p}_1, \tilde{p}_4) = 1$ and $\tilde{p}_4 \tilde{p}_1 \neq 0$). Hence, we only have to compute the implicit equation of that surface by applying, for instance, the method that is presented in [44].

Reasoning, similarly, $(r_1, r_2, p_i/p_4)$ can be seen as a parametrization of the surface defined by the irreducible polynomial $\tilde{p}_i(x_1, x_2) - x_3 \tilde{p}_4(x_1, x_2) \in \mathbb{K}[x_1, x_2, x_3]$, for $i = 2, 3$. Summarizing, we have the following proposition.

Proposition 1. *Under the conditions that are stated above, it holds that the implicit equation of the parametrization $(r_1, r_2, p_i/p_4)$ is given as $\tilde{p}_i(x_1, x_2) - x_3 \tilde{p}_4(x_1, x_2)$, for $i = 1, 2, 3$.*

In Remark 3, we apply the same idea that is stated in Proposition 1, but for the particular case of curves.

Given a rational projective parametrization $\mathcal{N}(\bar{t})$ of a surface over $\mathbb{K}$, in Definition 2 we introduce some auxiliary parametrizations over $\mathbb{K}(t_i)$ that are defined from $\mathcal{N}$.

Definition 2. *Let $\mathcal{N}(\bar{t})$ be a parametrization with coefficients in $\mathbb{K}$. We define the partial parametrizations associated to $\mathcal{N}$ as the parametrizations $\mathcal{N}_i(t_j) := \mathcal{N}(\bar{t})$ with coefficients in $\mathbb{K}[t_i]$ (i.e., $\mathcal{N}_i$ is defined over $\mathbb{K}[t_i]$), for $i, j \in \{1, 2\}$ and $i \neq j$.*

We note that the partial parametrization $\mathcal{N}_i(t_j)$ ($i \neq j$) determines a space curve in $\overline{\mathbb{K}(t_i)}^3$, where $\overline{\mathbb{K}(t_i)}$ is the algebraic closure of $\mathbb{K}[t_i]$. In addition, we note that Definition 2 can also be stated for any $N(\bar{t}) \in \mathbb{K}(\bar{t})^2$. That is, given $N(\bar{t}) \in \mathbb{K}(\bar{t})^2$, one may consider $N_i(t_j) := N(\bar{t}) \in (\mathbb{K}[t_i])(t_j)^2$ (i.e., N is seen defined over $\mathbb{K}[t_i]$ and in the variable t_j), for $i, j \in \{1, 2\}$ and $i \neq j$. Similarly, one also may adapt Definition 2 for any polynomial $n(\bar{t}) \in \mathbb{K}(\bar{t})$.

The properness of the input parametrization $\mathcal{P}$ of a surface $\mathcal{V}$ can be characterized by means of the properness of its partial parametrizations. In particular, it is proved that $\mathcal{P}$ is birational if and only if its associated partial parametrizations, $\mathcal{P}_i$, are proper and $\mathcal{P}_i^{-1} \in \mathbb{K}(\bar{x}) \setminus \mathbb{K}(t_i)$, where $\bar{x} = (x_1, x_2, x_3, x_4)$ (see [32]).

In the following, given a rational affine parametrization $\mathcal{P}(\bar{t})$ of a surface $\mathcal{V}$, we develop an algorithm that computes a rational parametrization $\mathcal{Q}(\bar{t})$ of $\mathcal{V}$, and $R(\bar{t}) \in \mathbb{K}(\bar{t})^2$, such that $\mathcal{P}(\bar{t}) = \mathcal{Q}(R(\bar{t}))$. The algorithm is based on the computation of polynomial gcds and univariate resultants whose computing time performance is very satisfactory.

We prove that the partial parametrizations that correspond to the output parametrization, $\mathcal{Q}(\bar{t})$, are proper (see Theorem 5), and we get properties relating the degree of $\phi_\mathcal{P}$ with the degree of the rational map $\phi_\mathcal{Q}$, and the degree of $R(\bar{t})$ (see Theorem 6). More precisely, we prove that

$$\deg(\phi_\mathcal{P}) = \deg(\phi_\mathcal{Q})\deg_{t_1}(S)\deg_{t_2}(T)$$

where

$$R(\bar{t}) = (S(\bar{t}), T(S(\bar{t}), t_2)), \quad S, T \in \mathbb{K}(\bar{t}).$$

In Corollaries 2 and 3, we analyze in which conditions $\deg(\phi_\mathcal{Q}) = 1$ or, otherwise, the degree of the rational map induced by $\mathcal{Q}(\bar{t})$ is lower than the degree that is induced by the input parametrization $\mathcal{P}(\bar{t})$.

In Theorem 5, we have that the partial parametrizations associated to the output parametrization, $\mathcal{Q}(\bar{t})$, are proper (see [32]) but we can not ensure that $\mathcal{Q}$ is proper.

Theorem 5. *The partial parametrizations $\mathcal{Q}_1(t_2)$ and $\mathcal{Q}_2(t_1)$ associated to the parametrization $\mathcal{Q}$ computed by Algorithm 2 are proper.*

Algorithm 2 Proper Reparametrization for Surfaces using μ-Basis

---

Input a rational parametrization $\mathcal{P}(\bar{t}) = (\wp_1(\bar{t}) : \wp_2(\bar{t}) : \wp_3(\bar{t}) : \wp_4(\bar{t}))$ of an algebraic surface $\mathcal{V}$.
Output a rational parametrization $\mathcal{Q}(\bar{t})$ of $\mathcal{V}$, and $R(\bar{t}) \in (\mathbb{K}(\bar{t}) \setminus \mathbb{K})^2$ such that $\mathcal{P}(\bar{t}) = \mathcal{Q}(R(\bar{t}))$.
Steps

1. Compute a μ-basis of $\mathcal{P}$. Let $\mathbf{p}(\bar{t})$, $\mathbf{q}(\bar{t})$, $\mathbf{r}(\bar{t})$ be this μ-basis.
2. Apply Algorithm 1 to $\mathcal{P}_2(t_1)$. If $\mathcal{P}_2$ is not proper, then the algorithm returns the proper parametrization $\mathcal{M}_2(t_1)$, and $S_2(t_1) \in (\mathbb{K}[t_2])(t_1)$ $(S_2(t_1) = S(t_1, t_2)$ seen with coefficients in $\mathbb{K}[t_2]$), such that $\mathcal{P}_2(t_1) = \mathcal{M}_2(S_2(t_1))$. Otherwise, the algorithm returns $\mathcal{M}(\bar{t}) = \mathcal{P}(\bar{t})$ (i.e $\mathcal{M}_2(t_1) = \mathcal{P}_2(t_1)$, and $S_2(t_1) = t_1$).
3. Apply Algorithm 1 to $\mathcal{M}_1(t_2)$. If $\mathcal{M}_1$ is not proper, the algorithm returns the proper parametrization $\mathcal{Q}_1(t_2)$, and $T_1(t_2) \in (\mathbb{K}[t_1])(t_2)$ $(T_1(t_2) = T(t_1, t_2)$ seen with coefficients in $\mathbb{K}[t_2]$) such that $\mathcal{M}_1(t_2) = \mathcal{Q}_1(T_1(t_2))$. Otherwise, the algorithm returns $\mathcal{Q}(\bar{t}) = \mathcal{M}(\bar{t})$ (i.e $\mathcal{Q}_1(t_2) = \mathcal{M}_1(t_2)$, and $T_1(t_2) = t_2$). Then,

$$\mathcal{P}(\bar{t}) = \mathcal{M}(S(\bar{t}), t_2) = \mathcal{Q}(t_1, T(\bar{t}))(S(\bar{t}), t_2) = \mathcal{Q}(S(\bar{t}), T(S(\bar{t}), t_2)).$$

4. Return the rational parametrization $\mathcal{Q}(\bar{t})$ of the surface $\mathcal{V}$, and

$$R(\bar{t}) = (S(\bar{t}), T(S(\bar{t}), t_2)) \in \mathbb{K}(\bar{t})^2.$$

From Algorithm 2, and while using that the degree of a rational map is multiplicative under composition, we deduce some properties that relate the degree the rational map $\phi_\mathcal{P}$ with the degree of the rational maps $\phi_\mathcal{Q}$, $\phi_\mathcal{M}$, $\phi_{\mathcal{P}_i}$, $i = 1, 2$, and with $\deg(R)$, where $R(\bar{t}) = (S(\bar{t}), T(S(\bar{t}), t_2))$. One reasons, similarly as in [32].

Theorem 6. *It holds that*

$$\deg(\phi_\mathcal{P}) = \deg(\phi_\mathcal{Q})\deg_{t_1}(S(\bar{t}))\deg_{t_2}(T(\bar{t})), \quad \text{and}$$

$$\deg(\phi_{\mathcal{P}_2}) = \deg_{t_1}(S(\bar{t})), \quad \deg(\phi_{\mathcal{M}_1}) = \deg_{t_2}(T(\bar{t})).$$

In addition,

$$\deg(\phi_{\mathcal{P}}) = \deg(\phi_{\mathcal{M}})\deg_{t_1}(S(\bar{t})), \quad \deg(\phi_{\mathcal{M}}) = \deg(\phi_{\mathcal{Q}})\deg_{t_2}(T(\bar{t})).$$

Corollary 2. *The following statements are equivalent:*
1. $\mathcal{Q}$ *is proper.*
2. $\deg(\phi_{\mathcal{M}}) = \deg_{t_2}(T).$
3. $\deg(\phi_{\mathcal{P}}) = \deg_{t_1}(S)\deg_{t_2}(T).$

Finally, in Corollary 3, we show in which conditions Algorithm 2 does not return a better reparametrization than the input one (in the sense of the degree of the rational map that is induced by the rational parametrization).

Corollary 3. *It holds that* $\deg(\phi_{\mathcal{Q}}) = \deg(\phi_{\mathcal{P}})$ *if and only if* $\deg(\phi_{\mathcal{P}_2}) = \deg(\phi_{\mathcal{M}_1}) = 1$. *In particular, if* $\deg(\phi_{\mathcal{P}_i}) = 1$ *for* $i = 1, 2$, *then* $\deg(\phi_{\mathcal{Q}}) = \deg(\phi_{\mathcal{P}})$.

We observe that, while using previous results, one may easily analyze whether some families of surfaces can be properly reparametrized using the approach presented in this paper. For instance, if $\deg(\phi_{\mathcal{P}_i}) \neq 1$ for some $i = 1, 2$, and $\deg(\phi_{\mathcal{P}}) = n$, where n is a prime number, then $\deg(\phi_{\mathcal{Q}}) = 1$.

To finish this section, we illustrate Algorithm 2 with one example. The times of our implementation performance is similar to the times that were obtained in [32].

Example 4. *Let* $\mathcal{V}$ *be the rational surface that is defined by the parametrization*

$$\mathcal{P}(\bar{t}) = (\wp_4(\bar{t})) : \wp_2(\bar{t}) : \wp_3(\bar{t}) : \wp_4(\bar{t}) =$$

$$\left(t_2^4 t_1^2 - t_1^4 : -t_2^2 + t_2^6 + t_2^2 t_1^4 : -t_1^2 + t_2^2 t_1^2 + t_2^4 t_1^2 - t_1^4 : -t_1^2 + t_2^4 t_1^2 + t_1^4\right).$$

For this purpose, in Step 1 of Algorithm 2, *we compute a μ-basis of* $\mathcal{P}$ *and we get that is given by*

$\mathbf{p}(\bar{t}) = (t_2^2 t_1^2 + 2t_1^2 + 4t_2^8 + 6t_2^6 - 4t_2^2 - 4, -2t_1^2, -2t_2^2 t_1^2 - t_1^2 - 4t_2^8 - 4t_2^6 + 4t_2^2 + 2, t_2^2 t_1^2 + t_1^2 + 2t_2^6 - 2)$

$\mathbf{q}(\bar{t}) = (-2t_2^2 t_1^2 - 3t_1^2 + t_2^2 + 1, 0, 2t_2^2 t_1^2 + 2t_1^2 - t_2^2 - 1, -t_1^2 + 1)$

$\mathbf{r}(\bar{t}) = (-2t_2^4 - t_2^2 + 2, 0, 2t_2^4 - 1, -t_2^2 + 1).$

Using Theorem 4, *one gets that* $\deg(\phi_{\mathcal{P}}) = 4$. *Now, we apply Algorithm* 1 *to* $\mathcal{P}_2(t_1)$. *We obtain that*

$$G^{\mathcal{P}_2}(t_1, s_1) = s_1^2 - t_1^2 \in (\mathbb{K}[t_2])[t_1, s_1],$$

and $S_2(t_1) = -t_1^2 \in (\mathbb{K}[t_2])[t_1]$ *(remind that* $S_2(t_1) = S(t_1, t_2)$ *is seen as a polynomial in the variable* t_1 *and with coefficients in* $\mathbb{K}[t_2]$). *Subsequently, we determine the polynomials*

$$L_i(s_1, x_i) = \mathrm{Res}_{t_1}(x_i\wp_4(\bar{t}) - \wp_i(\bar{t}), s_1 - S_2(t_1)) = (m_{i,2}(s_1)x_i - m_{i,1}(s_1))^{\deg_{t_1}(S)}$$

for $i = 1, 2, 3$. *We obtain that*

$$\mathcal{M}(\bar{t}) = ((-t_2^4 - t_1)t_1 : t_2^2(t_2^4 - 1 + t_1^2) : (1 - t_2^2 - t_2^4 - t_1)t_1 : t_1(1 - t_2^4 + t_1)).$$

Now, in Step 3 of the algorithm, we apply Algorithm 1 *to* $\mathcal{M}_1(t_2)$. *For this purpose, we first compute a μ-basis of* $\mathcal{M}$ *and we get that it is given by (see Remark* 3)

$\mathbf{p}_{\mathcal{M}}(\bar{t}) = (4t_2^2 + 4 - 4t_2^8 - 6t_2^6 + t_2^2 t_1 + 2t_1, -2t_1, -4t_2^2 - 2 + 4t_2^8 + 4t_2^6 - 2t_2^2 t_1 - t_1, -2t_2^6 + 2 + t_2^2 t_1 + t_1)$

$$\mathbf{q}_{\mathcal{M}}(\bar{t}) = (t_2^2 + 1 + 2t_2^2 t_1 + 3t_1, 0, -t_2^2 - 1 - 2t_2^2 t_1 - 2t_1, 1 + t_1)$$
$$\mathbf{r}_{\mathcal{M}}(\bar{t}) = (-2 + 2t_2^4 + t_2^2, 0, -2t_2^4 + 1, -1 + t_2^2).$$

We obtain that
$$G^{\mathcal{M}_1}(t_2, s_2) = s_2^2 - t_2^2 \in (\mathbb{K}[t_1])[t_2, s_2]$$

that is, $\mathcal{M}_1$ is not proper. Afterwards, we compute $T_1(t_2) = -t_2^2 \in (\mathbb{K}[t_1])[t_2]$, and the polynomials

$$L_i(s_2, x_i) = \text{Res}_{t_2}(x_i m_4(\bar{t}) - m_i(\bar{t}), s_2 - T_1(t_2)) = (q_{i,2}(s_2)x_i - q_{i,1}(s_2))^{\deg_{t_2}(T)}$$

for $i = 1, 2, 3$ (remind that $T_1(t_2) = T(t_1, t_2)$ is seen as a polynomial in the variable t_2 and with coefficients in $\mathbb{K}[t_1]$). We obtain that

$$\mathcal{Q}(\bar{t}) = \left((-t_1 - t_2^2)t_1 : -t_2(t_2^2 - 1 + t_1^2) : (-t_1 + 1 + t_2 - t_2^2)t_1 : t_1(1 + t_1 - t_2^2) \right).$$

In Step 4, the algorithm returns the parametrization $\mathcal{Q}(\bar{t})$, and $R(\bar{t}) = (S(\bar{t}), T(S(\bar{t}), t_2)) = (-t_1^2, -t_2^2)$. We observe that

$$\deg_{t_2}(T) = 2, \quad \text{and} \quad \deg_{t_1}(S) = 2.$$

Thus, since $\deg(\phi_{\mathcal{P}}) = 4$, by Theorem 6, we conclude that $\deg(\phi_{\mathcal{Q}}) = 1$ and, hence, $\mathcal{Q}$ is proper.

Remark 3. Using Lemma 1, we may compute a μ-basis, $\mathbf{p}_{\mathcal{M}}, \mathbf{q}_{\mathcal{M}}, \mathbf{r}_{\mathcal{M}}$, of $\mathcal{M}$ from the μ-basis, $\mathbf{p}, \mathbf{q}, \mathbf{r}$, of $\mathcal{P}$. Remind that $\mathcal{P}(\bar{t}) = \mathcal{M}(S_2(t_1), t_2)$, and $S_2(t_1) = S(t_1, t_2)$ is seen as a polynomial in the variable t_1 with coefficients in $\mathbb{K}[t_2]$. Thus, we have a particular case (case of curves) of the reasoning that is presented in Proposition 1 (which is stated for surfaces). More precisely, we write $\mathbf{p} = (p_1, p_2, p_3, p_4)$ and $\mathbf{p}_{\mathcal{M}} = (p_{m_1}, p_{m_2}, p_{m_3}, p_{m_4})$. Observe that the implicit equation of the parametrization $(S_2(t_1), p_i/p_4)$ (seen with coefficients in $\mathbb{K}(t_2)$ and in the variable t_1) is given by the polynomial $p_{m_i}(x_1, t_2) - x_2 p_{m_4}(x_1, t_2) \in (\mathbb{K}(t_2))[x_1, x_2]$ for $i = 1, 2, 3$ (i.e., the coefficients of the the implicit equation are in $\mathbb{K}(t_2)$). In order to compute this implicit equation, we may use that

$$\text{Res}_{t_1}(x_i p_4(\bar{t}) - p_i(\bar{t}), x_1 - S_2(t_1)) = (p_{m_4}(x_1, t_2)x_i - p_{m_i}(x_1, t_2))^{\deg_{t_1}(S)}, \quad i = 1, 2, 3$$

(see, e.g., [44]). Similarly one reasons to get $\mathbf{q}_{\mathcal{M}}$ from $\mathbf{q}$ and $\mathbf{r}_{\mathcal{M}}$ from $\mathbf{r}$. Observe that this is a particular case of the result presented in Proposition 1 (we apply the same idea stated in Proposition 1, but for the particular case of curves).

5. Implicitization Using μ-Basis

In the following, we assume that we are in the affine space (i.e., $x_4 = 1$; this simplifies the time on the computations), and we consider the polynomials

$$G_1^{\mathcal{P}}(\bar{t}, \bar{x}) := p^{\mathcal{P}}(\bar{t}, \bar{x})r_3(\bar{t}) - r^{\mathcal{P}}(\bar{t}, \bar{x})p_3(\bar{t}) \in \mathbb{K}[\bar{t}, x_1, x_2]$$
$$G_2^{\mathcal{P}}(\bar{t}, \bar{x}) := q^{\mathcal{P}}(\bar{t}, \bar{x})r_3(\bar{t}) - r^{\mathcal{P}}(\bar{t}, \bar{x})q_3(\bar{t}) \in \mathbb{K}[\bar{t}, x_1, x_2]$$
$$G_3^{\mathcal{P}}(\bar{t}, \bar{x}) := r^{\mathcal{P}}(\bar{t}, \bar{x}) \in \mathbb{K}[\bar{t}, \bar{x}], \qquad \bar{x} = (x_1, x_2, x_3).$$

Note that, then, we may write

$$G_1^{\mathcal{P}}(\bar{t}, \bar{x}) := x_1(p_1 r_3 - p_3 r_1) + x_2(p_2 r_3 - p_3 r_2) + (p_4 r_3 - p_3 r_4)$$
$$G_2^{\mathcal{P}}(\bar{t}, \bar{x}) := x_1(q_1 r_3 - q_3 r_1) + x_2(q_2 r_3 - q_3 r_2) + (q_4 r_3 - q_3 r_4)$$
$$G_3^{\mathcal{P}}(\bar{t}, \bar{x}) := x_1 r_1 + x_2 r_2 + x_3 r_3 + r_4.$$

In addition, let
$$S_{12}^{\mathcal{P}}(t_1, \bar{x}) = \text{pp}_{\bar{x}}(\text{Res}_{t_2}(G_1^{\mathcal{P}}, G_2^{\mathcal{P}})) \in \mathbb{K}[t_1, x_1, x_2],$$

$$T_{12}^{\mathcal{P}}(t_2, \overline{x}) = \text{pp}_{\overline{x}}(\text{Res}_{t_1}(G_1^{\mathcal{P}}, G_2^{\mathcal{P}})) \in \mathbb{K}[t_2, x_1, x_2].$$

Finally, $\mathcal{F}_{\mathcal{P}_{12}}(x_1, x_2)$ denotes the fiber of a point $Q_{12} := \pi_{12}(Q) = (x_1, x_2)$, where $Q = (x_1, x_2, x_3) \in \mathcal{V}_a$ and $\pi_{12}(\mathcal{V}_a)$ is the $(1,2)$-projection of $\mathcal{V}_a$. That is

$$\mathcal{F}_{\mathcal{P}_{12}}(Q_{12}) = \mathcal{P}_{12}^{-1}(Q_{12}) = \{\overline{t} \in \mathbb{K}^2 \mid \mathcal{P}_{12}(\overline{t}) = Q_{12}\},$$

where $\mathcal{P}_{12} := (\wp_1/\wp_4, \wp_2/\wp_4) := \pi_{12}(\mathcal{P}_a)$.

Lemma 2. *It holds that*

$$\deg_{t_1}(S_{12}^{\mathcal{P}}) = \deg_{t_2}(T_{12}^{\mathcal{P}}) = \text{Card}(\mathcal{F}_{\mathcal{P}_{12}}(x_1, x_2)).$$

Proof. Because $\mathbf{p}, \mathbf{q}, \mathbf{r}$ is a μ-basis of $\mathcal{P}(\overline{t})$, we have

$$\begin{aligned} p_1\wp_1 + p_2\wp_2 + p_3\wp_3 + p_4\wp_4 &= 0, \\ q_1\wp_1 + q_2\wp_2 + q_3\wp_3 + q_4\wp_4 &= 0, \\ r_1\wp_1 + r_2\wp_2 + r_3\wp_3 + r_4\wp_4 &= 0. \end{aligned} \qquad (1)$$

Consider a generic point $Q = (x_1, x_2, x_3)$ on the variety generated by $(\wp_1 : \wp_2 : \wp_4)$ and the associated polynomials

$$p^{\mathcal{P}}(\overline{t}, \overline{x}) = \mathbf{p}(\overline{t}) \cdot Q, \text{ and } q^{\mathcal{P}}(\overline{t}, \overline{x}) = \mathbf{q}(\overline{t}) \cdot Q,$$

where $\mathbf{p}(\overline{t}) = (p_1 r_3 - p_3 r_1, p_2 r_3 - p_3 r_2, p_4 r_3 - p_3 r_4)$, and $\mathbf{q}(\overline{t}) = (q_1 r_3 - q_3 r_1, q_2 r_3 - q_3 r_2, q_4 r_3 - q_3 r_4)$. It holds that $p^{\mathcal{P}}(\overline{t}, \mathcal{P}(\overline{t})) = q^{\mathcal{P}}(\overline{t}, \mathcal{P}(\overline{t})) = 0$. In fact, $p^{\mathcal{P}}(\overline{t}, \mathcal{P}(\overline{t})) = (p_1 r_3 - p_3 r_1)\wp_1 + (p_2 r_3 - p_3 r_2)\wp_2 + (p_4 r_3 - p_3 r_4)\wp_4 = 0$ is derived by eliminating $\wp_3$ from the first and third equations in (1). Similarly, to find $q^{\mathcal{P}}(\overline{t}, \mathcal{P}(\overline{t})) = 0$ from the last two equations in (1).

Thus, one may reason as in Theorem 4 and Corollary 1 (also see Remark 2) to get that

$$\deg_{t_1}(S_{12}^{\mathcal{P}}(t_1, \overline{x})) = \deg_{t_2}(T_{12}^{\mathcal{P}}(t_2, \overline{x})) = \text{Card}(\mathcal{F}_{\mathcal{P}_{12}}(x_1, x_2))$$

(remind that $\mathcal{P}_{12} := (\wp_1/\wp_4, \wp_2/\wp_4) = \pi_{12}(\mathcal{P}_a)$). □

Theorem 7. *Let $\mathbf{p}(\overline{t}), \mathbf{q}(\overline{t})$ and $\mathbf{r}(\overline{t})$ a μ-basis for $\mathcal{P}(\overline{t})$. It holds that*

$$\text{pp}_{x_3}(h(\overline{x})) = f(\overline{x})^{\deg(\phi_{\mathcal{P}})}$$

where

$$h(\overline{x}) = \text{Cont}_{\{Z,W\}}(\text{Res}_{t_2}(T_{12}^{\mathcal{P}}(t_2, \overline{x}), K(t_2, Z, W, \overline{x}))) \in \mathbb{K}[\overline{x}],$$

$$K(t_2, Z, W, \overline{x}) = \text{Res}_{t_1}(S_{12}^{\mathcal{P}}(t_1, \overline{x}), H^{\mathcal{P}}(\overline{t}, Z, W, \overline{x})) \in \mathbb{K}[t_2, Z, W, \overline{x}],$$

and

$$H^{\mathcal{P}}(\overline{t}, Z, W, \overline{x}) = G_3^{\mathcal{P}}(\overline{t}, \overline{x}) + Z G_1^{\mathcal{P}}(\overline{t}, \overline{x}) + W G_2^{\mathcal{P}}(\overline{t}, \overline{x}) \in \mathbb{K}[\overline{t}, Z, W, \overline{x}].$$

Proof. First, we recall that

$$\deg_{t_1}(S_{12}^{\mathcal{P}}) = \deg_{t_2}(T_{12}^{\mathcal{P}}) = \text{Card}(\mathcal{F}_{\mathcal{P}_{12}}(\overline{x})).$$

Let d_{12} be this quantity. Clearly, $d_{12} \geq 1$. In addition, let $m = \deg_{t_1}(H^{\mathcal{P}})$ and $k = \deg_{t_2}(H^{\mathcal{P}})$. Regarding $S_{12}^{\mathcal{P}}$ and $H^{\mathcal{P}}$ as polynomials in $\mathbb{K}(t_2, Z, W, \overline{x})[t_1]$, and using that the resultant of two univariate polynomials is the product of the evaluations of one of them in the roots of the other, we get

$$K(t_2, Z, W, \bar{x}) = \operatorname{Res}_{t_1}(S_{12}^{\mathcal{P}}, H^{\mathcal{P}}) = A(\bar{x})^m \prod_{i=1}^{d_{12}} H^{\mathcal{P}}(\alpha_i, t_2, Z, W, \bar{x}),$$

where A is the leading coefficient of $S_{12}^{\mathcal{P}}$ w.r.t. t_1, and where $\{\alpha_1, \ldots, \alpha_{d_{12}}\}$ are the roots of $S_{12}^{\mathcal{P}}$ in the algebraic closure $\overline{\mathbb{K}(x_1, x_2)}$ of $\mathbb{K}(x_1, x_2)$ (we regard $S_{12}^{\mathcal{P}}$ as an univariate polynomial in t_1). Similarly,

$$\operatorname{Res}_{t_2}(T_{12}^{\mathcal{P}}, K) = B(\bar{x})^k \prod_{j=1}^{d_{12}} K(\beta_j, Z, W, \bar{x}),$$

where B is the leading coefficient of $T_{12}^{\mathcal{P}}$ w.r.t. t_2, and $\{\beta_1, \ldots, \beta_{d_{12}}\}$ are the roots of $T_{12}^{\mathcal{P}}$ in $\overline{\mathbb{K}(x_1, x_2)}$ (we regard $T_{12}^{\mathcal{P}}$ as a univariate polynomial in t_2). Therefore,

$$\operatorname{Res}_{t_2}(T_{12}^{\mathcal{P}}, K) = B^k A^{m \cdot d_{12}} \prod_{i=1}^{d_{12}} \prod_{j=1}^{d_{12}} H^{\mathcal{P}}(\alpha_i, \beta_j, Z, W, \bar{x}).$$

By Lemma 2, there exist d_{12} pairs of points (α_i, β_j) being in $\mathcal{F}_{\mathcal{P}_{12}}(x_1, x_2)$, and for each $U(x_1, x_2) \in \mathcal{F}_{\mathcal{P}_{12}}(x_1, x_2)$ it holds that $G_1^{\mathcal{P}}(U, \bar{x}) = G_2^{\mathcal{P}}(U, \bar{x}) = 0$. Thus,

$$\operatorname{Res}_{t_2}(T_{12}^{\mathcal{P}}, K) = B^k A^{m \cdot d_{12}} Q(\bar{x}, Z, W) \prod_{U \in \mathcal{F}_{\mathcal{P}_{12}}(x_1, x_2)} G_3^{\mathcal{P}}(U, \bar{x}),$$

where

$$Q(\bar{x}, Z, W) = \prod_{\substack{1 \leq i,j \leq d_{12} \\ (\alpha_i, \beta_j) \notin \mathcal{F}_{\mathcal{P}_{12}}(x_1, x_2)}} H(\alpha_i, \beta_j, Z, W, \bar{x}).$$

Note that for each root α_i there exists a unique b_j satisfying that the pair (α_i, β_j) is in the fiber. Furthermore, for $(\alpha_i, \beta_j) \notin \mathcal{F}_{\mathcal{P}_{12}}(x_1, x_2)$, either $G_1^{\mathcal{P}}(\alpha_i, \beta_j, \bar{x}) \neq 0$ or $G_2^{\mathcal{P}}(\alpha_i, \beta_j, \bar{x}) \neq 0$ (see Lemma 2). Hence, $Q(\bar{x}, Z, W)$ depends on Z or W. In addition, each $H^{\mathcal{P}}(\alpha_i, \beta_j, Z, W, \bar{x})$ does depend on Z or W.

Next, we show that $Q(\bar{x}, Z, W)$, regarded as polynomial in $\overline{\mathbb{K}[\bar{x}]}[\bar{x}, Z, W]$, is primitive w.r.t. the variables $\{Z, W\}$. For this purpose, we denote, by $N(x_3) \in \overline{\mathbb{K}[x_1, x_2]}[x_3]$, the content of Q w.r.t. $\{Z, W\}$. Thus, there exists $(\alpha_i, \beta_j) \notin \mathcal{F}_{\mathcal{P}_{12}}(x_1, x_2)$ satisfying that the polynomial N divides $H(\alpha_i, \beta_j, Z, W, \bar{x})$; that is, $N(x_3)$ divides $G_3^{\mathcal{P}}(\alpha_i, \beta_i, \bar{x}) + ZG_1^{\mathcal{P}}(\alpha_i, \beta_j, \bar{x}) + WG_2^{\mathcal{P}}(\alpha_i, \beta_j, \bar{x})$ and, then, $N(x_3)$ divides $G_1^{\mathcal{P}}(\alpha_i, \beta_j, x_1, x_2)$ and $G_2^{\mathcal{P}}(\alpha_i, \beta_j, x_1, x_2)$. Taking into account that at least one of them is not zero, we get that $N \in \overline{\mathbb{K}[x_1, x_2]}$ and, thus, Q is primitive w.r.t. $\{Z, W\}$. Now, using that

$$h(\bar{x}) = \operatorname{Cont}_{\{Z, W\}}(\operatorname{Res}_{t_2}(T_{12}^{\mathcal{P}}, K)),$$

we obtain that

$$h(\bar{x}) = B^k A^{m \cdot d_{12}} \cdot N(x_1, x_2) \cdot \prod_{U \in \mathcal{F}_{\mathcal{P}_{12}}(x_1, x_2)} G_3^{\mathcal{P}}(U, \bar{x}),$$

where $N \in \overline{\mathbb{K}[x_1, x_2]}$. Thus,

$$\operatorname{PP}_{x_3}(h(\bar{x})) = \operatorname{PP}_{x_3}\left(\prod_{U \in \mathcal{F}_{\mathcal{P}_{12}}(x_1, x_2)} G_3^{\mathcal{P}}(U(\bar{x}), \bar{x})\right).$$

Under these conditions, it holds that $\deg_{x_3}(\text{pp}_{x_3}(h(\overline{x}))) = d_{12}$. Indeed, clearly one has $\deg_{x_3}(\text{pp}_{x_3}(h(\overline{x}))) \leq d_{12}$. If $\deg_{x_3}(\text{pp}_{x_3}(h(\overline{x}))) < d_{12}$, thus, there exists $U \in \mathcal{F}_{\mathcal{P}_{12}}(x_1, x_2)$, such that $r_3(U) = 0$. Moreover, $U \in \mathbb{K}^2$, which is impossible since $U \in \mathcal{F}_{\mathcal{P}_{12}}(x_1, x_2)$.

Now, we prove that

$$\text{pp}_{x_3}\left(\prod_{U \in \mathcal{F}_{\mathcal{P}_{12}}(x_1,x_2)} G_3^{\mathcal{P}}(U(\overline{x}), \overline{x})\right) = f(\overline{x})^r.$$

Indeed, clearly one has that

$$\text{pp}_{x_3}\left(\prod_{U \in \mathcal{F}_{\mathcal{P}_{12}}(x_1,x_2)} G_3^{\mathcal{P}}(U(\overline{x}), \overline{x})\right) = f(\overline{x})^r g(\overline{x}).$$

Furthermore, $r \geq \deg(\phi_\mathcal{P})$, since $G_3^{\mathcal{P}}(U(\overline{x}), \overline{x}) = G_3^{\mathcal{P}}(V(\overline{x}), \overline{x})$ for $U, V \in \mathcal{F}_\mathcal{P}(\overline{x})$ (observe that $\mathcal{F}_\mathcal{P}(\overline{x}) \subseteq \mathcal{F}_{\mathcal{P}_{12}}(x_1, x_2)$)). Thus, since $\deg_{x_3}(f) = d_{12}/\deg(\phi_\mathcal{P})$ (see [44]) and $\deg_{x_3}(\text{pp}_{x_3}(h(\overline{x}))) = d_{12}$, we get that

$$d_{12}/\deg(\phi_\mathcal{P}) \cdot r + \deg(g) = d_{12}$$

, which implies that $d_{12}(1 - r/\deg(\phi_\mathcal{P})) = \deg(g)$ and, hence, $r \leq \deg(\phi_\mathcal{P})$. Because $r \geq \deg(\phi_\mathcal{P})$, we conclude that $\deg(g) = 0$ and $r = \deg(\phi_\mathcal{P})$. □

In the following examples, we illustrate the above theorem. These examples are taken from [5].

Example 5. *Let $\mathcal{V}$ be the rational surface that is defined by the parametrization*

$$\mathcal{P}(\overline{t}) = (t_2^2 t_1 - t_1^2 : -t_2 + t_2^3 + t_2 t_1^2 : -t_1 + t_2 t_1 + t_2^2 t_1 - t_1^2 : -t_1 + t_2^2 t_1 + t_1^2).$$

We determine the polynomials $p^\mathcal{P}(\overline{t}, \overline{x}) = \mathbf{p}(\overline{t}) \cdot \overline{x}$, $q^\mathcal{P}(\overline{t}, \overline{x}) = \mathbf{q}(\overline{t}) \cdot \overline{x}$, $r^\mathcal{P}(\overline{t}, \overline{x}) = \mathbf{r}(\overline{t}) \cdot \overline{x}$, where the μ-basis is given by

$\mathbf{p}(\overline{t}) = (t_2 t_1 + 2t_1 + 4t_2^4 + 6t_2^3 - 4t_2 - 4, -2t_1, -2t_2 t_1 - t_1 - 4t_2^4 - 4t_2^3 + 4t_2 + 2, t_2 t_1 + t_1 + 2t_2^3 - 2)$

$\mathbf{q}(\overline{t}) = (-2t_2 t_1 - 3t_1 + t_2 + 1, 0, 2t_2 t_1 + 2t_1 - t_2 - 1, -t_1 + 1)$

$\mathbf{r}(\overline{t}) = (-2t_2^2 - t_2 + 2, 0, 2t_2^2 - 1, -t_2 + 1).$

Now, we determine

$$G_1^{\mathcal{P}}(\overline{t}, \overline{x}) := p^{\mathcal{P}}(\overline{t}, \overline{x})r_3(\overline{t}) - r^{\mathcal{P}}(\overline{t}, \overline{x})p_3(\overline{t}) \in \mathbb{K}[\overline{t}, x_1, x_2]$$

$$G_2^{\mathcal{P}}(\overline{t}, \overline{x}) := q^{\mathcal{P}}(\overline{t}, \overline{x})r_3(\overline{t}) - r^{\mathcal{P}}(\overline{t}, \overline{x})q_3(\overline{t}) \in \mathbb{K}[\overline{t}, x_1, x_2]$$

$$G_3^{\mathcal{P}}(\overline{t}, \overline{x}) := r^{\mathcal{P}}(\overline{t}, \overline{x}) \in \mathbb{K}[\overline{t}, \overline{x}], \quad \overline{x} = (x_1, x_2, x_3)$$

and we compute

$S_{12}^{\mathcal{P}}(t_1, \overline{x}) = \text{pp}_{\overline{x}}(\text{Res}(G_1^{\mathcal{P}}, G_2^{\mathcal{P}}, t_2)) = -t_1 + 2t_1^2 + x_1 + 2x_1^2 - x_2^2 t_1^2 x_1^2 + x_2^2 t_1^2 x_1 - 4x_2^2 t_1^3 x_1 + 4x_2^2 t_1^4 x_1 + 4x_2^2 t_1^3 x_1^2 - x_2^2 t_1^2 x_1^3 - 4x_2^2 t_1^4 x_1^3 + 4x_2^2 t_1^3 x_1^3 - 4x_2^2 t_1^4 x_1^2 + 4x_2^2 t_1^4 - 4x_2^2 t_1^3 + x_2^2 t_1^2 - 5x_1 t_1^4 - t_1^5 x_1 - 2x_1^4 t_1 + 2x_1^3 t_1^4 + 2x_1^5 t_1^5 + 2x_1^2 t_1^5 + x_1^5 t_1^2 + 3x_1^5 t_1^4 - x_1^4 t_1^5 - 3x_1^5 t_1^3 - 7x_1^3 t_1 + x_1 t_1^3 - x_1^5 t_1^5 + 16x_1^3 t_1^2 - 4x_1^2 t_1^4 - 6x_1^2 t_1^3 + 8x_1^4 t_1^2 - 5x_1 t_1 + 14x_1^2 t_1^2 - 9x_1^2 t_1 + 7x_1 t_1^2 + 6x_1^4 t_1^4 - 11x_1^4 t_1^3 - 14x_1^3 t_1^3 - t_1^5 + x_1^3 - 2t_1^4 + t_1^3,$

$$T_{12}^{\mathcal{P}}(t_2, \overline{x}) = \text{pp}_{\overline{x}}(\text{Res}(G_1^{\mathcal{P}}, G_2^{\mathcal{P}}, t_1)) = -2t_2^5 x_1 + 4x_1 t_2^3 - 2t_2^4 x_2 + t_2^2 x_2 + t_2^5 + x_1^2 t_2^5 - x_1^2 t_2^3 + 2x_1 t_2^4 x_2 - 3x_1 t_2^2 x_2 + x_1 x_2 - t_2 - 2x_1 t_2 + t_2^3.$$

Also, let

$$H^{\mathcal{P}}(\overline{t}, Z, W, \overline{x}) = G_3^{\mathcal{P}}(\overline{t}, \overline{x}) + ZG_1^{\mathcal{P}}(\overline{t}, \overline{x}) + WG_2^{\mathcal{P}}(\overline{t}, \overline{x}) \in \mathbb{K}[\overline{t}, Z, W, \overline{x}]$$

and

$$K(t_2, Z, W, \overline{x}) = \text{Res}_{t_1}(S_{12}^{\mathcal{P}}(t_1, \overline{x}), H^{\mathcal{P}}(\overline{t}, Z, W, \overline{x})) \in \mathbb{K}[t_2, Z, W, \overline{x}].$$

Finally, we compute

$$h(\overline{x}) = \text{Cont}_{\{Z,W\}}(\text{Res}_{t_2}(T_{12}^{\mathcal{P}}(t_2, \overline{x}), K(t_2, Z, W, \overline{x}))) \in \mathbb{K}[\overline{x}],$$

and

$$\begin{aligned}\text{pp}_{x_3}(h(\overline{x})) = f(\overline{x})^{\deg(\phi_{\mathcal{P}})} &= -1 - 46x_1^2 x_2 x_3 + 38x_1 x_2 x_3^2 - 8x_1 x_2 x_3 + 4x_2^2 x_1^2 x_3 - 12x_2^2 x_1 x_3 \\ &+ 10x_1^3 x_2 x_3 + x_1^2 x_2 x_3^2 - 10x_1 x_2 x_3^3 + 4x_1^2 x_2^2 x_3 - 10x_2^3 x_3 + 8x_2^3 x_3^2 - 4x_2^2 x_3 + 5x_2 x_3^2 - 5x_1^4 x_2 \\ &+ 19x_1^3 x_2 - 4x_2^2 x_3^3 + 4x_2 x_3^4 + 2x_1^2 x_2 - 2x_2^2 x_3^3 + 47x_3^2 x_1^2 - 12x_1^4 x_3 + 14x_1 x_3^3 + 11x_1^2 x_3^3 - 6x_1 x_3^4 \\ &- 32x_1 x_3^3 - 22x_3 x_1^3 - 12x_3 x_1^2 + 4x_3 x_1 - 4x_1^3 x_2^2 + 4x_1^2 x_2^2 + 4x_1 x_2^2 - x_1 x_2 + 7x_3^4 - 5x_3^3 - 5x_3^2 \\ &+ 8x_1^5 + 2x_1^2 + 2x_1^3 + x_3^5 + 3x_3 - 2x_1 + x_2.\end{aligned}$$

Observe that we may conclude that $\deg(\phi_{\mathcal{P}}) = 1$ and, thus, $\mathcal{P}(\overline{t})$ is a proper rational parametrization.

We have implemented this method while using Maple 2016 on a Lenovo ThinkPad Intel(R) Core(TM) i7-7500U CPU @ 2.70 GHz 2.90 GHz and 16 GB of RAM, OS-Windows 10 Pro. The time, in CPU seconds, for this example is 10.907 and using Gröbner basis, we get 0.187.

Example 6. Let $\mathcal{V}$ be the rational surface defined by the parametrization

$$\mathcal{P}(\overline{t}) = (t_1^2 + t_2 t_1 + 2t_2^2 - 2t_2^2 t_1 : t_1^2 + 2t_2 t_1 + t_2 t_1^2 + 2t_2^2 - t_2^2 t_1 + 2t_2^2 t_1^2 : -t_1^2 + t_2 t_1 + 2t_2 t_1^2 + 2t_2^2 - t_2^2 t_1 - 2t_2^2 t_1^2 : 2t_2 t_1 - 2t_2 t_1^2 - 2t_2^2 t_1 - t_2^2 t_1^2).$$

We determine the polynomials $p^{\mathcal{P}}(\overline{t}, \overline{x}) = \mathbf{p}(\overline{t}) \cdot \overline{x}$, $q^{\mathcal{P}}(\overline{t}, \overline{x}) = \mathbf{q}(\overline{t}) \cdot \overline{x}$, $r^{\mathcal{P}}(\overline{t}, \overline{x}) = \mathbf{r}(\overline{t}) \cdot \overline{x}$, where the μ-basis is given by

$\mathbf{p}(\overline{t}) = (-1344390 t_2^2 t_1 + 34075368 t_2 t_1 - 22657890 t_1 - 5710808 t_2^3 - 181563 t_2^2 - 23392736 t_2 - 4984080, 1344390 t_2^2 t_1 - 25195836 t_2 t_1 + 10711400 t_1 + 3569255 t_2^3 + 1074194 t_2^2 + 18408656 t_2 + 4984080, 1344390 t_2^2 t_1 - 17483628 t_2 t_1 - 11946490 t_1 + 2141553 t_2^3 - 892631 t_2^2 + 4984080 t_2, 9704572 t_2 - 11391246 t_2 t_1 + 6590790 t_1 + 2855404 t_2^3 + 6075203 t_2^2 - 2492040)$

$\mathbf{q}(\overline{t}) = (-229530 t_2 t_1 - 50278 t_1 + 139288 t_2^2 - 174717 t_2 + 194136, 131160 t_2 t_1 + 155206 t_1 - 87055 t_2^2 + 85766 t_2 - 194136, 65580 t_2 t_1 + 104928 t_1 - 52233 t_2^2 + 88951 t_2, 131160 t_2 t_1 + 100556 t_1 - 69644 t_2^2 - 58603 t_2 + 97068)$

$\mathbf{r}(\overline{t}) = (-8t_2^3 + 11t_2^2 - 4t_2 + 4, 5t_2^3 - 6t_2^2 + 8t_2 - 4, 3t_2^3 - 5t_2^2 - 4t_2, 4t_2^3 + t_2^2 + 2).$

Now, we determine

$$G_1^{\mathcal{P}}(\overline{t}, \overline{x}) := p^{\mathcal{P}}(\overline{t}, \overline{x}) r_3(\overline{t}) - r^{\mathcal{P}}(\overline{t}, \overline{x}) p_3(\overline{t}) \in \mathbb{K}[\overline{t}, x_1, x_2]$$

$$G_2^{\mathcal{P}}(\overline{t}, \overline{x}) := q^{\mathcal{P}}(\overline{t}, \overline{x}) r_3(\overline{t}) - r^{\mathcal{P}}(\overline{t}, \overline{x}) q_3(\overline{t}) \in \mathbb{K}[\overline{t}, x_1, x_2]$$

$$G_3^{\mathcal{P}}(\overline{t}, \overline{x}) := r^{\mathcal{P}}(\overline{t}, \overline{x}) \in \mathbb{K}[\overline{t}, \overline{x}], \quad \overline{x} := (x_1, x_2, x_3)$$

and we compute $S_{12}^{\mathcal{P}}(t_1, \overline{x})$ and $T_{12}^{\mathcal{P}}(t_2, \overline{x})$. Additionally, let

$$H^{\mathcal{P}}(\overline{t}, Z, W, \overline{x}) = G_3^{\mathcal{P}}(\overline{t}, \overline{x}) + Z G_1^{\mathcal{P}}(\overline{t}, \overline{x}) + W G_2^{\mathcal{P}}(\overline{t}, \overline{x}) \in \mathbb{K}[\overline{t}, Z, W, \overline{x}]$$

and

$$K(t_2, Z, W, \overline{x}) = \mathrm{Res}_{t_1}(S_{12}^{\mathcal{P}}(t_1, \overline{x}), H^{\mathcal{P}}(\overline{t}, Z, W, \overline{x})) \in \mathbb{K}[t_2, Z, W, \overline{x}].$$

Finally, we compute

$$h(\overline{x}) = \mathrm{Cont}_{\{Z,W\}}(\mathrm{Res}_{t_2}(T_{12}^{\mathcal{P}}(t_2, \overline{x}), K(t_2, Z, W, \overline{x}))) \in \mathbb{K}[\overline{x}],$$

and from $\mathrm{pp}_{x_3}(h(\overline{x}))$, we get that

$f(\overline{x}) = -449792 + 51270879 x_3 x_1 x_2^2 + 13092929 x_3^2 x_1 x_2 - 3482416 x_3 x_1^2 x_2 + 22904376 x_3 x_1 x_2 + 675054 x_1 x_3^6 - 862596 x_2 x_3^6 - 29225028 x_1^2 x_2 x_3^3 + 11830146 x_1 x_2 x_3^3 + 32231373 x_1 x_2^2 x_3^3 + 110760512 x_1^5 x_2 x_3 - 90099948 x_1^4 x_2^2 x_3 - 129717124 x_1^4 x_2 x_3^2 + 40844546 x_1^3 x_2^3 x_3 + 124810810 x_1^3 x_2^2 x_3^2 + 74702726 x_1^3 x_2 x_3^3 - 19905824 x_1^2 x_2^4 x_3 - 64077866 x_1^2 x_2^3 x_3^2 - 75736662 x_1^2 x_2^2 x_3^3 + 18645124 x_1 x_2^4 x_3^2 + 9481980 x_1 x_2^5 x_3 + 33323142 x_1 x_2^3 x_3^3 - 40980736 x_1^6 x_2 - 50645760 x_1^6 x_3 + 34182816 x_1^5 x_2^2 + 54875936 x_1^5 x_3^2 - 24633612 x_1^4 x_2^3 - 26906244 x_1^4 x_3^3 + 30171008 x_1^3 x_2^4 - 25238190 x_1^2 x_2^5 + 9961900 x_1 x_2^6 - 1918400 x_2^6 x_3 + 30499314 x_1 x_2^4 x_3 - 28601154 x_3^2 x_1^2 x_2^2 + 22849453 x_3^2 x_1 x_2^3 + 27085291 x_3^2 x_1^3 x_2 - 2534290 x_2^5 x_3^2 - 5374808 x_2^4 x_3^3 + 3235863 x_1^3 x_3^3 - 9477510 x_3^3 x_2^3 + 2815992 x_1^2 x_3^3 - 8545002 x_2^2 x_3^3 + 3831717 x_1 x_3^3 - 5284450 x_2 x_3^3 - 15996836 x_3^2 x_1^3 + 119044501 x_1 x_2^4 - 25566784 x_1^5 x_2 + 9070776 x_1^4 x_2^2 + 80443033 x_1^3 x_2^2 + 9622080 x_1^5 x_3 - 109224707 x_1^2 x_2^4 + 53021640 x_1 x_2^5 - 9557440 x_2^5 x_3 - 19184968 x_2^3 x_1^4 - 6794107 x_3^2 x_2^4 + 16282880 x_1^4 x_2 + 119771644 x_1^3 x_2^2 - 205506824 x_1^2 x_2^3 + 49430528 x_3 x_1^4 + 41274000 x_1^4 x_3 x_2 - 63080603 x_1^3 x_3 x_2^2 - 3378276 x_1^2 x_3 x_2^3 - 96074840 x_3 x_1^3 x_2 + 9101736 x_3 x_1^2 x_2^2 + 53646214 x_3 x_1 x_2^3 + 10909032 x_3^2 x_1^2 x_2 - 21171376 x_3 x_2^4 - 10777540 x_2^3 x_3^2 + 7054782 x_2^4 x_3^3 - 6762492 x_2^2 x_3^4 + 2808045 x_1^2 x_3^4 + 16635132 x_3^2 x_1 x_2^2 - 1041561 x_1 x_3^4 - 2701863 x_2 x_3^4 - 5920710 x_1^4 x_3^3 - 3201822 x_3^5 x_2^2 - 367011 x_3^5 x_1 - 1776762 x_3^5 x_2 - 2007234 x_3^5 x_1^2 + 10217583 x_2 x_1 x_3^4 - 26748144 x_3^4 x_2 x_1^2 + 23463306 x_3^2 x_2^2 x_1 + 6779862 x_3^2 x_2 x_1 + 4466880 x_1 - 3813952 x_2 - 520768 x_3 - 16392064 x_1^2 - 13928144 x_2^2 - 857264 x_3^2 + 23714880 x_1^3 - 28078300 x_2^3 - 573568 x_1^4 - 33492106 x_2^4 - 24141568 x_1^5 - 23550085 x_2^5 - 196096 x_1^6 - 9046200 x_2^6 + 39690 x_3^5 - 638436 x_3^3 - 1440780 x_3^3 + 3591536 x_3 x_1 - 2849856 x_3 x_1^2 + 4138196 x_3^2 x_1 + 31582352 x_1 x_2 - 4613120 x_3 x_2 - 90566848 x_1^2 x_2 + 92152884 x_1 x_2^2 - 15211892 x_3 x_2^2 + 88688752 x_1^3 x_2 - 195019560 x_1^2 x_2^2 + 140970299 x_1 x_2^3 - 25290352 x_3 x_1^3 - 24626550 x_3 x_2^3 - 5042468 x_2^2 x_2 + 1087832 x_3^2 x_1^3 - 10534970 x_3^2 x_2^2 + 18348032 x_1^7 - 1475500 x_2^7 - 119313 x_3^6 - 96066 x_{3}^7,$

and $\deg(\phi_{\mathcal{P}}) = 1$. That is, $\mathcal{P}(\overline{t})$ is a proper rational parametrization.

The time, in CPU seconds, for this example is 71.703, and using Gröbner basis, we get a time that is > 5000.

Remark 4. In order to improve the time of computations, one may compute the polynomial

$$h(\overline{x}) = \mathrm{Cont}_{\{Z,W\}}(\mathrm{Res}_{t_2}(T_{12}^{\mathcal{P}}(t_2, \overline{x}), K(t_2, Z, W, \overline{x})))$$

as $\gcd(R_1, R_2)$, where

$$R_i = \mathrm{Res}_{t_2}(T_{12}^{\mathcal{P}}(t_2, \overline{x}), K(t_2, a_i, b_i, \overline{x})), \; i = 1, 2$$

and $a_i, b_i \in \mathbb{K}$ are random constants. The answer is correct with a probability of almost one, since, taking into account the behavior of the gcd under specializations, this property holds in an open Zariski subset (see e.g., Lemmas 7 and 8 in [45]).

6. Conclusions

The μ-basis has shown as a bridge tool between the parametric form and the implicit form of curves and surfaces. Moreover, the μ-basis has also been introduced into applications in singularities analysis and collision detections. The μ-basis theory of curves are more complete than that of surfaces, but surfaces would certainly attract more attention, although the discussion is more difficult. We study the μ-basis further for improper rational surfaces. The results are essential to the theoretical completeness of the μ-basis of surface.

We show how the μ-basis allows for computing the inversion of a given proper parametrization $\mathcal{P}(\overline{t})$ of an algebraic surface. If $\mathcal{P}(\overline{t})$ is not proper, we show how the degree of the rational map that is induced by $\mathcal{P}(\bar{t})$ can be computed as well as the elements of the fiber. Furthermore and directly from $\mathcal{P}(\bar{t})$, we propose a method to find a μ-basis for a proper reparametrization $\mathcal{Q}(\bar{t})$ with some assumptions. If $\mathcal{P}(\bar{t})$ is improper, we give some partial results in finding a proper reparametrization of $\mathcal{V}$. Finally, we show how the μ-basis of a given not being necessarily proper parametrization also allows for computing the implicit equation of a given surface bysubsequence univariate resultants.

As the further work, the numerical consideration could be an interesting extension of the μ-basis theory. One possible way would consist in generalizing the symbolic computation to numerical situation using the ideas and techniques that have already been implemented in some other problems, such as the numerical proper reparametrization of surfaces [46].

Author Contributions: Writing—original draft, S.P.-D. and L.-Y.S. All authors have read and agreed to the published version of the manuscript.

Funding: This work has been partially supported by Beijing Natural Science Foundation under Grant Z190004, NSFC under Grant 61872332, the Fundamental Research Funds for the Central University and FEDER/Ministerio de Ciencia, Innovación y Universidades-Agencia Estatal de Investigación/MTM2017-88796-P (Symbolic Computation: new challenges in Algebra and Geometry together with its applications).

Institutional Review Board Statement: Not applicable.

Informed Consent Statement: Not applicable.

Data Availability Statement: Not applicable.

Acknowledgments: First author belongs to the Research Group ASYNACS (Ref. CT-CE2019/683).

Conflicts of Interest: The authors declare no conflict of interest.

References

1. Adkins, W.A.; Hoffman, J.W.; Wang, H.H. Equations of parametric surfaces with base points via syzygies. *J. Symb. Comput.* **2005**, *39*, 73–101. [CrossRef]
2. Buchberger, B. Groebner-Bases: An Algorithmic Method in Polynomial Ideal Theory. In *Multidimensional Systems Theory-Progress, Directions and Open Problems in Multidimensional Systems*; Bose, N., Ed.; Reidel Publishing Company: Dordrecht, The Netherlands, 1985; chapter 6, pp. 184–232.
3. Busé, L.; Chardin, M.; Jouanolou, J.P. Torsion of the Symmetric Algebra and Implicitization. *Proc. Am. Math. Soc.* **2009**, *137*, 1855–1865. [CrossRef]
4. Busé, L.; Cox, D.A.; D'andrea, C. Implicitization of surfaces in $\mathbb{P}^3$ in the presence of base points. *J. Algebra Its Appl.* **2003**, *2*, 189–214. [CrossRef]
5. Chen, F.; Cox, D.A.; Liu, Y. The μ-basis and implicitization of a rational parametric surface. *J. Symb. Comput.* **2005**, *39*, 689–706. [CrossRef]
6. Cox, D.A.; Little, J.B.; O'Shea, D. *Ideals, Varieties, and Algorithms: An Introduction to Computational Algebraic Geometry and Commutative Algebra*, 4th ed.; Springer: New York, NY, USA, 2015.
7. D'Andrea, C.; Emiris, I.Z. Sparse Resultant Perturbations. In *Algebra, Geometry and Software Systems*; Joswig, M., Takayama, N., Eds.; Springer: Berlin/Heidelberg, Germany, 2003; pp. 93–107. [CrossRef]
8. Emiris, I.Z.; Kotsireas, I. Implicitization exploiting sparseness. In *Geometric and Algorithmic Aspects of Computer-Aided Design and Manufacturing*; Dutta, D., Smid, M., Janardan, R., Eds.; DIMACS Series in Discrete Mathematics and Theoretical Computer Science 67; American Mathematical Society, Providence RI: Berlin/ Heidelberg, Germany, 2005; pp. 281–298.
9. Manocha, D.; Canny, J.F. Algorithm for implicitizing rational parametric surfaces. *Comput. Aided Geom. Des.* **1992**, *9*, 25–50. [CrossRef]
10. Sederberg, T.W.; Chen, F. Implicitization Using Moving Curves and Surfaces. In Proceedings of the 22nd Annual Conference on Computer Graphics and Interactive Techniques, SIGGRAPH '95, Los Angeles, CA, USA, 6–11 August 1995; ACM: New York, NY, USA, 1995; pp. 301–308. [CrossRef]
11. Shen, L.Y.; Goldman, R. Implicitizing Rational Tensor Product Surfaces Using the Resultant of Three Moving Planes. *ACM Trans. Graph.* **2017**, *36*, 167:1–167:14. [CrossRef]
12. Shen, L.Y.; Goldman, R. Combining complementary methods for implicitizing rational tensor product surfaces. *Comput.-Aided Des.* **2018**, *104*, 100–112. [CrossRef]
13. Shen, L.Y.; Pérez-Díaz, S.; Goldman, R.; Feng, Y. Representing rational curve segments and surface patches using semi-algebraic sets. *Comput. Aided Geom. Des.* **2019**, *74*, 101770. [CrossRef]

14. Shen, L.Y.; Yuan, C.M. Implicitization using univariate resultants. *J. Syst. Sci. Complex.* **2010**, *23*, 804–814. [CrossRef]
15. Busé, L.; Mantzaflaris, A.; Tsigaridas, E. Matrix formulae for resultants and discriminants of bivariate tensor-product polynomials. *J. Symb. Comput.* **2020**, *98*, 65–83. [CrossRef]
16. Chionh, E.W.; Zhang, M.; Goldman, R. Fast Computation of the Bezout and Dixon Resultant Matrices. *J. Symb. Comput.* **2002**, *33*, 13–29. [CrossRef]
17. Cox, D.A.; Sederberg, T.W.; Chen, F. The moving line ideal basis of planar rational curves. *Comput. Aided Geom. Des.* **1998**, *15*, 803–827. [CrossRef]
18. Dickenstein, A.; Emiris, I.Z. Multihomogeneous resultant formulae by means of complexes. *J. Symb. Comput.* **2003**, *36*, 317–342. [CrossRef]
19. Emiris, I.Z.; Mantzaflaris, A. Multihomogeneous resultant formulae for systems with scaled support. *J. Symb. Comput.* **2012**, *47*, 820–842. [CrossRef]
20. Shi, X.; Jia, X.; Goldman, R. Using a bihomogeneous resultant to find the singularities of rational space curves. *J. Symb. Comput.* **2013**, *53*, 1–25. [CrossRef]
21. Shi, X.; Wang, X.; Goldman, R. Using μ-bases to implicitize rational surfaces with a pair of orthogonal directrices. *Comput. Aided Geom. Des.* **2012**, *29*, 541–554. [CrossRef]
22. Busé, L. Implicit matrix representations of rational Bézier curves and surfaces. *Comput.-Aided Des.* **2014**, *46*, 14–24. [CrossRef]
23. Chen, F.; Wang, W. The μ-basis of a planar rational curve—Properties and computation. *Graph. Model.* **2002**, *64*, 368–381. [CrossRef]
24. Lai, Y.; Chen, F.; Shi, X. Implicitizing rational surfaces without base points by moving planes and moving quadrics. *Comput. Aided Geom. Des.* **2019**, *70*, 1–15. [CrossRef]
25. Shen, L.Y.; Goldman, R. Strong μ-bases for Rational Tensor Product Surfaces and Extraneous Factors Associated to Bad Base Points and Anomalies at Infinity. *SIAM J. Appl. Algebra Geom.* **2017**, *1*, 328–351. [CrossRef]
26. Zhang, M.; Goldman, R.; Chionh, E.W. Efficient Implicitization of Rational Surfaces by Moving Planes. In *Proceedings of the ASCM*; Gao, X.-S., Wang, D., Eds.; World Scientific: Singapore, 2000; pp. 142–151.
27. Zheng, J.; Sederberg, T.W.; Chionh, E.W.; Cox, D.A. Implicitizing rational surfaces with base points using the method of moving surfaces. In *Topics in Algebraic Geometry and Geometric Modeling*; American Mathematical Society: Providence, RI, USA, 2003; pp. 151–168. [CrossRef]
28. Walker, R.J. *Algebraic Curves*; Princeton University Press: Princeton, NJ, USA, 1950.
29. Gao, X.S.; Chou, S.C. On the Parameterization of Algebraic Curves. *Appl. Algebra Eng. Commun. Comput.* **1992**, *3*, 27–38. [CrossRef]
30. Pérez-Díaz, S. On the problem of proper reparametrization for rational curves and surfaces. *Comput. Aided Geom. Des.* **2006**, *23*, 307–323. [CrossRef]
31. Sederberg, T.W. Improperly parametrized rational curves. *Comput. Aided Geom. Des.* **1986**, *3*, 67–75. [CrossRef]
32. Pérez-Díaz, S. A partial solution to the problem of proper reparametrization for rational surfaces. *Comput. Aided Geom. Des.* **2013**, *30*, 743–759. [CrossRef]
33. Jia, X.; Shi, X.; Chen, F. Survey on the theory and applications of μ-bases for rational curves and surfaces. *J. Comput. Appl. Math.* **2018**, *329*, 2–23. [CrossRef]
34. Zheng, J.; Sederberg, T.W. A Direct Approach to Computing the μ-basis of Planar Rational Curves. *J. Symb. Comput.* **2001**, *31*, 619–629. [CrossRef]
35. Deng, J.; Chen, F.; Shen, L. Computing μ-bases of Rational Curves and Surfaces Using Polynomial Matrix Factorization. In Proceedings of the 2005 International Symposium on Symbolic and Algebraic Computation, ISSAC '05, Beijing, China, 24–27 July 2005; ACM: New York, NY, USA, 2005; pp. 132–139. [CrossRef]
36. Fabiańska, A.; Quadrat, A. Applications of the Quillen-Suslin theorem to multidimensional systems theory. In *Gröbner Bases in Control Theory and Signal Processing*; Park, H., Regensburger, G., Eds.; Radon Series on Computation and Applied Mathematics 3, de Gruyter Publisher: Berlin, Germany, 2007; pp. 23–106.
37. Fabiańska, A.; Quadrat, A. QuillenSuslin: A Maple Implementation of a Constructive Version of the Quillen-Suslin Theorem. 2007. Available online: https://who.rocq.inria.fr/Alban.Quadrat/QuillenSuslin/ (accessed on 22 December 2020).
38. Shen, L.Y.; Goldman, R. Algorithms for computing strong μ-bases for rational tensor product surfaces. *Comput. Aided Geom. Des.* **2017**, *52–53*, 48–62. [CrossRef]
39. Pérez-Díaz, S.; Shen, L.Y. Inversion, degree, reparametrization and implicitization of improperly parametrized planar curves using μ-basis. *Comput. Aided Geom. Des.* **2021**, *84*, 101957. [CrossRef]
40. Cox, D.A.; Little, J.B.; O'Shea, D. *Using Algebraic Geometry*; Graduate Texts in Mathematics; Springer: New York, NY, USA, 1998.
41. Shafarevich, I.R. *Basic Algebraic Geometry Schemes; 1 Varieties in Projective Space*; Springer: Berlin/Heidelberg, Germany, 1994.
42. Harris, J. *Algebraic Geometry : A First Course*; Springer: Berlin/Heidelberg, Germany, 2000.
43. Pérez-Díaz, S.; Schicho, J.; Sendra, J. Properness and Inversion of Rational Parametrizations of Surfaces. *Appl. Algebra Eng. Commun. Comput.* **2002**, *13*, 29–51. [CrossRef]
44. Pérez-Díaz, S.; Sendra, J. A univariate resultant-based implicitization algorithm for surfaces. *J. Symb. Comput.* **2008**, *43*, 118–139. [CrossRef]

45. Pérez-Díaz, S.; Sendra, J. Computation of the degree of rational surface parametrizations. *J. Pure Appl. Algebra* **2004**, *193*, 99–121. [CrossRef]
46. Shen, L.Y.; Pérez-Díaz, S.; Yang, Z. Numerical Proper Reparametrization of Space Curves and Surfaces. *Comput.-Aided Des.* **2019**, *116*, 102732. [CrossRef]

Article

Vector Geometric Algebra in Power Systems: An Updated Formulation of Apparent Power under Non-Sinusoidal Conditions

Francisco G. Montoya [1,*], Raúl Baños [1], Alfredo Alcayde [1], Francisco Manuel Arrabal-Campos [1] and Javier Roldán-Pérez [2]

[1] Department of Engineering, University of Almeria, 04120 Almeria, Spain; rbanos@ual.es (R.B.); aalcayde@ual.es (A.A.); fmarrabal@ual.es (F.M.A.-C.)
[2] Electrical Systems Unit, IMDEA Energy Institute, 28935 Madrid, Spain; javier.roldan@imdea.org
* Correspondence: pagilm@ual.es; Tel.: +34-950-214-501

Citation: Montoya, F.G.; Baños, R.; Alcayde, A.; Arrabal-Campos, F.M.; Roldán-Pérez, J. Vector Geometric Algebra in Power Systems: An Updated Formulation of Apparent Power under Non-Sinusoidal Conditions. *Mathematics* **2021**, *9*, 1295. https://doi.org/10.3390/math9111295

Academic Editors: Sonia Pérez-Díaz and Yang-Hui He

Received: 11 May 2021
Accepted: 2 June 2021
Published: 4 June 2021

Publisher's Note: MDPI stays neutral with regard to jurisdictional claims in published maps and institutional affiliations.

Copyright: © 2021 by the authors. Licensee MDPI, Basel, Switzerland. This article is an open access article distributed under the terms and conditions of the Creative Commons Attribution (CC BY) license (https://creativecommons.org/licenses/by/4.0/).

Abstract: Traditional electrical power theories and one of their most important concepts—apparent power—are still a source of debate, because they present several flaws that misinterpret the power-transfer and energy-balance phenomena under distorted grid conditions. In recent years, advanced mathematical tools such as geometric algebra (GA) have been introduced to address these issues. However, the application of GA to electrical circuits requires more consensus, improvements and refinement. In this paper, electrical power theories for single-phase systems based on GA were revisited. Several drawbacks and inconsistencies of previous works were identified, and some amendments were introduced. An alternative expression is presented for the electric power in the geometric domain. Its norm is compatible with the traditional apparent power defined as the product of the RMS voltage and current. The use of this expression simplifies calculations such as those required for current decomposition. This proposal is valid even for distorted currents and voltages. Concepts are presented in a simple way so that a strong background on GA is not required. The paper included some examples and experimental results in which measurements from a utility supply were analysed.

Keywords: geometric algebra; non-sinusoidal power; Clifford algebra; power theory

1. Introduction

The full understanding of power flows in electrical and electronic systems has been a topic of interest during the last century. It is of paramount relevance because of the increasing energy losses in transmission systems, as well as the the poor power quality on electrical devices. In particular, the reduction of power losses involves the reduction of CO_2 emissions. This aim requires the installation of smart metering systems in the grid to collect a large amount of electrical power data [1]. To accomplish this task, relevant efforts have been carried out in the frequency domain for systems operating in steady state [2] and in the time domain by using both instantaneous and averaged approaches [3–5]. The outcomes from these studies were sometimes inconsistent and even contradictory. For example, the well-known instantaneous power theory can yield incoherent results under specific conditions [6]. Similar controversial results have been found for well-established regulations such as the standard IEEE 1459 [7]. Traditional techniques that are commonly applied for analysing power flows are based on linear algebra tools such as complex numbers, matrices, tensors, etc., and they have been proven to be useful from the application point of view [8]. However, none of them provide a clear overview of power flows under disported and unbalanced grid conditions, and this point is still an open discussion [9]. Geometric algebra (GA) is a mathematical tool developed by W. K. Clifford and H. Grassmann at the end of the 19th Century. It has been rediscovered and refined in the last few decades by Hestenes [10] and others [11]. This tool has brought new possibilities to important fields such as physics [10], computer graphics [12] and rendering problems [13], producing compact and generalised formulations [14]. Furthermore, it can be easily used to manipulate integral and differential equations in multi-component systems [15,16]. Even though GA is not widely known by the scientific community, it has a great potential and has attracted a lot of interest in recent publications [17]. GA has already been introduced to reformulate the apparent power as the geometric product between voltage and current, which is commonly written as M [18–21]. Compared to the traditional definition of apparent power $S = VI$, the use of M has several advantages. A relevant one is that M is conservative in spite of S, and this is of interest for its application in distorted environments [22]. The use of a different letter M instead of the traditional one S is therefore justified. It also has important implications for the definition of reactive power and its compensation, and this is a topic of interest for the power community [23].

Contributions

The main contributions and novelties of this paper are based on the following points:

- GA power theories proposed by different authors were briefly reviewed in order to analyse some of the inconsistencies raised so far, while additional ones not yet found in the literature were also discussed [7,18,24]. Menti's pioneering expression for geometric electric power was recovered because it has several advantages and benefits over other proposals for power computations. For example, one of the most relevant is that its norm equals the product between the norms of geometric voltage and current, thus retaining the traditional approach in the apparent power definition. It should be emphasized that this approach is different from those already published and based on k-blades or complex-vectors;
- A new mapping between the Fourier basis for periodic time functions and the Euclidean basis was introduced, accounting for harmonics, inter- and sub-harmonics and DC components. Because no additional restrictions were imposed on the waveforms, the developed methodology is valid even in the case of distorted currents and voltages. Furthermore, the relevant features of GA for power and circuit analysis and power calculations were maintained: electrical circuits can be easily solved, and the principle of energy conservation was still satisfied;
- Another relevant contribution was the formulation by means of vectors in GA for some of the most important laws in basic circuit theory, i.e., Kirchoff's laws or Ohm's law, to mention a few. This is a crucial issue when solving steady-state AC circuits in GA without the use of complex phasors. The concept of geometrical impedance as a bivector was also introduced;
- Another very relevant aspect is the current decomposition proposal based on the use of the inverse of the voltage vector, which has important implications in the use of active filters and current compensation. It was shown that the use of this approach allowed a comprehensive current decomposition for optimal passive/active filtering based on the concept of the vector inverse, not discussed previously in the literature.

Numerical and experimental results were included in order to validate the main contributions of this work. A brief introduction to GA and its terminology was included in order to make the paper self-contained.

2. Geometric Algebra for Power Flow Analysis

The geometric product was introduced by Clifford at the end of the 19th Century, and it includes the exterior (Grassmann) and interior (dot) products of vectors. Suppose a Euclidean two-dimensional vector space $\mathbb{R}^2$ spanned by the basis $\{\sigma_1, \sigma_2\}$ and elements such as $a = \alpha_1 \sigma_1 + \alpha_2 \sigma_2$ and $b = \beta_1 \sigma_1 + \beta_2 \sigma_2$ with $\alpha_i, \beta_i \in \mathbb{R}$. Their interior product can be calculated as follows:

$$a \cdot b = \|a\| \|b\| \cos \varphi = \sum \alpha_i \beta_i \qquad (1)$$

while the exterior product is:

$$a \wedge b = \|a\| \|b\| \sin \varphi \, \sigma_1 \sigma_2 \tag{2}$$

The operation in Equation (2) does not exist in traditional linear algebra, and its result is not a scalar, nor a vector, but a new entity that is commonly known as a bivector [10]. Bivectors play a key role in calculations related to non-active power, as is shown later. The exterior product is anticommutative, i.e., $a \wedge b = -b \wedge a$. The fundamental operation in GA is the geometric product:

$$M = ab = (\alpha_1 \sigma_1 + \alpha_2 \sigma_2)(\beta_1 \sigma_1 + \beta_2 \sigma_2) = \underbrace{(\alpha_1 \beta_1 + \alpha_2 \beta_2)}_{\langle M \rangle_0} + \underbrace{(\alpha_1 \beta_2 - \alpha_2 \beta_1)}_{\langle M \rangle_2} \sigma_1 \sigma_2 \tag{3}$$

where M consists of two elements. As these elements are of a different nature, M is commonly referred to as a multivector. The operation $\langle X \rangle_k$ refers to the extraction of the k-grade component of a multivector X. In Equation (3), the term $\langle M \rangle_0$ is a scalar, while the term $\langle M \rangle_2$ is a bivector. Multivectors are classified according to their degree: scalars have degree zero, vectors one, bivectors two, etc. The norm of a multivector is:

$$\|M\| = \sqrt{\langle M^\dagger M \rangle_0} \tag{4}$$

where $M^\dagger$ is the reverse of M (see [10] for details). Note that in the rest of this work, vectors are represented by bold lower case letters while bivectors and multivectors are represented by bold upper case letters. Considering a single-phase system operating under perfect periodic sinusoidal conditions, it is possible to select an orthonormal basis in the space vector of Fourier functions:

$$\varphi = \{1, \sqrt{2} \cos \omega t, \sqrt{2} \sin \omega t\} \tag{5}$$

This basis also belongs to the L^2 Hilbert space [25] of integrable and finite energy functions, equipped with a norm:

$$\|x(t)\| = \sqrt{\frac{1}{T} \int_0^T x^2(t) dt} \tag{6}$$

Note that $x(t)$ can be expressed as a linear combination of the orthonormal elements φ_j with $j = 1, 2, 3$ as in $x(t) = \sum_{j=1}^{3} x_j \varphi_j(t)$, which, in general, can be represented in a Euclidean vector space as a vector $x = \sum_{j=1}^{3} x_j \sigma_j$ as defined previously. The above can be readily extended to non-sinusoidal signals by updating the basis to $\varphi = \{1, \sqrt{2} \cos k\omega t, \sqrt{2} \sin k\omega t\}_{k=1}^{n}$ where n is the number of harmonics under consideration. For simplicity, the DC term will not be considered at this moment (but it can be added without problems). Therefore, voltages and currents are transformed to the proposed Euclidean vector space as follows:

$$\begin{aligned} u(t) &\longrightarrow & u = \alpha_1 \sigma_1 + \alpha_2 \sigma_2 \\ i(t) &\longrightarrow & i = \beta_1 \sigma_1 + \beta_2 \sigma_2 \end{aligned} \tag{7}$$

The geometric product defined in Equation (3) can be used to calculate the geometric power:

$$M = ui = \underbrace{(\alpha_1 \beta_1 + \alpha_2 \beta_2)}_{P} + \underbrace{(\alpha_1 \beta_2 - \alpha_2 \beta_1)}_{Q} \sigma_1 \sigma_2 \tag{8}$$

This expression consists of two terms of a different nature that can be clearly identified: P is a scalar, and Q is a bivector. Note that $\|Q\|$ is the traditional reactive power Q.

For convenience, $\sigma_1\sigma_2$ is often written as σ_{12}. This result is extended to non-sinusoidal conditions in later sections. The geometric power fulfils:

$$\|M\|^2 = \langle M\rangle_0^2 + \langle M\rangle_2^2 = P^2 + Q^2 = \|u\|^2\|i\|^2 \tag{9}$$

3. GA-Based Power Theories: Overview

In this section, the main power theories based on GA are briefly and critically discussed so that the main contributions of this paper can be better understood.

- Menti: This theory was developed by Anthoula Menti et al. in 2007 [18]. This was the first application of GA to electrical circuits. The apparent power multivector was defined by multiplying the voltage and current in the geometric domain:

$$S = ui = u \cdot i + u \wedge i = \langle S\rangle_0 + \langle S\rangle_2 \tag{10}$$

The scalar part matches the active power P, while the bivector part represents power components with zero mean value. Unfortunately, the theory did not establish a general framework for the resolution of electrical circuits under distorted conditions. Furthermore, the proposal was not applied to decompose currents (for non-linear load compensation, for example), and it was not extended to multi-phase systems.

- Castilla–Bravo: This theory was developed by Castilla and Bravo in 2008 [19]. The authors introduced the concept of generalised complex geometric algebra. Vector-phasors were defined for both voltage and current:

$$\tilde{U}_p = U_p e^{j\alpha_p}\sigma_p = \tilde{U}_p\sigma_p, \quad \tilde{I}_q = I_q e^{j\beta_q}\sigma_q = \tilde{I}_q\sigma_q \tag{11}$$

Geometric power results from multiplying the harmonic voltage and conjugated harmonic current vector-phasors:

$$\tilde{S} = \sum_{\substack{p\in N\cup L \\ q\in N\cup M}} \tilde{U}_p\tilde{I}_q^* = \tilde{P} + j\tilde{Q} + \tilde{D} \tag{12}$$

This proposal is able to capture the multicomponent nature of apparent power through the so-called complex scalar $\tilde{P} + j\tilde{Q}$ and the complex bivector $\tilde{D}$. However, this formulation requires the use of complex numbers, which could have been avoided by using appropriate bivectors [14]. Furthermore, only definitions of powers were presented, and it was not extended to multi-phase systems.

- Lev-Ari: This theory was developed by Lev-Ari [20,26], and it was the first application of GA to multi-phase systems in the time domain. However, this work did not contain examples, nor fundamentals for load compensation. Furthermore, practical aspects required to solve electrical circuits were not explained.

- Castro-Núñez: This theory was developed by Castro-Núñez in the year 2010 [27] and then extended and refined in further works [7,28]. A relevant contribution of this work consisted of the resolution of electrical circuits by using GA (without requiring complex numbers). Furthermore, a multivector called geometric power that is conservative and fulfils the Tellegen theorem was defined [29]. As in the Menti and Castilla–Bravo proposals, the results were presented only for single-phase systems. Another contribution was the definition of a transformation based on k-blades, i.e., objects that can be expressed as the exterior product of k basis vectors. They form an orthonormal base. However, this basis presents some drawbacks. The main one is the definition of the geometric power [30]. In particular, active power calculations did not match with those obtained by using classical theories. Therefore, the authors needed to include an ad hoc corrective coefficient [7]. Finally, the definition of geometric power norm did not follow the traditional expression as a product or voltage and current norms (i.e., RMS in the complex domain) due to the proposed axiomatic transformation.

- Montoya: This framework was proposed by Montoya et al. [30], and it is an upgrade of the Menti and Castro-Núñez theories [7,18]. It establishes a general framework for power calculations in the frequency domain. Since it was the most recent work, it provided solutions to some problems detected so far in other proposals, and the formulation was more compact and efficient. However, this framework was based on the use of k-blades, and therefore, drawbacks related to the non-standardised definition of apparent power were inherited from previous theories.

The most relevant contributions of this work compared to existing proposals are presented in Table 1.

Table 1. Comparison of the main contributions of GA power theories.

Feature	Menti [18]	Castilla and Bravo [19]	Lev-Ari [20]	Castro-Núñez [27]	Montoya [21]	This Work
Based on	vectors	complex-vectors	vectors	k-blades	k-blades	vectors
GA power definition	$S = ui$	$\check{S} = \sum \check{U}_p \check{s} \check{I}_q^*$	$\mathbb{S} = \mathbf{vi}$	$M = VI$	$M = VI^\dagger$	$M = ui$
Power norm	$\|S\| = \|u\|\|i\|$	$\|\check{S}\|^2 = \|\check{U}\|^2 \|\check{I}\|^2$	$\|\mathbb{S}\| = \|\mathbf{v}\|\|\mathbf{i}\|$	$\|M\| \neq \|V\|\|I\|$	$\|M\| \neq \|V\|\|I^\dagger\|$	$\|M\| = \|u\|\|i\|$
Circuit theory ready	No	No	No	Yes	Yes	Yes
Current decomposition	No	No	No	No	Not Always	Yes
Interharmonic handling	No	No	No	No	Yes	Yes
Impedance definition	No	No	No	Yes	Yes	Yes

4. GA Framework and Methodology

4.1. Circuit Analysis by Means of GA

In this theory, different approaches already available in the literature are unified and enhanced in order to analyse electrical circuits in the geometric domain. The proposed modifications are deemed to remain consistent with the physical basic principles observed in electrical circuits. An orthonormal basis $\sigma = \{\sigma_1, \sigma_2, \ldots, \sigma_n\}$ is used in order to represent the multi-component nature of periodic signals with finite energy. Consider a voltage signal $u(t)$:

$$u(t) = U_0 + \sqrt{2}\sum_{k=1}^{n} U_k \cos(k\omega t + \varphi_k) + \sqrt{2}\sum_{l \in L} U_l \cos(l\omega t + \varphi_l) \quad (13)$$

where U_0 is the DC component, while U_k and φ_k are the RMS and phase of the k-th harmonic, respectively. The set L represents sub- and inter-harmonics included in the signal [31]. As in traditional circuit analysis based on complex variables, a sinusoidal and steady-state signal can be considered as a part of a rotating vector $n(t)$ (in a similar fashion to $e^{j\omega t}$). It was demonstrated that this facilitated the analysis in the geometric domain. In addition, thanks to the linear properties of GA, it was possible to define a single multivector that included all the harmonic frequencies present in the signal (this was not possible by using the traditional complex analysis). This rotating vector $n(t)$ in a two-dimensional geometric space $\mathcal{G}_2$ can be obtained as follows [32]:

$$n(t) = e^{\frac{1}{2}\omega t \sigma_{12}} N e^{-\frac{1}{2}\omega t \sigma_{12}} = RNR^\dagger = e^{\omega t \sigma_{12}} N = R^2 N = NR^{\dagger 2} \quad (14)$$

where $R = e^{\frac{1}{2}\omega t \sigma_{12}}$ is a geometric *rotor* (or simply a rotor) [33] and N is a vector. In Equation (14), left-multiplying produces opposite effects compared to right-multiplying. Figure 1 shows a graphical representation of a vector N left-multiplied by a rotor $R' = e^{\varphi \sigma_{12}}$ with positive angle φ. This operation produces a rotated vector (in green) in clockwise direction. Similarly, the same vector N right-multiplied by the same rotor R' produces a rotation of φ degrees in the counter-clockwise direction (vector in red).

In order to maintain the commonly accepted convention of signs in electrical engineering, vectors are always left-multiplied by rotors. Therefore, a positive sign in a phase angle refers to the clockwise direction. This implies that an inductor reactance will have positive phase angles, while a capacitor will have negative phase angles. However, the phase lead and lag now change its role in the geometric domain: lag implies rotation in the counter-clockwise direction and lead in the clockwise direction (see Figure 1).

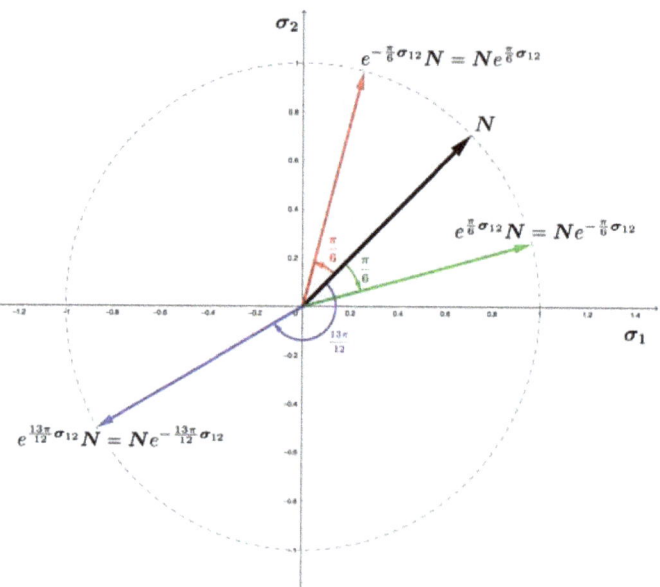

Figure 1. A vector multiplied by a rotor $e^{\varphi \sigma_{12}}$ (with positive angle φ) rotates in the clockwise or counter-clockwise direction depending whether the multiplication is performed by the left or by the right, respectively.

It can be proven that the projection of a rotating vector voltage $u_1(t)$ over the basis σ_1 yields the original voltage waveform, i.e., $u_1(t) = \sqrt{2}(\alpha_1 \cos \omega t + \alpha_2 \sin \omega t)$. This resembles the extraction of the real part of a complex rotating phasor, i.e., $\mathcal{R}e\{\sqrt{2}\bar{V}e^{j\omega t}\}$. By using the Euclidean orthonormal basis $\sigma = \{\sigma_1, \sigma_2\}$ isomorphic to the Fourier basis in Equation (5), then the original time signal $u_1(t)$ is transformed into the vector $\boldsymbol{u}_1 = \alpha_1 \sigma_1 + \alpha_2 \sigma_2$; therefore:

$$\boldsymbol{u}_1(t) = e^{\omega t \sigma_{12}} \boldsymbol{u}_1 = (\cos \omega t + \sin \omega t \sigma_{12})(\alpha_1 \sigma_1 + \alpha_2 \sigma_2) =$$
$$= (\alpha_1 \cos \omega t + \alpha_2 \sin \omega t)\sigma_1 + (\alpha_2 \cos \omega t - \alpha_1 \sin \omega t)\sigma_2$$
$$= \frac{1}{\sqrt{2}}(u_1(t)\sigma_1 - \mathcal{H}[u_1(t)]\sigma_2) \tag{15}$$

where $\mathcal{H}$ refers to the Hilbert transform of a signal [34]. Hence, $u_1(t) = \text{proj}_{\sigma_1}[\sqrt{2}\boldsymbol{u}_1(t)] = \sqrt{2}\boldsymbol{u}_1(t) \cdot \sigma_1$ can be recovered as the scalar product, i.e., the projection of a rotating vector $\boldsymbol{u}_1(t)$ onto σ_1. It is worth pointing out that the rotating vector $\boldsymbol{u}_1(t)$ is not the original time domain voltage waveform, $u_1(t)$. This is a different interpretation compared to that of other authors [7,22]. This discrepancy was analysed by using the simple RLC circuit depicted in Figure 2. Its solution is well known in both the time and complex domain (for the steady state), but it is presented here to highlight that the proposed framework can be applied to the most basic electrical circuits. The time-domain equation that governs the circuit dynamics is:

$$u_1(t) = Ri(t) + L\frac{di(t)}{dt} + \frac{1}{C}\int i(t)dt \tag{16}$$

If we perform the derivative and the integral [32] of the rotating vector defined in Equation (15), we obtain:

$$\frac{d\boldsymbol{u}_1(t)}{dt} = \omega \sigma_{12} \boldsymbol{u}_1(t)$$
$$\int \boldsymbol{u}_1(t)dt = -\frac{\sigma_{12}}{\omega}\boldsymbol{u}_1(t) \tag{17}$$

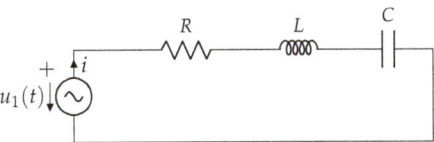

Figure 2. RLC circuit used in Example 1.

Because the source is sinusoidal and assuming that the circuit is operating in the steady state, Equation (15) can be substituted in Equation (16),

$$\text{proj}_{\sigma_1}\left[\sqrt{2}e^{\omega t\sigma_{12}}u_1\right] = \sqrt{2}\,\text{proj}_{\sigma_1}\left[R\,e^{\omega t\sigma_{12}}i + L\omega\sigma_{12}e^{\omega t\sigma_{12}}i - \frac{\sigma_{12}}{C\omega}e^{\omega t\sigma_{12}}i\right] \qquad (18)$$

so, applying the same rationale used in Equation (15), Equation (18) can be simplified, yielding:

$$u_1 = Ri + L\omega\sigma_{12}i - \frac{\sigma_{12}}{C\omega}i \qquad (19)$$

Rotors such as $e^{\omega t\sigma_{12}}$ are cancelled out because they commute with σ_{12}. Therefore, it is not necessary to set any specific time instant t_0 after performing the derivative, as suggested by Castro-Núñez [27]. The result is an algebraic equation where only vectors such as u_1 and i are present. Right-multiplying Equation (19) by the inverse of the current results in a generalised Ohm's law, where the geometric impedance can be defined as:

$$Z = u_1 i^{-1} = R + \left(L\omega - \frac{1}{C\omega}\right)\sigma_{12} = R + X\sigma_{12} \qquad (20)$$

The geometric admittance can be defined as the inverse of the geometric impedance:

$$Y = Z^{-1} = \frac{Z^\dagger}{Z^\dagger Z} = \frac{Z^\dagger}{\|Z\|^2} = G + B\sigma_{12} \qquad (21)$$

Both elements have similar definitions to those of impedance/admittance in the complex domain (the complex algebra is already a subalgebra of $\mathbb{G}^2$). However, now they are multivectors because they consist of a scalar part plus a bivector (this kind of multivector is commonly known as a spinor). The use of this criterion allows overcoming the drawbacks of other theories in which inductive reactance is negative, while capacitive reactance is positive [27]. In order to transform the voltage signal in Equation (13) from the Fourier to the geometric domain, a new Euclidean basis was proposed based on Equation (5). This proposal was supported by the principle of isomorphism among vector spaces. Let V and W be vector spaces over the same field F, which preserves the addition and scalar multiplication of elements in both spaces. Then, for all vectors u and v in V and all scalars $c \in F$, a transformation $T: V \rightarrow W$ exists such as:

$$T(\mathbf{u}+\mathbf{v}) = T(\mathbf{u}) + T(\mathbf{v}) \text{ and } T(c\mathbf{v}) = cT(\mathbf{v}) \qquad (22)$$

This was a major contribution of this work, not previously reported in the literature. This isomorphism is then defined as:

$$\begin{aligned}
\varphi_{DC} &= 1 &&\longleftrightarrow& \sigma_0 \\
\varphi_{c1}(t) &= \sqrt{2}\cos\omega t &&\longleftrightarrow& \sigma_1 \\
\varphi_{s1}(t) &= \sqrt{2}\sin\omega t &&\longleftrightarrow& \sigma_2 \\
&\vdots \\
\varphi_{cn}(t) &= \sqrt{2}\cos n\omega t &&\longleftrightarrow& \sigma_{2n-1} \\
\varphi_{sn}(t) &= \sqrt{2}\sin n\omega t &&\longleftrightarrow& \sigma_{2n}
\end{aligned} \qquad (23)$$

In addition, l sub- and inter-harmonics can be added by increasing the number of elements in the basis by $2l$ after the highest order harmonic (n) [31]. Now, $u(t)$ can be completely transferred to the geometric domain as:

$$\begin{aligned} u &= U_0\sigma_0 + \sum_{k=1}^{n} U_k e^{\varphi_k \sigma_{(2k-1)(2k)}} \sigma_{(2k-1)} + \sum_{m=1}^{l} U_m e^{\varphi_m \sigma_{(2n+2m-1)(2n+2m)}} \sigma_{(2n+2m-1)} \\ &= U_0 + \sum_{k=1}^{n} U_{k1}\sigma_{(2k-1)} + U_{k2}\sigma_{(2k)} + \sum_{m=1}^{l} U_{m1}\sigma_{(2m-1)} + U_{m2}\sigma_{(2m)} \end{aligned} \quad (24)$$

where $U_{k1} = U_k \cos\varphi_k$ and $U_{k2} = U_k \sin\varphi_k$. The same transformation can be applied to $i(t)$ in order to calculate the geometric current $\boldsymbol{i}$. It is worth noting that $\boldsymbol{i}$ may include harmonics not present in the voltage. By using the same rationale presented in Equations (16)–(19), the geometric impedance can be defined for each harmonic as:

$$Z_k = u_k i_k^{-1} = R + \left(kL\omega - \frac{1}{kC\omega}\right)\sigma_{(2k-1)(2k)} \quad (25)$$

where $\boldsymbol{u}_k$ and $\boldsymbol{i}_k$ are the vector representation for the harmonic k in the geometric domain. This proposal overcomes some drawbacks of previous GA-based power theories. First, it can accommodate DC components in voltages and currents. Second, the traditional idea behind the definition of apparent power based on the product of the RMS voltage and current is preserved, and this does not happens in other proposals [7]. These are the contributions of this work.

4.2. Power definitions in GA

There exist different definitions for apparent power in other power theories based on GA. Menti and Castro-Núñez chose $S = UI$ and $M = UI$, respectively, while Castilla–Bravo used $S = UI^*$. All of them are compatible with the energy conservation principle due to the multi-component nature of GA [28]. However, the results might be inconsistent if the orthonormal basis that spans the geometric space is not carefully chosen. For example, in the proposal of Castro-Núñez, k-blades were used for the basis [27]. As a result, the geometric power calculation should be corrected so that power components are computed in accordance with physics principles, as already mentioned in Section 3. Furthermore, non-active power calculations can lead to erroneous results since the geometric power is not calculated accordingly. In [30], the correction $M = UI^\dagger$ was proposed as a solution. In this work, the geometric power was defined as:

$$M = ui = u \cdot i + u \wedge i \quad (26)$$

In Equation (26), several terms of engineering interest can be identified. On the one hand, the scalar term:

$$\langle M \rangle_0 = u \cdot i = \sum_{k=1}^{n} u_k i_k \quad (27)$$

matches the well-known active power P, and it will be referred to as active geometric power (or just active power). It corresponds to the mean value of the instantaneous active power $p(t)$ that is converted into useful work in power systems. On the other hand, the bivector term $\langle M \rangle_2 = u \wedge i$ is the so-called non-active geometric power or M_N. It can be decomposed into other terms with engineering significance:

$$M_N = M_Q + M_D = \sum_{k=1}^{n}(u_{(2k-1)}i_{(2k)} - u_{(2k)}i_{(2k-1)})\sigma_{(2k-1)(2k)} + \sum_{\substack{k,l \\ k \neq l \\ k < l+1}}^{n}(u_k i_l - u_l i_k)\sigma_{kl} \quad (28)$$

The term M_Q is a bivector with coordinates representing the classical reactive power generated by harmonics in the Budeanu sense Q_k. Finally, the term M_D is a new concept not existing in complex algebra approaches. It stands for cross-products between a voltage and current of different frequencies.

Note that the units of the geometrical quantities previously introduced are volts (V) for voltage u, amperes (A) for current i, volt-amperes (VA) for power M, M_N, M_Q and M_N. Obviously, the active power P is in watts (W). It can be readily checked that the norm of M is the product of the voltage and current norms, provided that u and i are vectors:

$$\|M\| = \sqrt{\langle M^\dagger M \rangle_0} = \sqrt{\langle (ui)^\dagger (ui) \rangle_0} = \sqrt{\langle (i^\dagger u^\dagger)(ui) \rangle_0} = \sqrt{\|u\|^2 \|i\|^2} = \|u\| \|i\| \quad (29)$$

where the property $a^\dagger = a$ was applied for vectors. The application of this property is the key to overcoming a definition based on the complex conjugate current. This feature cannot be applied in other power theories based on GA, since, in general, $A^\dagger \neq A$ for any k-blade A with $k > 1$ [27]. The power triangle also holds for the geometric power:

$$\|M\|^2 = P^2 + \|M_N\|^2 = P^2 + \|M_Q\|^2 + \|M_D\|^2 \quad (30)$$

The above expression does not hold for other GA power theories such as that proposed by Castro-Núñez because of the use of k-blades in the definition of voltage and current.

4.3. Current Decomposition in GA

In this section, the current demanded by a load is decomposed by using the proposed power theory. Simplifying Equation (26) and taking into account that for any given vector $a^{-1} = a/\|a\|^2$, the following result is obtained:

$$M = ui \longrightarrow u^{-1}M = \underbrace{u^{-1}u}_{1} i = i$$

$$i = u^{-1}M = \frac{u}{\|u\|^2}(M_a + M_N) = i_a + i_N \quad (31)$$

where i_a is the geometric counterpart of the active or Fryze current [35], while i_N is the non-active current. This decomposition procedure has not been used before for GA power theories in the frequency domain, and it was a novel contribution of this work. Furthermore, in previous power theories based on GA current decomposition were not guaranteed since multivectors might not have an inverse, and in any case, its calculation is not straightforward [36]. Each of the currents presented above has a well-established engineering meaning. The current i_a is the minimum current required to produce the same active power to that consumed by the load, while the non-active current i_N is the current that does not affect the net active power. Therefore, the latter can be compensated by using either passive or active filters. For linear loads, the current i_N can be decomposed into two terms for practical engineering purposes. The first one is related to transient energy storage and leads to the reactive current. The second one does not include storage and leads to the scattered current introduced by Czarnecki [37]. In addition, by using Equations (21) and (24) and Ohm's law, the current i demanded by the linear load can be calculated as:

$$i = \sum_{k=1}^{n} Y_k u_k = \sum_{k=1}^{n} \left(G_k + B_k \sigma_{(2k-1)(2k)} \right) u_k = i_p + i_q \quad (32)$$

where i_p is commonly known as the parallel current while i_q as the quadrature current:

$$i_p = \sum_{k=1}^{n} G_k u_k, \quad i_q = \sum_{k=1}^{n} B_k \sigma_{(2k-1)(2k)} u_k \quad (33)$$

which is a geometric counterpart of the Shepherd and Zakikhani decomposition [38]. It can be demonstrated that they are orthogonal because of the term $\sigma_{(2k-1)(2k)}$ in i_q. Therefore, by comparing Equations (31) and (32), it follows that:

$$i = i_a + i_N = i_p + i_q = i_a + i_s + i_q \tag{34}$$

where $i_s = i_p - i_a$ is the geometric counterpart of the scattered current [39], which can only be compensated by using active elements, while i_q can be compensated by using both passive and active elements [37]. There have been different attempts to give physical meaning to these current components. For that purpose, the scattered power could be defined as $M_s = ui_s$, while reactive power as $M_q = ui_q$. However, it has already been demonstrated that this decomposition has no real physical meaning (in the sense that these currents do not flow separately), even though it is useful for engineering practice [40,41]. In addition, for non-linear loads, the component i_G is included to model current components with frequencies that are not present in the voltage:

$$i = i_a + \underbrace{i_s + i_q + i_G}_{i_N} \tag{35}$$

The power factor can be defined in the geometric domain as:

$$pf = \frac{\langle M \rangle_0}{\|M\|} = \frac{P}{\|M\|} \tag{36}$$

5. Examples and Discussion

Two examples are given in order to validate the theoretical developments. The first one is the resolution of an RLC circuit under distorted conditions, while the other one consists of the analysis of experimental data. The results obtained with the proposed amendments were compared to those obtained by using other theories. Computations were performed by using the GA-Explorer library, which is available at https://github.com/ga-explorer (accessed on 1 May 2021) [42]. This library was chosen because it has a MATLAB connector. Furthermore, it performs calculations quickly and accurately. The Clifford Algebra toolbox was also used in some parts [43].

5.1. Example 1: Non-Sinusoidal Source

The RLC circuit presented in Figure 2 was used previously as an example and benchmark by different theories based on GA. Interestingly enough, the proposals by Menti, Castilla–Bravo and Lev-Ari cannot cope with the circuit analysis since they do not offer the right tools in the geometric domain. For these cases, it would be required to solve the circuit by using other techniques (such as complex algebra) and then transform the results to the geometric domain in order to analyse the power flow. Therefore, the circuit was only solved by using the theory proposed in this paper, CN [27] and CPC (Czarnecki) [2]. All of them allow a current decomposition into meaningful engineering terms.

In the circuit, $R = 1\ \Omega$, $L = 1/2$ H and $C = 2/3$ F. The source voltage is $u(t) = 100\sqrt{2}(\sin \omega t + \sin 3\omega t)$. The proposed theory was used to transform Equation (16) to the geometric domain:

$$u_1 + u_3 = R(i_1 + i_3) + L(\omega \sigma_{12} i_1 + 3\omega \sigma_{56} i_3) - \frac{1}{C}\left(\frac{\sigma_{12} i_1}{\omega} + \frac{\sigma_{56} i_3}{3\omega}\right) \tag{37}$$

It can be seen that the superposition theorem is embedded in the proposed formulation since all components are operated at the same time. This is a clear difference compared to theories based on complex numbers.

By using Equation (23), the geometric voltage turns into:

$$u = u_1 + u_3 = 100(\sigma_2 + \sigma_6) \tag{38}$$

while impedances and admittances are calculated with Equation (25):

$$\begin{aligned} Z_1 = 1 - \sigma_{12} &\longrightarrow Y_1 = 0.5 + 0.5\sigma_{12} \\ Z_3 = 1 + \sigma_{56} &\longrightarrow Y_3 = 0.5 - 0.5\sigma_{56} \end{aligned} \quad (39)$$

Therefore, the current becomes:

$$i = i_1 + i_3 = Y_1 u_1 + Y_3 u_3 = 50\sigma_1 + 50\sigma_2 - 50\sigma_5 + 50\sigma_6 \quad (40)$$

The geometric power is calculated by using Equation (26):

$$M = ui = \underbrace{10}_{M_a = P} \underbrace{- 5\sigma_{12} + 5\sigma_{56} - 5\sigma_{16} - 5\sigma_{25}}_{M_N}$$

The active power is a scalar with a value of 10 kW, while the other terms are the non-active power.

The reactive power (in the Budeanu sense) consumed by each harmonic is included in the $\sigma_{(2k-1)(2k)}$ terms.

Therefore, the reactive power of the first harmonic was $-5\sigma_{12}$, while that of the third one was $5\sigma_{56}$.

This result was in good agreement with traditional analyses in the frequency domain where the value for reactive power of each harmonic was identical, but of opposite sign. However, the term $-5\sigma_{16} - 5\sigma_{25}$ cannot be obtained by using complex algebra since it involves the cross-product between voltages and currents of different frequencies. This is one of the clear advantages of GA over complex numbers.

The norm (modulus) of the geometric power is:

$$\|M\| = \sqrt{\langle M^\dagger M \rangle_0} = \|u\| \|i\| = 141.42 \times 100 = 14,142 \text{ VA} \quad (41)$$

If the CN theory is applied, the geometric apparent power becomes:

$$\begin{aligned} M_{CN} &= 10 + 10\sigma_{12} + 10\sigma_{34} \text{ kVA} \\ \|M_{CN}\| &= 17,320 \text{ VA} \end{aligned} \quad (42)$$

The value of active power was 10 kW. However, the factor $f = (-1)^{k(k-1)/2}$ should be used for the calculations in order to obtain the right result. Furthermore, it can be seen that it was not possible to distinguish reactive power components generated by each harmonic since all of them were grouped into the term σ_{12}. Moreover, the CN proposal failed to provide the correct result as proven in [30]. Finally, it can be observed that $\|M_{CN}\| \neq \|u\| \|i\|$.

By using the CPC theory, it was not possible to generate a current vector in the frequency domain, nor a power multivector. Furthermore, the instantaneous value of currents should be used to describe independent terms of power. The results were:

$$\begin{aligned} P &= 10.000 \text{ W} & Q_r &= 10.000 \text{ VAr} \\ D_s &= 0 \text{ VA} & S &= 14.142 \text{ VA} \end{aligned}$$

The value of active power calculated by the CPC theory was, of course, correct. However, this theory cannot fully describe harmonic interactions between the voltage and current components. The norm of the total reactive power yielded 10 kVAr. However, it was not possible to calculate the individual contribution of each harmonic, nor its sign (sense).

Regarding current decomposition, by using Equation (31), it follows:

$$i = \underbrace{50\sigma_2 + 50\sigma_6}_{i_a} + \underbrace{50\sigma_1 - 50\sigma_5}_{i_N} \qquad (43)$$

Furthermore, if Equation (32) is applied, an identical result is obtained:

$$i = \underbrace{50\sigma_2 + 50\sigma_6}_{i_p} + \underbrace{50\sigma_1 - 50\sigma_5}_{i_q} \qquad (44)$$

If a harmonic compensator is to be designed, its susceptance at each harmonic would be the same as that of the load, but with the opposite sign:

$$B_{cp1} = -B_1 \quad B_{cp3} = -B_3 \qquad (45)$$

All the current would be compensated by using passive elements since no scattered current was present (see [44] for more details). This means that $i_a = i_p$. Therefore, i_N would be zero.

Consider now a value of $C = 2/7$ F in Figure 2. This set of parameters was used in other scientific works since power components cannot be distinguished if the classical concept of apparent power is applied [28,45]. For the voltage value presented in Equation (38), the current becomes:

$$i = 30\sigma_1 + 10\sigma_2 - 30\sigma_5 + 90\sigma_6 \qquad (46)$$

and the geometric power is:

$$M = 10 - 3\sigma_{12} + 3\sigma_{56} - 3\sigma_{16} - 3\sigma_{25} + 8\sigma_{26} \qquad (47)$$

Active power was the same as that obtained with other theories (10 kW). However, the rest of the terms were different. Reactive power consumption for each harmonic was reduced. The term $8\sigma_{26}$ appeared due to the interaction between in-phase components in the first voltage harmonic and the third current harmonic. This term highlights that the system cannot be fully compensated by using only passive elements. Despite the changes in various terms in the currents and powers, the norm of the geometric power remained unchanged:

$$\|M\| = \|u\| \|i\| = 141.42 \times 100 = 14.14 \text{ kVA} \qquad (48)$$

The current decomposition for this case is given in Table 2. If the CN theory is applied, the power becomes:

$$M = 10 + 6\sigma_{12} + 6\sigma_{34} + 8\sigma_{1234} \text{ kVA} \qquad (49)$$

where $\|M\| = 15.36$ kVA. This value differed from that obtained in the previous case, even though the voltages and currents did not change. Therefore, the proposed theory captured effects that others cannot.

Table 2. Current decomposition for the circuit in Figure 2 and $C = 2/7$ F.

	σ_1	σ_2	σ_3	σ_4	σ_5	σ_6	$\|\cdot\|$
i_a	0	50.00	0	0	0	50.00	70.71
i_s	0	−40.00	0	0	0	40.00	56.56
i_p	0	10.00	0	0	0	90.00	90.55
i_q	30.00	0	0	0	−30.00	0	42.42
i	30.00	10.00	0	0	−30.00	90.00	100.00

5.2. Example 2: Measurements Analysis

In this example, the voltage and current waveforms of a typical residential building in Almería (Spain) were analysed. The open-platform openZmeter (oZm) was used for the acquisition of the raw values of such waveforms [46]. Figure 3 shows voltage and current measurements in a time window of 200 ms, taken with a sampling frequency of 15.625 kHz (3125 samples). Several home appliances were on, such as a TV and LED lights, or electronic appliances, such as a router, satellite receiver and other devices in stand-by mode. The current waveform was highly distorted since the THDi was 88.3%, while the THDv was 6.63%.

Figure 4 shows the voltage and current spectrum for the first fifty harmonics (for the sake of clarity, the fundamental component is not shown). The fifth and seventh harmonic voltage components were prominent, while even harmonics were insignificant due to the half-wave symmetry of the waveform. From Table 3, it can be concluded that most of the energy was concentrated in the first five odd harmonics. The RMS value of the voltage was 234.011 V, while that of the current was 2.618A. Figure 5 shows the power waveform, as well as the value of P (359.15 W).

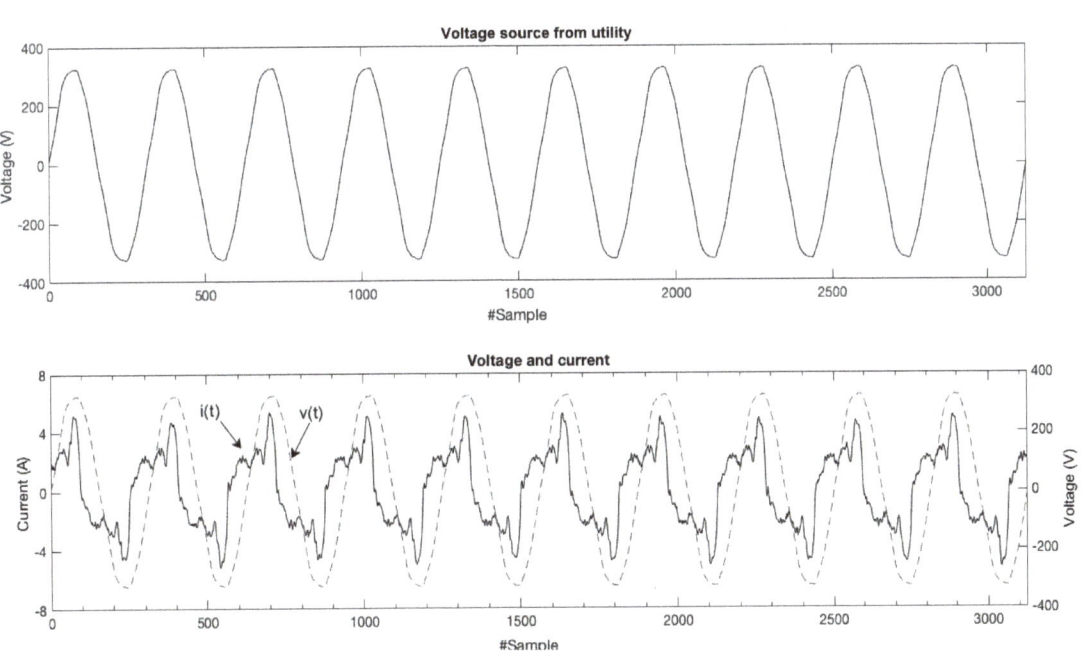

Figure 3. Voltage and current waveform measurements at a residential installation in Spain.

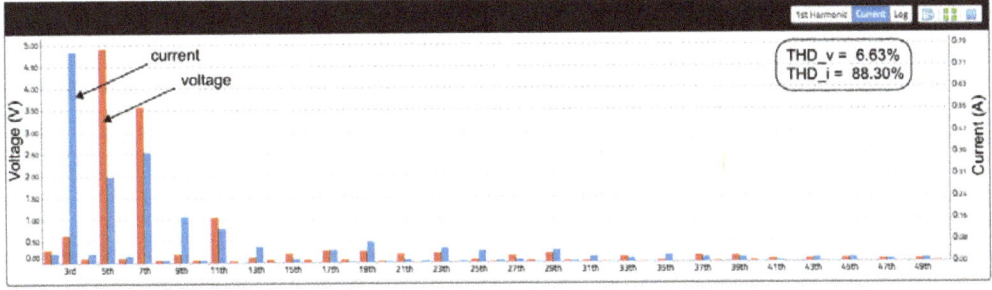

Figure 4. Voltage and current spectrum of the waveforms in Figure 3.

Table 3. Odd harmonics present in the waveforms of Example 2.

Order	Voltage		Current	
	$\|\|V\|\|$ (V)	φ_v (rad)	$\|\|I\|\|$ (A)	φ_i (rad)
fund	233.92	−1.57	2.33	−0.72
3rd	0.46	−2.61	0.93	1.85
5th	4.74	1.28	0.45	−1.69
7th	4.02	−0.07	0.49	1.70
9th	0.42	−2.60	0.16	−1.44

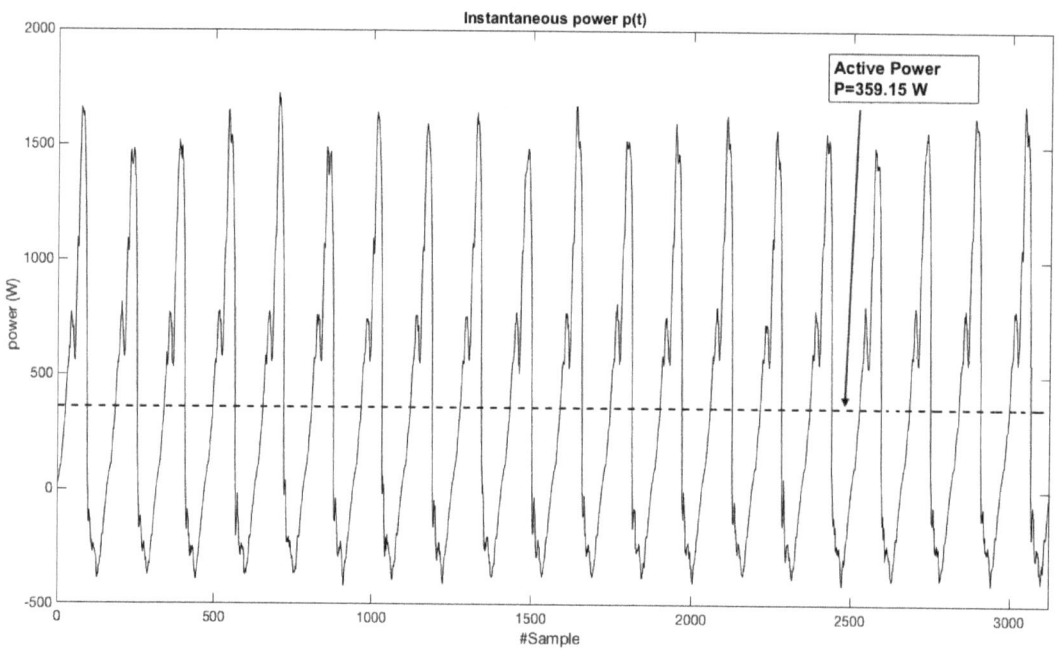

Figure 5. Instantaneous power waveform and active power P in Example 2.

A geometric vector of dimension 100 can be derived, but due to its length, only the five most energetic odd harmonics were selected (the fundamental component plus four odd harmonics), as shown in Table 3. It is worth pointing out that in the proposed theory, the dimension of the geometric space can be chosen according to specific requirements (e.g., the number of harmonics of interest). This is an advantage compared to other theories. In this case, the basis was $\sigma = \{\sigma_1, \ldots, \sigma_{10}\}$. The voltage and current expressions in polar form are:

$$u = 233.92e^{-1.57\sigma_{12}}\sigma_1 + 0.46e^{-2.61\sigma_{34}}\sigma_3 + 4.74e^{1.28\sigma_{56}}\sigma_5 + 4.02e^{-0.07\sigma_{78}}\sigma_7 + 0.42e^{-2.60\sigma_{(9)(10)}}\sigma_9$$

$$i = 2.33e^{-0.72\sigma_{12}}\sigma_1 + 0.93e^{1.85\sigma_{34}}\sigma_3 + 0.45e^{-1.69\sigma_{56}}\sigma_5 + 0.49e^{1.70\sigma_{78}}\sigma_7 + 0.16e^{-1.44\sigma_{(9)(10)}}\sigma_9$$

(50)

The most significant terms of the geometric power were those related to active and reactive power:

$$M = 359.14 - 408.56\sigma_{12} + 0.42\sigma_{34} + 0.34\sigma_{56} - 1.95\sigma_{78} - 0.06\sigma_{(9)(10)} + O \quad (51)$$

where O includes the rest of the bivectors that appeared due to the cross-frequency products and is not shown due to the lack of space. The norm was $\|M\| = 612.66$ VA, which was nearly the same as $\|u\|\|i\| = 234.011 \times 2.618 = 612.64$ VA. The value of M_a was 359.21 W, which was similar to that obtained by using the digital samples of voltages and currents. Results for the reactive power of each harmonic were also similar. Note that the reactive power of the fundamental frequency was encoded in the bivector term σ_{12}. In this case, it had a negative value, so it represented a capacitive behaviour. These values are also shown in Table 4. Table 5 shows the current components presented in Equation (31) for the five most energetic harmonics. In order to compute i_p and i_q according to Equation (25), the geometric impedances were calculated for each harmonic. The value of the total current was $\|i\| = 2.607$ A, while $\|i_a\| = 1.535$ A. Note that the norm of the total current differed slightly from the real one (2.618 A) because not all harmonics were included. It can be observed that $\|i_a\|$ was the minimum current that would produce the same active power. Figure 6 shows the waveforms of $i(t)$, $i_a(t)$ and $i_N(t)$.

Table 4. Harmonic active (W) and reactive (VAr) power measurements.

Order	P_i	Q_i	
	oZm	oZm	GA
fund	361.80	−408.56	−408.50
3rd	−0.102	0.426	0.425
5th	−2.134	0.346	0.346
7th	−0.408	−1.955	−1.955
9th	0.028	−0.063	−0.062
Total	359.15		

Table 5. Current components obtained from current measurements.

	i_p	i_a	i_s	i_q	i_N	i
σ_1	−0.007	−0.007	0.000	1.746	1.746	1.739
σ_2	1.547	1.534	0.012	0.008	0.020	1.555
σ_3	0.188	−0.003	0.190	−0.454	−0.263	−0.266
σ_4	−0.108	0.001	−0.109	−0.789	−0.898	−0.897
σ_5	−0.126	0.009	−0.135	0.070	−0.065	−0.056
σ_6	0.431	−0.030	0.461	0.020	0.482	0.452
σ_7	−0.101	0.026	−0.127	0.036	−0.091	−0.065
σ_8	−0.007	0.002	−0.010	−0.484	−0.494	−0.492
σ_9	−0.057	−0.002	−0.055	0.077	0.022	0.020
σ_{10}	0.034	0.001	0.033	0.129	0.162	0.163
$\|\cdot\|$	1.629	1.535	0.548	2.035	2.108	2.607

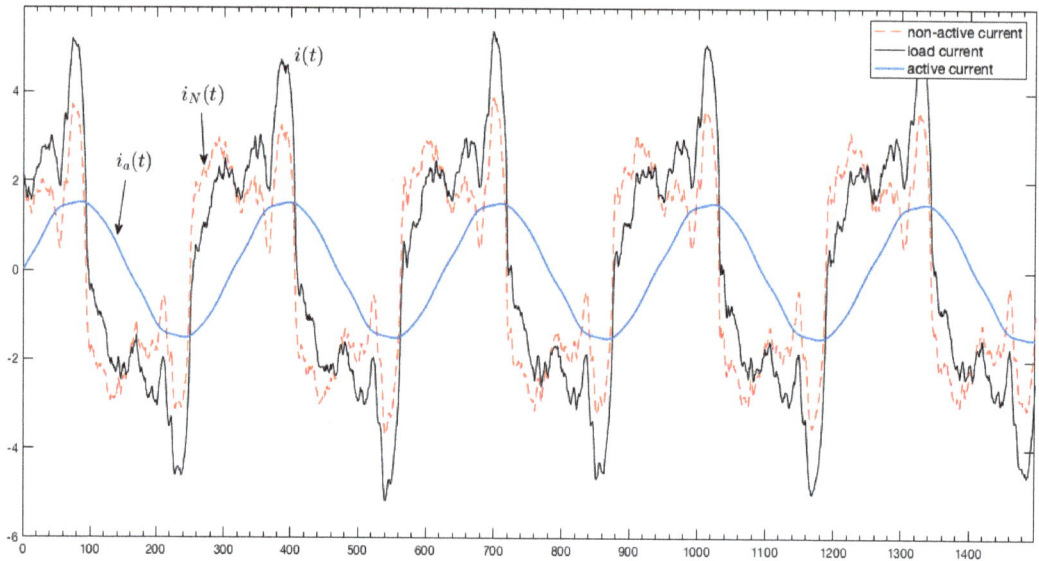

Figure 6. Total, active and non-active current for the measurements.

6. Conclusions

In this paper, an improved formulation of the power theory based on GA was presented. First, the main shortcomings of existing power theories based on GA were identified. It was shown that the use of k-blades as a basis for the geometric space led to an unclear definition of the geometric apparent power. Moreover, the energy conservation principle cannot be easily fulfilled without extra correction factors. Thus, Menti's proposal was recovered and favoured for power computations. A new comprehensive isomorphic transformation that accounted for harmonic, sub- and inter-harmonic and DC components' representation was presented. It simplified the power definitions in the frequency domain and provided a clear meaning to harmonic power. Furthermore, the norm of the geometric power was in good agreement with the traditional definition of apparent power based on the product of RMS voltage and current. Circuit theory analysis can also be performed in the steady state for AC circuits using geometric vectors. The concept of geometric impedance was also introduced with a similar meaning as the well-known complex algebra. It made it possible to analyse electrical circuits by using conventional techniques. Current decomposition for load compensation purposes can be easily carried out by means of the use of the inverse of the current vector. Through different examples, it was shown that the proposed framework overcame some limitations of existing GA-based power theories and provided a comprehensive tool for analysing and solving single-phase electrical circuits under distorted conditions. Moreover, new indices for power quality can be defined based on the suggested non-active power as a result of the cross-frequency products of the voltage and current. Harmonic and power factor correction can also benefit from the proposed approach. Future research is under way to extend this methodology to polyphase systems. This requires the use of orthogonal transformations such as the one derived from the application of the symmetric components. This fact can be addressed through a higher number of dimensions.

Author Contributions: Conceptualization, F.G.M. and F.M.A.-C.; Formal analysis, F.G.M. and J.R.-P.; Investigation, F.G.M. and J.R.-P.; Methodology, R.B.; Project administration, F.M.A.-C.; Software, A.A.; Validation, A.A.; Writing—original draft, F.G.M. and J.R.-P.; Writing—review & editing, R.B. All authors read and agreed to the published version of the manuscript.

Funding: This research was funded by Ministry of Science, Innovation and Universities Grant Number PGC2018-098813-B-C33.

Institutional Review Board Statement: Not applicable.

Informed Consent Statement: Not applicable.

Data Availability Statement: Not applicable.

Acknowledgments: This research was supported by the Ministry of Science, Innovation and Universities at the University of Almeria under the programme "Proyectos de I+D de Generacion de Conocimiento" of the national programme for the generation of scientific and technological knowledge and strengthening of the R+D+I system with Grant Number PGC2018-098813-B-C33.

Conflicts of Interest: The authors declare no conflict of interest.

References

1. Lee, R.P.; Lai, L.L.; Lai, C.S. Design and Application of Smart Metering System for Micro Grid. In Proceedings of the 2013 IEEE International Conference on Systems, Man, and Cybernetics, Manchester, UK, 13–16 October 2013; pp. 3203–3207.
2. Czarnecki, L.S. Currents' physical components (CPC) in circuits with nonsinusoidal voltages and currents. Part 1, Single-phase linear circuits. *Electr. Power Qual. Util.* **2005**, *11*, 3–14.
3. Akagi, H.; Watanabe, E.H.; Aredes, M. *Instantaneous Power Theory and Applications to Power Conditioning*; John Wiley & Sons: Hoboken, NJ, USA, 2017.
4. Staudt, V. Fryze-Buchholz-Depenbrock: A time-domain power theory. In Proceedings of the Nonsinusoidal Currents and Compensation, Lagow, Poland, 10–13 June 2008; pp. 1–12.
5. Salmerón, P.; Herrera, R.; Vazquez, J. Mapping matrices against vectorial frame in the instantaneous reactive power compensation. *IET Electr. Power Appl.* **2007**, *1*, 727–736. [CrossRef]
6. Czarnecki, L.S. On some misinterpretations of the instantaneous reactive power pq theory. *IEEE Trans. Power Electron.* **2004**, *19*, 828–836. [CrossRef]
7. Castro-Nuñez, M.; Castro-Puche, R. The IEEE Standard 1459, the CPC power theory, and geometric algebra in circuits with nonsinusoidal sources and linear loads. *IEEE Trans. Circuits Syst. Regul. Pap.* **2012**, *59*, 2980–2990. [CrossRef]
8. Chakraborty, S.; Simoes, M.G. Experimental evaluation of active filtering in a single-phase high-frequency AC microgrid. *IEEE Trans. Energy Convers.* **2009**, *24*, 673–682. [CrossRef]
9. Petroianu, A.I. A geometric algebra reformulation and interpretation of Steinmetz's symbolic method and his power expression in alternating current electrical circuits. *Elec. Eng.* **2015**, *97*, 175–180. [CrossRef]
10. Hestenes, D.; Sobczyk, G. *Clifford Algebra to Geometric Calculus: A Unified Language for Mathematics and Physics*; Springer Science & Business Media: Berlin/Heidelberg, Germany, 2012; Volume 5.
11. Ablamowicz, R. *Clifford Algebras: Applications to Mathematics, Physics, and Engineering*; Springer Science & Business Media: Berlin/Heidelberg, Germany, 2012; Volume 34.
12. Easter, R.B.; Hitzer, E. Conic and cyclidic sections in double conformal geometric algebra G 8, 2 with computing and visualization using Gaalop. *Math. Methods Appl. Sci.* **2020**, *43*, 334–357. [CrossRef]
13. Papaefthymiou, M.; Papagiannakis, G. Real-time rendering under distant illumination with conformal geometric algebra. *Math. Methods Appl. Sci.* **2018**, *41*, 4131–4147. [CrossRef]
14. Hestenes, D. *New Foundations for Classical Mechanics*; Springer Science & Business Media: Berlin/Heidelberg, Germany, 2012; Volume 15.
15. Dorst, L.; Fontijne, D.; Mann, S. *Geometric Algebra for Computer Science: An Object-Oriented Approach to Geometry*; Elsevier: Amsterdam, The Netherlands, 2010.
16. Chappell, J.M.; Drake, S.P.; Seidel, C.L.; Gunn, L.J.; Iqbal, A.; Allison, A.; Abbott, D. Geometric algebra for electrical and electronic engineers. *Proc. IEEE* **2014**, *102*, 1340–1363. [CrossRef]
17. Yao, H.; Li, Q.; Chen, Q.; Chai, X. Measuring the closeness to singularities of a planar parallel manipulator using geometric algebra. *Appl. Math. Model.* **2018**, *57*, 192–205. [CrossRef]
18. Menti, A.; Zacharias, T.; Milias-Argitis, J. Geometric algebra: A powerful tool for representing power under nonsinusoidal conditions. *IEEE Trans. Circuits Syst. Regul. Pap.* **2007**, *54*, 601–609. [CrossRef]
19. Castilla, M.; Bravo, J.C.; Ordonez, M.; Montaño, J.C. Clifford theory: A geometrical interpretation of multivectorial apparent power. *IEEE Trans. Circuits Syst. Regul. Pap.* **2008**, *55*, 3358–3367. [CrossRef]
20. Lev-Ari, H.; Stanković, A.M. A geometric algebra approach to decomposition of apparent power in general polyphase networks. In Proceedings of the 41st North American Power Symposium, Starkville, MS, USA, 4–6 October 2009; pp. 1–6.
21. Montoya, F.; Baños, R.; Alcayde, A.; Montoya, M.; Manzano-Agugliaro, F. Power Quality: Scientific Collaboration Networks and Research Trends. *Energies* **2018**, *11*, 2067. [CrossRef]
22. Castro-Núñez, M.; Londoño-Monsalve, D.; Castro-Puche, R. M, the conservative power quantity based on the flow of energy. *J. Eng.* **2016**, *2016*, 269–276. [CrossRef]

23. Wu, J.C. Novel circuit configuration for compensating for the reactive power of induction generator. *IEEE Trans. Energy Convers.* **2008**, *23*, 156–162.
24. Castilla, M.; Bravo, J.C.; Ordoñez, M. Geometric algebra: A multivectorial proof of Tellegen's theorem in multiterminal networks. *IET Circuits Devices Syst.* **2008**, *2*, 383–390. [CrossRef]
25. Weidmann, J. *Linear Operators in Hilbert Spaces*; Springer Science & Business Media: Berlin/Heidelberg, Germany, 2012; Volume 68.
26. Lev-Ari, H.; Stankovic, A.M. Instantaneous power quantities in polyphase systems—A geometric algebra approach. In Proceedings of the IEEE Energy Conversion Congress and Exposition, San Jose, CA, USA, 20–24 September 2009; pp. 592–596.
27. Castro-Núñez, M.; Castro-Puche, R.; Nowicki, E. The use of geometric algebra in circuit analysis and its impact on the definition of power. In Proceedings of the Nonsinusoidal Currents and Compensation (ISNCC), Lagow, Poland, 15–18 June 2010; pp. 89–95.
28. Castro-Nuñez, M.; Castro-Puche, R. Advantages of geometric algebra over complex numbers in the analysis of networks with nonsinusoidal sources and linear loads. *IEEE Trans. Circuits Syst. Regul. Pap.* **2012**, *59*, 2056–2064. [CrossRef]
29. Castro-Núñez, M.; Londoño-Monsalve, D.; Castro-Puche, R. Theorems of compensation and Tellegen in non-sinusoidal circuits via geometric algebra. *J. Eng.* **2019**, *2019*, 3409–3417. [CrossRef]
30. Montoya, F.G.; Baños, R.; Alcayde, A.; Arrabal-Campos, F.M. A new approach to single-phase systems under sinusoidal and non-sinusoidal supply using geometric algebra. *Electr. Power Syst. Res.* **2020**, *189*, 106605. [CrossRef]
31. Montoya, F.G.; Baños, R.; Alcayde, A.; Arrabal-Campos, F.M. Analysis of power flow under non-sinusoidal conditions in the presence of harmonics and interharmonics using geometric algebra. *Int. J. Elec. Power Energy Sys.* **2019**, *111*, 486–492. [CrossRef]
32. Jancewicz, B. *Multivectors and Clifford Algebra in Electrodynamics*; World Scientific: Singapore, 1989.
33. Hitzer, E. Introduction to Clifford's geometric algebra. *arXiv* **2013**, arXiv:1306.1660.
34. Lev-Ari, H.; Stankovic, A.M. A decomposition of apparent power in polyphase unbalanced networks in nonsinusoidal operation. *IEEE Trans. Power Sys.* **2006**, *21*, 438–440. [CrossRef]
35. Frize, S. Active reactive and apparent power in circuits with nonsinusoidal voltage and current. *Elektrotechnische Z.* **1932**, *53*, 596–599.
36. Hitzer, E.; Sangwine, S.J. Construction of Multivector Inverse for Clifford Algebras Over 2 m+ 12 m+ 1-Di mensional Vector Spaces fro m Multivector Inverse for Clifford Algebras Over 2 m-Di mensional Vector Spaces. *Adv. Appl. Clifford Algebr.* **2019**, *29*, 29. [CrossRef]
37. Czarnecki, L.S.; Pearce, S.E. Compensation objectives and Currents' Physical Components–based generation of reference signals for shunt switching compensator control. *IET Power Electron.* **2009**, *2*, 33–41. [CrossRef]
38. Shepherd, W.; Zakikhani, P. Suggested definition of reactive power for nonsinusoidal systems. *Proc. Inst. Electr. Eng. IET* **1972**, *119*, 1361–1362. [CrossRef]
39. Czarnecki, L.S. Considerations on the Reactive Power in Nonsinusoidal Situations. *IEEE Tran. Inst. Meas.* **1985**, *IM-34*, 399–404. [CrossRef]
40. Cohen, J.; De Leon, F.; Hernández, L.M. Physical time domain representation of powers in linear and nonlinear electrical circuits. *IEEE Trans. Power Deliv.* **1999**, *14*, 1240–1249. [CrossRef]
41. De Léon, F.; Cohen, J. AC power theory from Poynting theorem: Accurate identification of instantaneous power components in nonlinear-switched circuits. *IEEE Trans. Power Del.* **2010**, *25*, 2104–2112. [CrossRef]
42. Eid, A.H. An extended implementation framework for geometric algebra operations on systems of coordinate frames of arbitrary signature. *Adv. Appl. Clifford Algebr.* **2018**, *28*, 16. [CrossRef]
43. Sangwine, S.J.; Hitzer, E. Clifford multivector toolbox (for MATLAB). *Adv. Appl. Clifford Algebr.* **2017**, *27*, 539–558. [CrossRef]
44. Montoya, F.; Alcayde, A.; Arrabal-Campos, F.M.; Baños, R. Quadrature Current Compensation in Non-Sinusoidal Circuits Using Geometric Algebra and Evolutionary Algorithms. *Energies* **2019**, *12*, 692. [CrossRef]
45. Czarnecki, L. Budeanu and fryze: Two frameworks for interpreting power properties of circuits with nonsinusoidal voltages and currents. *Electr. Eng.* **1997**, *80*, 359–367. [CrossRef]
46. Viciana, E.; Alcayde, A.; Montoya, F.G.; Baños, R.; Arrabal-Campos, F.M.; Manzano-Agugliaro, F. An Open Hardware Design for Internet of Things Power Quality and Energy Saving Solutions. *Sensors* **2019**, *19*, 627. [CrossRef] [PubMed]

Article

A Topological View of Reed–Solomon Codes

Alberto Besana and Cristina Martínez *

Department of Physics and Mathematics, University of Alcalá, 28871 Madrid, Spain; albertobesana@gmail.com
* Correspondence: Cristina.martinezram@uah.es

Abstract: We studied a particular class of well known error-correcting codes known as Reed–Solomon codes. We constructed RS codes as algebraic-geometric codes from the normal rational curve. This approach allowed us to study some algebraic representations of RS codes through the study of the general linear group $GL(n,q)$. We characterized the coefficients that appear in the decompostion of an irreducible representation of the special linear group in terms of Gromov–Witten invariants of the Hilbert scheme of points in the plane. In addition, we classified all the algebraic codes defined over the normal rational curve, thereby providing an algorithm to compute a set of generators of the ideal associated with any algebraic code constructed on the rational normal curve (NRC) over an extension $\mathbb{F}_{q^n}$ of $\mathbb{F}_q$.

Keywords: 2000 mathematics subject classification; 05E10 (primary); 05A15 (secondary); algebraic code; symmetric group; partitions

1. Introduction

Let us denote by $\mathbb{F}_q$ the finite field of q elements with q a power of prime number p. One can consider field extensions $\mathbb{F}_q$ of $\mathbb{F}_p$ as q varies through powers of the prime p. Any $\mathbb{F}_{p^n}$ field extension of $\mathbb{F}_p$ is a vector space over $\mathbb{F}_p$ of dimension n and an $(n-1)$−dimensional projective space $PG(n-1, p)$.

Let V be an $n+1$ dimensional vector space over the field $\mathbb{F}_q$; we denote by $PG(n, q)$ or $\mathbb{P}(V)$ the n-dimensional projective space over V and by $\mathbb{P}^1$, the projective line. The set of all subspaces of dimension r in V is a Grassmannian, and it is denoted by $\mathcal{G}_{r,n}(\mathbb{F}_q)$ or by $PG^r(n, q)$. The dual of an r−space in $PG(n, q)$ is an $(n-r-1)$−space.

Consider the $\mathbb{F}_q$−rational points of $\mathcal{G}_{r,n}(\mathbb{F}_q)$ as a projective system; we obtain a q-ary linear code, called the Grassmann code, which we denote $[n, r]_q$ code. The length l and the dimension k of $G(r, n)$ are given by the q binomial coefficient $l = \begin{bmatrix} n \\ r \end{bmatrix}_q = \frac{(q^{n+1}-1)(q^{n+1}-q)\dots(q^{n+1}-q^r)}{(q^{r+1}-1)(q^{r+1}-q)\dots(q^{r+1}-q^r)}$, and $k = \binom{n}{r}$, respectively.

We study the relation between codes constructed from vector bundles and the representation theory of the general linear group $GL(n, \mathbb{F}_q)$. Following [1], we consider the right action of the general linear group $GL(n, \mathbb{F}_q)$ on $\mathcal{G}_{k,n}(\mathbb{F}_q)$:

$$\mathcal{G}_{k,n}(\mathbb{F}_q) \times GL(n, \mathbb{F}_q) \to \mathcal{G}_{k,n}(\mathbb{F}_q) \quad (1)$$
$$(\mathcal{U}, A) \to \mathcal{U}A.$$

Observe that the action is defined independently of the choice of the representation matrix $\mathcal{U} \in \mathbb{F}_q^{k \times n}$.

Let $\mathcal{U} \in \mathcal{G}_{k,n}(\mathbb{F}_q)$ and $G < GL(n, \mathbb{F}_q)$ be a subgroup; then $C = \{\mathcal{U}A \mid A \in G\}$ is an orbit in $\mathcal{G}_{k,n}(\mathbb{F}_q)$ of the induced action.

In order to classify all the orbits, we need to classify all the conjugacy classes of subgroups of $GL(n, \mathbb{F}_q)$. In [2], we studied cyclic coverings of the projective line that correspond to orbits defined by a cyclic subgroup of order p as the multiplicative group of

p^{th} roots of unity or the additive group of integers modulo p for some prime number p. In particular, we showed that any irreducible cyclic plane cover of the projective line can be given by a prime ideal

$$(y^m - (x-a_1)^{d_1} \ldots (x-a_n)^{d_n}) \subset \mathbb{F}_q[x,y].$$

This ideal defines an affine curve in $\mathbb{A}^2(\mathbb{F}_q)$ which has singularities, if $d_k > 1$ for some $1 \leq k \leq n$. There exists a unique projective curve birationally equivalent to this affine curve obtained by homogenization of the polynomial. Here we study the connection between ideal sheaves on $\mathbb{F}_q[x,y]$ and its numerical invariants together with the combinatorics of partitions of n and the representation theory of the general linear group $GL(\mathbb{F}_q, n)$. In other words, we want to understand which subspaces are invariant by the action of elements of the general linear group or finite subgroups of $GL(n, \mathbb{F}_q)$ and how the $GL(n, \mathbb{F}_q)$ group's action on the Grassmannian changes the Grassmann code, as this action simply permutes basis elements of the Grassmann code.

When one considers as an alphabet a set $\mathcal{P} = \{P_1, \ldots, P_N\}$ of $\mathbb{F}_q$-rational points lying on a smooth projective curve defined over a finite field, algebraic codes are constructed by evaluation of the global sections of a line bundle or a vector bundle on the curve. Any cyclic cover of $\mathbb{P}^1$ which is simply ramified corresponds to an unordered tuple of n points on $\mathbb{P}^1$. More generally, in Section 4 we consider configurations of n points in a d-dimensional projective space $PG(d,q)$ which generically lies on a rational normal curve (NRC) and we study the algebraic codes defined on it, providing a complete classification in terms of divisors defined over the NRC; see Theorem 2. These are the so called Reed–Solomon codes. Moreover, in the last section as an application of the Horn problem, we provide a set of generators of the ideal associated with any algebraic code constructed on the NRC over an extension $\mathbb{F}_q^n$ of $\mathbb{F}_q$.

From now on, $\mathbb{F}_q$ will be a field with $q = p^n$ elements and $\mathcal{C}$ a non-singular, projective, irreducible curve defined over $\mathbb{F}_q$ with q elements.

Notation

For d a positive integer, $\alpha = (\alpha_1, \ldots, \alpha_m)$ is a partition of d into m parts if the α_i are positive and decreasing integers summing to n. We will denote as $\mathcal{P}(d)$ the set of all partitions of d. We set $l(\alpha) = m$ for the length of α, that is, the number of cycles in α, and l_i for the length of α_i. The notation $(a_1, \ldots, a_k)$ stands for a permutation in S_d that sends a_i to a_{i+1}. For $\lambda \in \mathcal{P}(d)$, we write $[\lambda]$ for the corresponding character of S_n. We write $PGL(2,k) = GL(2,k)/k^*$, where k is field of arbitrary characteristic and elements of $PGL(2,k)$, which will be represented by equivalence classes of matrices $\begin{pmatrix} a & b \\ c & d \end{pmatrix}$, with $ad - bc \neq 0$. A q-ary constant weight code of length n, distance d and weight w will be denoted as an $[n,d,w]_q$ code.

2. Horn Problem: An Application to Convolutional Codes

In this section we present a description of the Horn problem, i.e., the study of the eigenvalues of the sum $C = A + B$ of two matrices, given the spectrum of A and B, in the context of polynomial matrices with polynomial entries associated with torsion modules or dually submodules of a polynomial ring with coefficients in a field. Next, we introduce some important matrices that define a linear error-correcting code.

Let R be any complete valued field R with a closed coefficient field k of an arbitrary characteristic, for example, a finite field or the ring $R = \mathbb{C}\{x\}$ of convergent power series. If $f \in R$ is a nonzero divisor, then we define the encoder A as the matrix associated with the corresponding torsion module R/fR. The matrix A can be diagonalized by elementary row and column operations with diagonal entries $x^{\alpha_1}, x^{\alpha_2}, \ldots, x^{\alpha_n}$, for unique non-negative integers $\alpha_1 \geq \ldots \geq \alpha_n$. More precisely, these matrices are in correspondence with endomorphisms of R^n, with cokernels being torsion modules with at most n generators. Such a module is isomorphic to a direct sum

$$R/x^{\alpha_1}R \oplus R/x^{\alpha_2}R \oplus \ldots \oplus R/x^{\alpha_n}R, \quad \alpha_1 \geq \ldots \geq \alpha_n.$$

The set $(\alpha_1, \ldots, \alpha_n)$ of invariant factors of A defines a partition α of size $d = |\lambda|$. Reciprocally, when $R = \mathbb{C}\{x\}$ is the ring of convergent power series, any partition λ defines a rank one torsion-free sheaf on $\mathbb{C}$ by setting $\mathcal{I}_\lambda = (x^{\lambda_1}, x^{\lambda_2}, x^{\lambda_3}, \ldots, x^{\lambda_n})$. In particular, the ideal sheaf corresponding to the identity partition $(1)^n$, defines a maximal ideal $\mathcal{I}_{(1)^n} = (x, \overbrace{\ldots}^{n \text{ times}}, x)$ in $\mathbb{C}[x]$. The Horn problem is then equivalent to the following question: which partitions α, β, γ can be the invariant factors of matrices A, B and C if $C = A \cdot B$?

In the case of convergent power series, this problem was proposed by I. Gohberg and M.A. Kaashoek. By denoting the cokernels of A, B and C as $\mathcal{A}, \mathcal{B}$ and $\mathcal{C}$, respectively, one has a short exact sequence:

$$0 \to \mathcal{A} \to \mathcal{B} \to \mathcal{C} \to 0,$$

i.e., $\mathcal{B}$ is a submodule of $\mathcal{C}$ with $\mathcal{C}/\mathcal{B} \cong \mathcal{A}$; such an exact sequence corresponds to matrices A, B and C with $A \cdot B = C$.

If we specialize C to be the identity matrix I, by the correspondence between partitions and ideal sheaves above, the invariant factors of the identity matrix are defined by the partition $(1)^n$, then the question becomes: which partitions α, β can be the invariant factors of matrices A, B if $A \cdot B = I$? The case of interest for us will be the case in which R is an $\mathbb{F}_q[x]$-module with q a prime power of p.

Duly, the code can be defined as an R−submodule of R^n, where $R = \mathbb{F}[z]$ is a polynomial ring with coefficients in a field $\mathbb{F}$ and z is a uniformizing parameter in R (see [3]). When $\mathbb{F}$ is a finite field, these are known as convolutional codes which have been very well studied; see, for example, [4]. A full row rank matrix $G(z) \in \mathbb{F}[z]^{k \times n}$ with the property that

$$\mathcal{C} = \text{Im}_{\mathbb{F}[z]} G(z) = \{f(z)g(z) : f(z) \in [\mathbb{F}^k(z)]\}$$

is called a generator matrix. The degree d of a convolutional code $\mathcal{C}$ is the maximum of the degrees of the determinants of the $k \times k$ submatrices of one, and hence any generator matrix of $\mathcal{C}$. The main difference between block and convolutional codes is that at the encoder, in a convolutional code we may have different states. Linear block codes may be considered as a particular case of convolutional codes with only one state. In next section we describe an example of block codes known as Reed–Solomon codes.

Remark 1. *The set of convolutional codes of a fixed degree is parametrized by the Grothendieck Quot scheme of degree d, rank $n - k$ coherent sheaf quotients of $\mathcal{O}^n$ on a curve X defined over $\mathbb{F}$. If the degree is zero, these schemes describe a Grassmann variety and constitute the so called class of block codes of parameters (n,k). Namely, the space of all matrix divisors $\mathcal{D}_k(r,d)$ of rank r and degree d can be identified with the set of rational points of $\text{Quot}^m_{\mathcal{O}_X(D)^n/X/k}$ parametrizing torsion quotients of $\mathcal{O}_X(D)^n$ and having degree $m = r \cdot \deg D - d$, see [5].*

An Example with Algebraic-Geometric Codes: Reed–Solomon Codes

Let X be a smooth projective curve defined over a finite field $\mathbb{F}_q$ with q elements. The classical algebraic-geometric (AG) code due to Goppa is defined by evaluating rational functions associated with a divisor D at a finite set of $\mathbb{F}_q$-rational points. From another point of view, we are considering the evaluation of sections of the corresponding line bundle $\mathcal{O}_X(D)$ on X. Namely, let $\mathcal{P} := \{P_1, \ldots, P_n\}$ be a configuration of distinct $\mathbb{F}_q$-

rational points of X, the usual algebraic-geometric code is defined to be the image of the evaluation map:

$$\varphi_D : L(D) \to \mathbb{F}_q^n \qquad (2)$$
$$f \mapsto (f(P_1), \ldots, f(P_n)),$$

where $L(D)$ denotes the vector space of sections associated with the line bundle $\mathcal{O}_X$. The parameters of these codes, the length n, the dimension k and the minimum distance d are determined by the theorem of Riemann-Roch and it is easy to see that they satisfy the following bound $k + d \geq n + 1 - g$, where g is the genus of the curve X. Using this definition, the notion of AG codes is easily generalized for varieties of higher dimension.

Namely, let E be a vector bundle of rank r on X defined over $\mathbb{F}_q$. The Goppa code $C(X, D, G)$ takes as input a divisor D supported on the finite set $\mathcal{P}$ of $\mathbb{F}_q$-rational points and a divisor G associated with the vector bundle E and evaluates each section $\sigma \in \mathcal{L}(G)$ in the linear series attached to the divisor G:

$$C(X, D, G) = \{(\sigma(P_i))_{i=1}^n : \sigma \in \mathcal{L}(G)\} \subseteq \mathbb{F}_q^n.$$

Observe that $C(X, \mathcal{P}, E)$ is an $\mathbb{F}_q$-linear subspace of $\mathbb{F}_{q^r}^n$ and thus a point of the Grassmannian $\mathcal{G}_{r,n}(\mathbb{F}_q)$. Moreover, for the same subset of evaluation points and any $r \leq k$, we have $G(r, n) \subseteq G(k, n) \subseteq \mathbb{F}_q^n$, where $r \leq k$. Further, we get a partial flag of $\mathbb{F}_q$-vector spaces $\{0\} = E^k \subset E^{k-1} \subset \ldots \subset E^1 \subset E^0 = \mathbb{F}_q^n$ such that $\dim(E^{i-1}/E^i) = \lambda_i$, to which we associate the partition $\lambda = (\lambda_1, \ldots, \lambda_r)$ of n. In this way, each partition λ of n determines a variety $\mathcal{F}_\lambda = \mathcal{F}_\lambda(\mathbb{F}_q)$ of partial flags of $\mathbb{F}_q$-vector spaces.

The representation theory of the special linear group $SL(n, \mathbb{F}_q)$ can be viewed as a form of Gale duality first proven by Goppa in the context of algebraic coding theory.

Let D and G be effective divisors supported over a smooth projective curve X defined over $\mathbb{F}_q$ such that $\text{Supp}(G) \cap \text{Supp}(D) = \emptyset$, then the geometric Goppa code associated with the divisors D and G is defined by

$$\mathcal{C}(D, G) = \{(x(P_1), \ldots, x(P_n)), x \in \mathcal{L}(G)\} \subseteq \mathbb{F}_{q^r}^n,$$

where $\mathcal{L}(G)$ denotes the linear system associated with the divisor G.

Definition 1. *Let C_1 and C_2 be the corresponding codes obtained by evaluating non-constant rational functions $f(x)$ and $g(x)$ with non common roots on X over the support of the divisor D. We define the quotient code of C_1 and C_2 to be the code associated with the quotient rational function $\varphi = f/g$.*

Since f and g take the value ∞, they are defined by non constant polynomials $f(x)$ and $g(x)$ in $\overline{\mathbb{F}_q}[x]$. Here $\overline{\mathbb{F}_q}$ denotes the algebraic closure of $\mathbb{F}_q$. The degree of φ is defined to be $\deg(\varphi) = \max\{\deg(f), \deg(g)\}$.

As φ is a finite morphism, one may associate to each rational point $x \in X(\mathbb{F}_q)$ a local degree or multiplicity $m_\varphi(x)$ defined as:

$$m_\varphi(x) = \text{ord}_{z=0} \psi(z),$$

where $\psi = \sigma_2 \circ \varphi \circ \sigma_1$, $y = \varphi(x)$, and $\sigma_1, \sigma_2 \in PGL(2, \mathbb{F}_q)$ such that $\sigma_1(0) = x$ and $\sigma_2(y) = 0$.

With each non-constant rational function φ over X, one can associate a matrix A with entries in the ring $\mathbb{F}_q[x]$. Namely, let us call $f_0 := f(x)$ and call f_1 the divisor polynomial $g(x)$, and f_2 the remainder polynomial; then by repeated use of the Euclid's algorithm, we construct a sequence of polynomials $f_0, f_1, \ldots, f_k$, and quotients $q_1, \ldots q_k$, $K \leq n$. Then the quotient matrix A is defined to be the diagonal matrix with entries $q_1, \ldots, q_k$ corresponding to the continued fraction expansion of the rational function φ.

Here we include a SAGE code [6] which implements the algorithm.

```
def euclid(f, g):
r = f % g
q = f // g
while r.degree() >= 0:
yield q
f = g
g = r
r = f % g
q = f // g
```

Let λ_i be the partition of the integer k, defining the degree multiplicities of the polynomial q_i. Then the Horn problem applied to this situation reads:

Which partitions α, β and γ can be the degree multiplicities of polynomials q_A, q_B and q_C such that the corresponding diagonal matrices A, B, and C satisfy $C = A \cdot B$?

Another important family of Goppa codes is obtained considering the normal rational curve $\mathcal{C}^n$ defined over $\mathbb{F}_q$:

$$\mathcal{C}^n := \{\mathbb{F}_q(1, \alpha, \ldots, \alpha^n) : \alpha \in \mathbb{F}_q \cup \{\infty\}\}.$$

The points are distinct elements of $\mathbb{F}_q$ and L is the vector space of polynomials of degree at most $k - 1$ and with coefficients in $\mathbb{F}_q$. Such polynomials have at most $k - 1$ zeros, so nonzero codewords have at least $n - k + 1$ non-zeros. Hence, this is a $[n, k, n - k + 1]_q$ code whenever $k \leq n$. Any codeword $(c_0, c_1, \ldots, c_{n-1})$ can be expressed into a q-ary k-vector with respect to the basis $\{1, \alpha, \ldots, \alpha^{k-1}\}$. These codes are just generalized Reed–Solomon codes of parameters $[n, k, d]_q$ over $\mathbb{F}_q$ with parity check polynomial $h(x) = \prod_{i=1}^{k-1}(x - \alpha^i)$ where α is a primitive root of $\mathbb{F}_q$ such that $\alpha^k = \alpha + 1$. In other words, the GRS code is an ideal in the ring $\mathbb{F}_q[x]/(x^k - x - 1)$ generated by a polynomial $g(x)$ with roots in the splitting field $\mathbb{F}_q^l$ of $x^k - x - 1$, where $k | q^l - 1$. Since the NRC is a genus 0 curve, it is easy to see that these codes satisfy the Singleton bound $d \geq n - k + 1$.

Construction of Reed–Solomon codes over $\mathbb{F}_q$ only employs elements of $\mathbb{F}_q$, hence their lengths are at most q. In order to get longer codes, one can make use of elements of an extension of $\mathbb{F}_q$, for instance considering subfield subcodes of Reed–Solomon codes.

As in [2], where we considered a variant of the Horn problem in the context of cyclic coverings of the projective line defined over an arbitrary field k, the problem is reduced to study the representation theory of the general linear group $GL(n, \mathbb{F}_q)$.

3. Representation Theory of $GL(n, \mathbb{F}_q)$

We focus on Grassman codes $\mathcal{G}_{k,n}(\mathbb{F}_q)$ that were described in the introduction as $[n, k]_q$–codes by considering an action (1) of the general linear group $GL(n, q)$ on the Grassmannian. The study of the representation theory of $GL(n, q)$ will allow us to understand better the orbits of this action that will be characterized in Section 5.

The multiplication in the finite field $\mathbb{F}_{q^n}$ is a bilinear map from $\mathbb{F}_{q^n} \times \mathbb{F}_{q^n}$ into $\mathbb{F}_{q^n}$. Thus it corresponds to a linear map from the tensor product $m : \mathbb{F}_{q^n} \otimes \mathbb{F}_{q^n} \to \mathbb{F}_{q^n}$. The symmetric group S_n acts on $\mathbb{F}_{q^n}$ via the permutation matrix:

$$\sigma \cdot v_i = v_{\sigma(i)}, \quad v_i \in \mathbb{F}_{q^n}. \tag{3}$$

The d-Veronese embedding of $\mathbb{P}^n(\mathbb{F}_q)$ maps the line spanned by the vector $v \in \mathbb{F}_{q^n}$ to the line spanned by $v^{\otimes d} = v \otimes \ldots \otimes v$. Thus the symmetric group S_n acts diagonally on the basis of simple tensors of $\mathbb{F}_{q^n}$.

$$\sigma \cdot (v_{i_1} \otimes \ldots \otimes v_{i_r}) = v_{\sigma(i_1)} \otimes \ldots \otimes v_{\sigma(i_r)}. \tag{4}$$

For each partition $\lambda = (\lambda_1, \ldots, \lambda_k)$ we consider its Young diagram. The diagram of λ is an array of boxes, lined up at the left, with λ_i boxes in the ith row, with rows arranged from top to botton. For example,

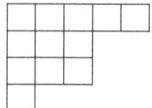

is the Young diagram of the partition $\lambda = (5,3,3,1)$ with $l(\lambda) = 4$ and $|\lambda| = 12$. We define the Schur projection:

$$c_\lambda : \bigotimes^d \mathbb{F}_{q^n} \to \bigotimes^d \mathbb{F}_{q^n}.$$

Let S_n be the symmetric group of permutations over d elements. Any permutation $\sigma \in S_n$ acts on a given Young diagram by permuting the boxes. Let $R_\lambda \subseteq S_n$ be the subgroup of permutations preserving each row. Let $C_\lambda \subseteq S_n$ be the subgroup of permutations preserving each column, let $c_\lambda = \sum_{\sigma \in R_\lambda} \sum_{\tau \in C_\lambda} \epsilon(\tau) \sigma \tau$.

The image of c_λ is an irreducible $GL(n, \mathbb{F}_q)$-module, which is nonzero iff the number of rows is less or equal than dimV_λ. All irreducible $GL(n, \mathbb{F}_q)$-modules can be obtained in this way. Every $GL(n, \mathbb{F}_q)$-module is a sum of irreducible ones.

In terms of irreducible representations of $GL(n, \mathbb{F}_q)$, a partition η corresponds to a finite irreducible representation that we denote as $V(\eta)$. Since $GL(n, \mathbb{F}_q)$ is reductive, any finite dimensional representation decomposes into a direct sum of irreducible representations, and the structure constant $c^\eta_{\lambda,\mu}$ is the number of times that a given irreducible representation $V(\eta)$ appears in an irreducible decomposition of the tensor product of the representations $V(\lambda) \otimes V(\mu)$. These are known as Littlewood–Richardson coefficients, since they were the first to give a combinatorial formula encoding these numbers (see [7]). In terms of the Hopf algebra Λ of Schur functions, let s_λ be the Schur function indexed by the partition λ, we have $s_\lambda \cdot s_\mu = \sum_\nu k^\nu_{\lambda \mu} s_\nu$ for the product and we get the coefficients $k^\eta_{\lambda \mu}$ as the structure constants of the dual Hopf algebra Λ^*. These are known as Kronecker coefficients (see [8,9]) since they appear as expansion coefficients in the Kronecker product $[\lambda][\mu] = \sum_\nu k_{\lambda \mu}[\nu]$ of characters of the symmetric group S_n, as the authors proved in Proposition 4.3 of [2]. Recall that the Schur function s_λ attached to the partition $\lambda = (\lambda_1, \ldots, \lambda_n)$ of length less or equal than n is defined by the quotient:

$$s_\lambda(x_1, \ldots, x_n) = \frac{det(x_i^{\lambda_i+n-j})_{1 \leq i,j, \leq n}}{det(x_i^{n-j})_{1 \leq i,j \leq 0}}.$$

It is a homogeneous polynomial of degree $|\lambda|$ in $x_1, \ldots, x_n$. It easily seen that $s_\lambda(x_1, \ldots, x_n, 0) = s_\lambda(x_1, \ldots, x_n)$. Moreover we can define the Schur function s_λ as the unique symmetric function with this property for all $n \geq l(\lambda)$. It is well known that the Schur functions constitute a basis for the ring Λ of symmetric functions. In addition, there are at least other three well known bases for the ring Λ of symmetric functions. The basis e_k of k-elementary symmetric functions, the h_k complete homogeneous symmetric functions of degree k and the power sums $p_k = z_1^k + z_2^k + \ldots$. This has been applied in Reed–Solomon coding, that is, for AG codes defined on the projective line $\mathbb{P}^1$, as a way to encode information words. Namely, for each codeword $a = (a_0, a_1, \ldots, a_n)$, $a_i \in \mathbb{F}_q$, let us define $a_{n+1} = \sum_{i=1}^n a_i \in \mathbb{F}_q$ which is nothing but the first elementary symmetric function e_1. If we consider the variables $x_1, \ldots, x_r$ as a fixed list of nonzero elements in $\mathbb{F}_q$, then the information word a can be encoded into the codeword $d = (d_1, \ldots, d_r)$, where $d_i = \sum_{j=1}^n a_j x_i^j$. The secret is $a_0 = -\sum_{i=1}^r d_i$, while the pieces of the secret are the d_is.

3.1. Relation between Littlewood–Richardson Coefficients and Kronecker Coefficients

One can stack Littlewood–Richardson coefficients $c_{\lambda\mu}^{\nu}$ in a 3D matrix or 3-dimensional matrix. Intuitively a 3D matrix is a stacking of boxes in the corner of a room. The elements of the principal diagonal are called rectangular coefficients and are indexed by triples $(\lambda, \mu, \nu) = ((i^n), (i^n), (i^n))$ of partitions (i^n) with all their parts equal to the same integer $1 \leq i \leq n$.

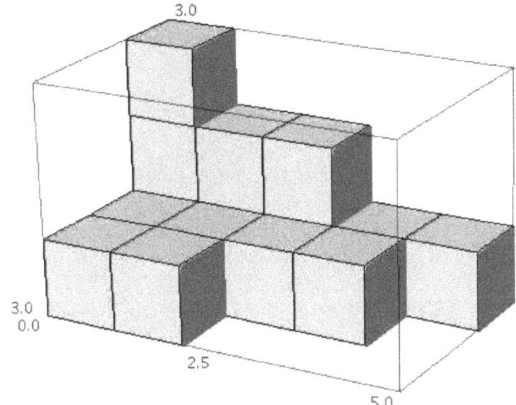

Consider $\mathcal{B}$ and $\mathcal{C}$, two 3D matrices, then we define the product matrix $\mathcal{B} \cdot \mathcal{C}$ as the 3D matrix

$$\mathcal{B} \cdot \mathcal{C} = \prod_{\nu \in \mathcal{P}(n), B^{\nu}, C^{\nu} \in M_{p(n) \times p(n)}(\mathbb{Q})} B^{\nu} \cdot C^{\nu}.$$

Namely, for each index ν fixed, λ and μ run over all partitions $\mathcal{P}(n)$ of n. Thus the coefficients $\left(c_{\lambda,\mu}^{\nu}\right)_{\lambda,\mu \in \mathcal{P}(n)}$ are encoded in a matrix of order $p(n) \times p(n)$, where $p(n)$ denotes the number of unrestricted partitions of n, that is, the number of ways of writing the integer n as a sum of positive integers without regard to order. Thus the product matrix $B^{\nu} \cdot C^{\nu}$ is the standard product of square matrices in $M_{p(n) \times p(n)}(\mathbb{Q})$. In particular, the property of associativity follows easily from the associativity in the vector space $M_{p(n) \times p(n)}(\mathbb{Q})$.

Proposition 1. *Let $\mathcal{C}$ be the 3D matrix whose entries are the Littlewood–Richardson coefficients, and $\mathcal{K}$ the 3D matrix of Kronecker coefficients. Then the matrices are inverse one to each other.*

Proof. Since $c_{\lambda\mu}^{\nu}$ and $k_{\lambda\mu}^{\nu}$ correspond to the structure constants of the Hopf algebra of Schur functions and its dual one respectively (see Proposition 4.3 of [2]), and the Hopf algebra of Schur functions is self-dual (see [9]), one gets that the product matrix $\mathcal{C} \cdot \mathcal{K}$ is the identity 3D matrix $\mathcal{I}$, that is, the matrix whose rectangular coefficients are identically 1. Thus both matrices are inverse one to each other. □

3.2. The Polytope of Triples (λ, μ, η) for Which $c_{\lambda,\mu}^{\eta}$ Is Positive

The convex hull in $\mathbb{R}^3$ of all triples (λ, μ, ν) with $c_{\lambda,\mu}^{\nu} > 0$ is the Newton polytope of $f(x, y, z) = \sum_{\lambda,\mu,\nu} c_{\lambda,\mu}^{\nu} x^{\lambda} y^{\mu} z^{\nu} \in \mathbb{C}[x, y, z]$. Here x^{λ} denotes the monomial $x^{\lambda_1} \cdots x^{\lambda_n}$ of partition degree λ. In particular, when $\lambda = (1^r)$, we have $x^{(1)^r} = e_r = \sum_{i_1 < \ldots < i_r} x_{i_1} \ldots x_{i_r}$, the r-th elementary symmetric function. At the other extreme, when $\lambda = (r)$ we have $x^{(r)} = p_r = \sum x_i^r$, the r-power sum. As we have seen in the previous section, it is clear that every symmetric function $f \in \Lambda$ is uniquely expressible as a finite linear combination of the $(x^{\lambda})_{\lambda \in \mathcal{P}}$. Moreover, the following theorem shows that f is the the generating series for the Gromov–Witten invariant $N_{d,g}(\lambda, \mu, \nu)$ counting irreducible plane curves of given degree and genus g passing through a generic configuration of $3d - 1 + g$ points on $\mathbb{P}^2(\mathbb{C})$ with ramification type at $0, \infty$ and 1 described by the partitions λ, μ and ν and simple ramification over other specified points with $|\lambda| + |\mu| + |\nu| = d$, and these have been computed by Fomin and Mikhalkin in [10].

Theorem 1. *The power series $f(x, y, z) = \sum_{\lambda, \mu, \nu} c^{\nu}_{\lambda, \mu} x^{\lambda} y^{\mu} z^{\nu} \in \mathbb{C}[x, y, z]$, is the generating series for the Gromov–Witten invariants $N_{d,g}(\lambda, \mu, \nu)$, counting irreducible plane curves of given degree d and genus g passing through a generic configuration of $3d - 1 + g$ points on $\mathbb{P}^2(\mathbb{C})$ with ramification type at $0, \infty$ and 1 described by the partitions λ, μ and ν and simple ramification over other specified points with $|\lambda| + |\mu| + |\nu| = d$.*

Proof. Whenever the coefficient $c^{\nu}_{\lambda, \mu} > 0$ is positive consider the corresponding ideal sheaves $\mathcal{I}_{\lambda}, \mathcal{I}_{\mu}$ and $\mathcal{I}_{\nu}$ in $\mathbb{C}[x, y, z]$ associated with the partitions λ, μ and ν respectively. Each ideal sheaf determines a curve in $\mathbb{C}[x, y]$ via homogenization of the corresponding monomial ideals. Thus each coefficient represents the number of ideal sheaves on $\mathbb{C}^3$ of colength n and degree d equal to the size of the partition, that is the corresponding 3-point Gromov–Witten invariant $\langle \lambda, \mu, \nu \rangle_{0,3,d}$ of the Hilbert scheme Hilb_n of $n = 2d - 1 + |\nu| + |\mu| + |\lambda| + g$ distinct points in the plane, or the relative Gromov–Witten invariant $N_{d,g}(\lambda, \mu, \nu)$ counting irreducible plane curves of given degree d and genus g passing through a generic configuration of $3d - 1 + g$ points on $\mathbb{P}^2(\mathbb{C})$ with ramification type at $0, \infty$ and 1 respectively, described by the partitions λ, μ and ν of n (see section 4 of [2]). □

Remark 2. *The Euler characteristic of each ideal sheaf is fixed and coincides with the Euler characteristic χ of the polyhedra described in $\mathbb{R}^3$ by the convex hull of all triples (λ, μ, ν) with $c^{\nu}_{\lambda, \mu} > 0$, that is, the Newton polytope of $f(x, y, z) = \sum_{\lambda, \mu, \nu} c^{\nu}_{\lambda, \mu} x^{\lambda} y^{\mu} z^{\nu} \in \mathbb{R}[x, y, z]$. Thus each coefficient represents the number of ideal sheaves on $\mathbb{C}^3$ of fixed Euler characteristic $\chi = n$ and degree d equal to the size of the partition, that is the corresponding Donaldson-Thomas invariant of the blow-up of the plane $\mathbb{P}^1 \times (\mathbb{C}^2)$ with discrete invariants $\chi = n$ and degree d.*

Remark 3. *The Hilbert scheme Hilb_n of n points in the plane $\mathbb{C}^2$ parametrizing ideals $\mathcal{J} \subset \mathbb{C}[x, y]$ of colength n contains an open dense set in the Zariski topology parametrizing ideals associated with configurations of n distinct points. Moreover there is an isomorphism $\text{Hilb}_n \cong (\mathbb{C}^2)^n / S_n$. In particular, as we showed in [2], any conjugacy class in the symmetric group S_n determines a divisor class in the T-equivariant cohomology $H^{4n}_T(\text{Hilb}_n, \mathbb{Q})$, for the standard action of the torus $T = (\mathbb{C}^*)^2$ on $\mathbb{C}^2$. The T-equivariant cohomology of Hilb_n has a canonical Nakajima basis indexed by $\mathcal{P}(n)$. The map $\lambda \to \mathcal{J}_{\lambda}$ is a bijection between the set of partitions $\mathcal{P}(n)$ and the set of T-fixed points $\text{Hilb}_n^T \subset \text{Hilb}_n$.*

Denote the series $\langle \lambda, \mu, \nu \rangle^{\text{Hilb}_n}$ of 3-point invariants by a sum over curve degrees:

$$\langle \lambda, \mu, \nu \rangle^{\text{Hilb}_n} = \sum_{d \geq 0} q^d \langle \lambda, \mu, \nu \rangle^{\text{Hilb}_n}_{0,3,d}.$$

Corollary 1. *Let H be the divisor class in the Nakajima basis corresponding to the tautological rank n bundle $\mathcal{O}/\mathcal{J} \to \text{Hilb}_n$ with fiber $\mathbb{C}[x, y]/\mathcal{J}$ over $\mathcal{J} \in \text{Hilb}_n$ and ν the corresponding partition. Then we can recover inductively in the degree d, all the Littlewood–Richardson coefficients $(c^{\nu}_{\lambda, \mu})_{\lambda, \mu \in \mathcal{P}(n)}$.*

Proof. The non-negative degree of a curve class $\beta \in H_2(\text{Hilb}_n, \mathbb{Z})$ is defined by $d = \int_{\beta} H$. Then via the identification of $c^{\nu}_{\lambda, \mu}$ with the 3-point Gromov–Witten invariant $\langle \lambda, H, \mu \rangle^{\text{Hilb}_n}_{0,3,d}$ where $[\lambda], [\mu]$ are the corresponding classes in $H^{4n}_T(\text{Hilb}_n, \mathbb{Q})$ associated with the partitions λ and μ in $\mathcal{P}(n)$, we proceed by induction on the degree d as in section 3.6 of [11]. □

Remark 4. *If we choose the partition ν to be the empty partition $\emptyset$, we recover the relative Gromov–Witten invariants $N_{d,g}(\lambda,\mu)$ studied by Fomin and Mikhalkin in [10], and by Caporaso and Harris in [12].*

4. Configurations of Points over a Normal Rational Curve

In this section, we study codes defined from a linear series attached to a divisor on the normal rational curve NRC or equivalently Goppa codes on $\mathbb{P}^1$ and hence generalized Reed–Solomon codes. Assume V is a vector space of dimension $n+1$ over a field k equipped with a linear action, that is, G acts via a representation $G \to GL(V)$. We denote by $S^d V$ the d-th symmetric power of V.

Consider the d-Veronese embedding of $\mathbb{P}^n$

$$\mathbb{P}V^* \to \mathbb{P}S^d V^* \tag{5}$$
$$v \mapsto v^d,$$

mapping the line spanned by $v \in V^*$ to the line spanned by $v^d \in S^d V^*$. In coordinates, if we choose bases $\{\alpha, \beta\}$ for V and $\{[\frac{n!}{k!(n-k)!}]\alpha^k \beta^{d-k}\}$ for $S^d V^*$ and expanding out $(x\alpha + y\beta)^d$, we see that in coordinates this map may be given as

$$[x,y] \to [x^d, x^{d-1}y, x^{d-2}y^2, \ldots, xy^{d-1}, y^d].$$

In particular, the homogeneous coordinate ring for the natural projective embedding of the geometric invariant theory (GIT) quotient $(\mathbb{P}^d)^n // SL_{d+1}$ is the ring of invariants for n ordered points in the projective space up to, projectivity, i.e, if one considers the function field $k(x_1, \ldots, x_d)$ of the projective space $(\mathbb{P})^d$, the ring of invariants is defined by:

$$\{f \in k(x_1, \ldots, x_d) \mid \forall \sigma \in SL_{d+1}, \sigma \cdot f = f\}.$$

Generators for this ring are given by tableau functions, which appear in many areas of mathematics, particularly representation theory and Schubert calculus. Consider the hypersimplex:

$$\triangle(d+1,n) = \{(c_1,\ldots,c_n) \in \mathbb{Q}^n \mid 0 \leq c_i \leq 1, \sum c_i = d+1\},$$

for any $1 \leq d \leq n-3$ and choose a linearization $c \in \triangle(d+1,n)$, there is a morphism

$$\varphi : \bar{M}_{0,n} \to (\mathbb{P}^d)^n //_c SL_{d+1},$$

sending a configuration of distinct points on $\mathbb{P}^1$ to the corresponding configuration under the dth Veronese map.

The symmetric power $\text{Sym}^n C_d$ of the curve C_d is the quotient of the configuration space C_d^n of n unordered tuples of points on the normal rational curve C_d by the symmetric group S_n. Furthermore, we can identify the set of effective divisors of degree d on C_d with the set of k-rational points of the symmetric power $\text{Sym}^n C$, that is, $\text{Sym}^n C$ represents the functor of families of effective divisors of degree n on C.

Lee-Sullivan List-Decoding Algorithm of Reed–Solomon Codes

By definition, the rational normal curve C_d is the image by the d–Veronese embedding of $\mathbb{P}V^* = \mathbb{P}^1$ where V is a 2-dimensional vector space, therefore it is isomorphic to any curve of genus 0. The action of $PGL(2,k)$ on $\mathbb{P}^d$ preserves the rational normal curve C_d. Conversely, any automorphism of $\mathbb{P}^d$ fixing C_d pointwise is the identity. It follows that the group of automorphisms of $\mathbb{P}^d$ that preserves C_d is precisely $PGL(2,k)$. These codes are just generalized RS codes and they come with efficient decoding algorithms once we choose a metric consistent with channel errors and search of a set of vectors with given metric properties as a correcting code. In particular, these codes are consistent with the Hamming metric ([13,14]). Recall that given two vectors of length n, say U and V, the

Hamming distance $d_H(U, V)$ between U and V is the number of coordinates in which they differ.

Given a $[n, k]$ RS code C of length n and dimension k, we call d the minimum (Hamming distance) which attains the Singleton bound $n - k + 1$. We shall identify the code with the set of its codewords. A codeword of C is viewed as a polynomial $c_0 + c_1 x + \ldots + c_{n-1} x^{n-1}$ in the $\mathbb{F}$-vector space $\mathbb{F}[x]$, where $\mathbb{F}$ is a finite field. In the communication process, when a codeword is transmitted, it can be affected by errors and erasures. An error occurs when one codeword component is changed into another field element and an erasure occurs when the received component has an unknown value. The problem of minimum distance decoding is to find, for any given vector v, the set C_v of all codewords $c \in C$ at minimum distance from v. If C_v contains just one element c, then the sent codeword coincides with the received codeword and no decoding is needed. The codewords of minimum weight are the points lying in the intersection of any line and the curve. K. Lee and M.E. O'Sullivan in [15] describe a list decoding algorithm consisting of two steps: the interpolation step and the root-finding step. Starting with a set of generators of the module induced from the ideal for the n points $\{P_1, \ldots, P_n\}$, they convert the generators to a Gröbner basis of the module in which the minimal polynomial is found. This results in an efficient algorithm solving the interpolation problem.

Let v be the received vector, and fix n distinct points $\alpha_1, \ldots, \alpha_n$ from $\mathbb{F}$, for each $1 \leq i \leq n$, let P_i denote the point (α_i, v_i) by Lagrange interpolation we get the polynomial $h_v = \sum_{i=1}^n v_i h_i \in \mathbb{F}[x]_n$, where $h_i = \prod_{j=1}^n (x - \alpha_j)$, $j \neq i$ so that $h_i(\alpha_j) = 1$ if $j = 1$, and 0 otherwise. Now for $m \geq 1$, we define the ideal

$$I_{v,m} = \{f \in \mathbb{F}[x, y] | \, mult_{P_i}(f) \geq m \text{ for } 1 \leq i \leq n\} \cup \{0\}.$$

For $f \in \mathbb{F}[x, y]$ and $u \geq 1$, denoted by $deg_u(f)$, the $(1, u)$-weighted degree of f, that is, the variables x and y, are assigned weights 1 and u, respectively, and for a monomial $x^i y^j$, we define $deg_u(x^i y^j) = i + uj$.

The goal of the interpolation step is to find a polynomial in $I_{v,m}$ having the smallest $(1, k-1)$-weighted degree. The codewords of minimum weight are the points lying in the intersection of any line and the curve. Moreover if $wt(v - c) < n - \frac{w}{m}$, where $w = deg_{k-1}(f)$ and f is the polynomial representing the word c, then the polynomial h_c is a root of f as a polynomial in y over $\mathbb{F}[x]$. Moreover the set of polynomials $(y - h_c)^i \eta^{m-i}$, $0 \leq i \leq m$, where $\eta = \prod_{j=1}^n (x - \alpha_j)$ is a set of generators of $I_{v,m}$.

Let Q be the minimal polynomial of $I_{v,m}$ with respect to the monomial order $>_{k-1}$ of $\mathbb{F}[x, y]$. We can find Q by computing a Gröbner basis of $I_{v,m}$ with respect to $>_{k-1}$. In Appendix A, we provide Horn's algorithm to compute sets of indices which are admissible for the Horn problem. As a result, we provide a set of generators for the algebraic code induced on the NRC.

Proposition 2. *If we consider the set of orbits of $\mathcal{C}_d^n$ by the action of finite subgroups of the symmetric group S_n, we get all possible divisor classes in the group $\mathrm{Div}^n(\mathcal{C}_d)$ of degree n divisors on $\mathcal{C}_d$.*

Proof. Since the symmetric group S_n is generated by 3 elements, a reflection of order 2, a symmetry of order 3 and a rotation of order n, we get all the divisor classes by quotienting the configuration space $\mathcal{C}_d^n$ of n points on the normal rational curve, by the cyclic group generated by the rotation, or one of the triangle groups, the dihedral group D_n, the alternated groups A_4, A_5 or the symmetric group S_4. □

5. Notion of Collinearity on the Normal Rational Curve

A permutation matrix $\sigma \in GL(n, \mathbb{F}_q)$ acts on the Grassmannian by multiplication on the right of the corresponding representation matrix. In particular, we are interested in understanding the orbits by the action of any permutation matrix of $GL(n, \mathbb{F}_q)$ and moreover of any subgroup G contained in $GL(n, \mathbb{F}_q)$. Further, it is possible to count the orbits of the action in several cases and this is established by the correspondence given in Theorem 2 between sets of points satisfying certain geometrical conditions and partitions.

Definition 2. *An incidence structure $\mathcal{S}$ on V is a triple $(\mathcal{P}, \mathcal{B}, I)$, where $\mathcal{P}$ is a set whose elements are smooth, reduced points in V, $\mathcal{B}$ is a set whose elements are subsets of points called blocks (or lines in several specific cases) endowed with a relation of collinearity, and an incidence relation $I \subset \mathcal{P} \times \mathcal{B}$. If $(P, L) \in I$, then we say that P is incident with L or L is incident with P, or P lies in L or L contains P.*

When the collinearity relation is a symmetric ternary relation defined on triples $(p, q, r) \in \mathcal{P} \times \mathcal{P} \times \mathcal{P}$ by the geometric condition $(p, q, r) \in \mathcal{B}$ if either $p + q + r$ is the full intersection cycle of C_d with a k-line $l \subset \mathbb{P}^n(k)$ with the right multiplicities, or else if there exists a k-line $l \subset V$ such that, $p, q, r \in l$, then the triple (p, q, r) is called a plane section.

1. For any $(p, q) \in \mathcal{P}^2(V^*)$, there exists an $r \in \mathcal{P}(S^d V^*)$ such that $(p, q, r) \in I$. The triple (p, q, r) is strictly collinear if r is unique with this property, and p, q, r are pairwise distinct. The subset of strictly collinear triples is a symmetric ternary relation. When k is a field algebraically closed of characteristic 0, then r is unique with this property, and we recover the euclidean axioms.
2. Assume that $p \neq q$ and that there are two distinct $r_1, r_2 \in \mathcal{P}$ with $(p, q, r_1) \in \mathcal{B}$ and $(p, q, r_2) \in \mathcal{B}$. Denote by $l = l(p, q)$ the set of all such rs, then $l^3 \in \mathcal{B}$—that, is any triple (r_1, r_2, r_3) of points in which l is collinear. Such sets l are called lines in $\mathcal{B}$.

If V is a 3-dimensional vector space defined over the finite field $\mathbb{F}_p$, then the projective plane $\mathbb{P}^2(\mathbb{F}_p)$ on V is defined by the incidence structure $PG(2, p) = (\mathcal{P}(V), \mathcal{L}(V), I)$.

Definition 3.
1. *A $(k; r)$-arc $\mathcal{K}$ in $PG(2, p)$ is a set of k points such that some r, but not $r + 1$ of them are collinear. In other words, some line of the plane meets $\mathcal{K}$ in r points and no more than r points. A $(k; r)$-arc is complete if there is no $(k + 1; r)$ arc containing it.*
2. *A k-arc is a set of k points, such that, every subset of s points with $s \leq n$ points is linearly independent.*

Let q denote some power of the prime p and $PG(n, p)$ be the n-dimensional projective space $(\mathbb{F}_p)^{n+1} \cong \mathbb{F}_q$, where $n \geq 2$. The normal rational curve C is defined as:

$$\mathcal{V}_1^n := \left\{ \mathbb{F}_q(1, x, x^2, \ldots, x^n) \mid x \in \mathbb{F}_q \bigcup \{\infty\} \right\}.$$

If $q \geq n + 2$, the NRC is an example of a $(q + 1)$-arc. It contains $q + 1$ rational points, and every set of $n + 1$ points are linearly independent. For each $a \in (\mathbb{F}_p)^{n+1}$, the mapping:

$$\mathbb{F}_p(x_0, \ldots, x_n) \to \mathbb{F}_p(a^0 x_0, \ldots, a^n x_n),$$

describes an automorphic collineation of the NRC.

All invariant subspaces form a lattice with the operations of "join" and "meet".
For $j \in \mathbb{N}$, let $\Omega(j) = \{m \in \mathbb{N} \mid 0 \leq m \leq n, \binom{m}{j} \neq 0 \bmod p\}$. Given $J \subset \{0, 1, \ldots, n\}$, put $\Omega(J) = \bigcup_{j \in J} \Omega(j)$, $\Psi(J) := \bigcup_{j \in J} \{j, n - j\}$.

Both Ω and Ψ are closure operators on $\{0, 1, \ldots, n\}$. Likewise the projective collineation $\mathbb{F}_p(x_0, x_1, \ldots, x_n) \to \mathbb{F}_p(x_n, x_{n-1}, \ldots, x_0)$ leaves the NRC invariant whence Λ has to be closed with respect to Ψ. Any algebraic-geometric code constructed by evaluation of a function over the NRC with values in $\mathbb{F}_q$ is a generalized Reed–Solomon code of length at most q. In order to get longer codes, one needs to use elements from any finite extension $\mathbb{F}_q^r$ of $\mathbb{F}_q$.

Proposition 3. *Each subspace invariant under collineation of the NRC is indexed by a partition in $\mathcal{P}(t)$. If the ground field k is sufficiently large, then every subspace which is invariant under all collineations of the NRC is spanned by base points kc_λ, where $\lambda \in \mathcal{P}(t)$.*

Proof. Let
$$E_n^t := \{(e_0, e_1, \ldots, e_n) \in \mathbb{N}^{n+1} \mid e_0 + e_1 + \ldots + e_n = t\},$$
be the set of partitions of t of n parts and let Y be the $\binom{n}{t}$-dimensional vector space over $\mathbb{F}_p$ with basis
$$\{c_{e_0, e_1, \ldots, e_n} \in \mathbb{F}_q : (e_0, e_1, \ldots, e_n) \in E_n^t\}.$$
Let us call $\mathcal{V}_n^t$ the Veronese image under the Veronese mapping given by:
$$\mathbb{F}_p(\sum_{i=0}^n x_i b_i) \to \mathbb{F}_p(\sum_{E_n^t} c_{e_0, \ldots, e_n} x_1^{e_0} x_1^{e_1} \cdots c_n^{e_n}), \quad x_i \in \mathbb{F}_p.$$

The Veronese image of each r-dimensional subspace of $PG(n, p)$ is a sub-Veronesean variety $\mathcal{V}_r^t$ of $\mathcal{V}_n^t$, and all those subspaces are indexed by partitions in $\mathcal{P}(t)$. Thus by a Theorem due to Gmainer are invariant under the collineation group of the normal rational curve (see [16]).

The k-rational points $(p_0, p_1, \cdots, p_n)$ of the normal rational curve C correspond to collinear points on C that are defined over some Galois extension l of k and permuted by $\text{Gal}(l/k)$. □

5.1. An Application: Three-Point Codes on the Normal Rational Curve

As we showed in Proposition 3, each subspace invariant under collineation of the NRC is indexed by a partition $\lambda \in \mathcal{P}(d)$. Let us call the base point associated with the partition λ as P_λ. As we are considering that the ground field is $\mathbb{F}_q$, the $\mathbb{F}_q$-points might be defined over a finite extension $\mathbb{F}_{q^r}$ of $\mathbb{F}_q$. Observe that for any divisor r of n, one easily obtains a extension field of $\mathbb{F}_q$ of degree r. Namely, let ξ a non-trivial r-root of unity, one can consider the symbols $\xi^{q^r}, \ldots, \xi^q, \xi$ and the polynomial which has them as roots, $q(x) = \prod_{i=0}^{i=r}(1 - \xi^{q^i})$ gives an extension field of $\mathbb{F}_q$ of degree r.

Theorem 2. *Let $\sigma_1, \sigma_2, \sigma_3$ be three generators for the symmetric group S_d and let λ_1, λ_2 and λ_3 be the partitions of d indexing the corresponding irreducible representations in the special linear group $SL(n, \mathbb{F}_q)$. Then any algebraic code defined over the NRC is covered by a divisor defined as linear combination of the base points $(P_{\lambda_i})_{1 \leq i \leq 3}$ on the NRC, where the λ_i are LR coefficients.*

Proof. Consider the divisors associated with the rational maps $f(x, y, z) = nx + my + lz$ defined over the normal rational curve C_d defined over $\mathbb{F}_q$, with n, m and l integer numbers. In particular, if $d \mid q^2 - 1$, the points $P = (\alpha, 0, 0)$, $Q = (0, \beta, 0)$ and $R = (0, 0, \gamma)$ with $\alpha^d = 1$, $\beta^d = 1$ and $\gamma^d = 1$, are $\mathbb{F}_{q^2}$-rational points on C_d, and the divisors nP, mQ and lR define codes on it. Reciprocally, given a code on the NRC, by Proposition 2, the corresponding divisor defining the code is defined by a finite subgroup in the symmetric group. Since the symmetric group is generated by the 3 elements σ_1, σ_2 and σ_3, the divisor is a linear combination of the base points $(P_{\lambda_i})_{1 \leq i \leq 3}$ on the NRC. □

5.2. Conclusions

In [17], the authors considered a particular class of block codes known as quasi-cyclic codes as orbit codes in the Grassmannian parameterizing constant dimension codes. In the present paper we have focused on RS codes that can also be viewed as orbit codes in the Grassmannian through the action of $PGL(n, q)$, the collineation group of the NRC. This approach could be extended to study a wide class of codes, including convolutional codes with two states known as 2D finite support convolutional codes of rate $\frac{k}{n}$, which are defined as free $\mathbb{F}[z_1, z_2]$-submodules of $\mathbb{F}[z_1.z_2]^n$ with rank k.

Author Contributions: Writing—original draft, C.M.; Writing–review—editing, A.B. All authors have read and agreed to the published version of the manuscript.

Funding: This research received external funding from University of Alcalá.

Acknowledgments: We thank Diego Napp for very useful comments during the preparation of this manuscript and the referee for valuable observations.

Conflicts of Interest: The authors declare no conflict of interest.

Appendix A. Explicit Presentation of 3-Point Codes

In this section, we provide Horn's algorithm to compute sets of indices which are admissible for the Horn problem. As a result, we provide a set of generators for the algebraic code induced on the NRC. Given sets $I, J, K \subset \{0, 1, \ldots, n\}$, of cardinality r, we can associate to them partitions λ, μ and ν as follows. Let $I = \{i_1 < \ldots < i_r\} \subset \{1, \ldots, n\}$; then the corresponding partition is defined as $\lambda = (i_r - r, \ldots, i_1 - 1)$. We consider the corresponding codes defined by the base points c_λ, c_μ and c_ν, whenever the corresponding Littlewood–Richardson coefficient $c_{\lambda,\mu}^\nu$ is positive. Next, we give an algorithm to compute the Littlewood–Richardson coefficients $c_{\lambda,\mu}^\nu$. Horn defined sets of triples (I, J, K) by the following inductive procedure (see [7]):

$$U_r^n = \{(I, J, K) | \sum_{i \in I} + \sum_{j \in J} = \sum_{k \in K} k + r(r+1)/2\},$$

$$T_r^n = \{(I, J, K) \in U_r^n | \text{ for all } p < r \text{ and all } (F, G, H) \in T_p^r,$$

$$\sum_{f \in F} i_f + \sum_{g \in G} j_g \leq \sum_{h \in H} k_h + p(p+1)/2\}.$$

Note that Horn's algorithm produces all the triples from the lowest values. Even if it is possible to start with a random generator set I, you need first to compute the lower values. As a consequence of the classification Theorem 2, for any triple (I, J, K) of indices admissible for the Horn problem the polynomials defined by $f(x) = \prod_{j \in J}(x - \alpha^j)$, $g(x) = \prod_{i \in I}(x - \alpha^i)$, and $h(x) = \prod_{k \in K}(x - \alpha^k)$ where α is a primitive element of $\mathbb{F}_{q^m}$ and m is the least integer such that $n+1 | p^m - 1$ constitute a set of generators for the ideal of the corresponding algebraic code in the module of $n + 1$ $\mathbb{F}_{q^m}$-rational points lying on the NRC.

Here we present a Sage [6] code calculating the U_r^n and T_r^n index sets, followed by a table containing all the cases till $n = 4$ and $r = 3$. The algorithm is implemented using Python: this involves calculate and iterate through $r-$combination of $n-$element. The running time is $O(\binom{n}{r}^3)$.

```
from sage.combinat.subset import Subsets

def simple_cache(func):
cache = dict()
def cached_func(*args):
if args not in cache:
cache[args] = func(*args)
return cache[args]
cached_func.cache = cache
return cached_func
```

```
@simple_cache
def getUnr(n, r):
if r >= n:
raise ValueError(''r must be less than n: (n, r) =
(%d, %d)'' %(n, r))
s = Subsets(range(1, n + 1), r)
candidates = [(x, y, z) for x in s for y in s for z in s]
return [tuple(map(sorted, (x, y, z))) for (x, y, z) in candidates if (
sum(x) + sum(y)) == (sum(z) + r * (r + 1)/2)]


def index_filter(sub_index, index):
if max(sub_index) > len(index):
raise ValueError(''%s must be valid indexes for %s''
% (sub_index, index))
# our indexes lists start at 1
return [index[i - 1] for i in sub_index]


def condition((f, g, h), (i, j, k)):
p = len(f)
return sum(index_filter(f, i)) + sum(index_filter(g, j)) <= sum(
index_filter(h, k)) + p*(p + 1)/2

def genTillR(r):
return [getTnr(r, p) for p in range(1, r)]

@simple_cache
def getTnr(n, r):
if r == 1:
return getUnr(n, 1)
else:
return [(i, j, k) for (i, j, k) in getUnr(n, r) if all(
all(condition((f, g, h), (i, j, k)) for (f, g, h) in triplets)
for triplets in genTillR(r))]
```

Here we list code's remarks
- The sorted() mapping function in getUnr() is necessary because the order of elements in Subsets is unknown;
- There is a 1-offset between index in Python lists and index sets we use;
- The recursion in getTnr() is factored out in getTillR() call;
- The cache decorator mitigates the perils of performing the same calculation several times in a function that is already heavily recursive;
- Results are limited by constraints Python has on recursive function calls;
- The filtering performed on U_r^n to get T_r^n is implemented by two nested calls to all().

(n,r)	U_r^n	T_r^n
(2, 1)	({1}, {1}, {1}), ({1}, {2}, {2}), ({2}, {1}, {2})	({1}, {1}, {1}), ({1}, {2}, {2}), ({2}, {1}, {2})
(3, 1)	({1}, {1}, {1}), ({1}, {2}, {2}), ({1}, {3}, {3}), ({2}, {1}, {2}), ({2}, {2}, {3}), ({3}, {1}, {3})	({1}, {1}, {1}), ({1}, {2}, {2}), ({1}, {3}, {3}), ({2}, {1}, {2}), ({2}, {2}, {3}), ({3}, {1}, {3})
(3, 2)	({1,2}, {1,2}, {1,2}), ({1,2}, {1,3}, {1,3}), ({1,2}, {2,3}, {2,3}), ({1,3}, {1,2}, {1,3}), ({1,3}, {1,3}, {2,3}), ({2,3}, {1,2}, {2,3})	({1,2}, {1,2}, {1,2}), ({1,2}, {1,3}, {1,3}), ({1,2}, {2,3}, {2,3}), ({1,3}, {1,2}, {1,3}), ({1,3}, {1,3}, {2,3}), ({2,3}, {1,2}, {2,3})
(4, 1)	({1}, {1}, {1}), ({1}, {2}, {2}), ({1}, {3}, {3}), ({1}, {4}, {4}), ({2}, {1}, {2}), ({2}, {2}, {3}), ({2}, {3}, {4}), ({3}, {1}, {3}), ({3}, {2}, {4}), ({4}, {1}, {4})	({1}, {1}, {1}), ({1}, {2}, {2}), ({1}, {3}, {3}), ({1}, {4}, {4}), ({2}, {1}, {2}), ({2}, {2}, {3}), ({2}, {3}, {4}), ({3}, {1}, {3}), ({3}, {2}, {4}), ({4}, {1}, {4})
(4, 2)	({1,2}, {1,2}, {1,2}), ({1,2}, {1,3}, {1,3}), ({1,2}, {1,4}, {1,4}), ({1,2}, {1,4}, {2,3}), ({1,2}, {2,3}, {1,4}), ({1,2}, {2,3}, {2,3}), ({1,2}, {2,4}, {2,4}), ({1,2}, {3,4}, {3,4}), ({1,3}, {1,2}, {1,3}), ({1,3}, {1,3}, {1,4}), ({1,3}, {1,3}, {2,3}), ({1,3}, {1,4}, {2,4}), ({1,3}, {2,3}, {2,4}), ({1,3}, {2,4}, {3,4}), ({1,4}, {1,2}, {1,4}), ({1,4}, {1,2}, {2,3}), ({1,4}, {1,3}, {2,4}), ({1,4}, {1,4}, {3,4}), ({1,4}, {2,3}, {3,4}), ({2,3}, {1,2}, {1,4}), ({2,3}, {1,2}, {2,3}), ({2,3}, {1,3}, {2,4}), ({2,3}, {1,4}, {3,4}), ({2,3}, {2,3}, {3,4}), ({2,4}, {1,2}, {2,4}), ({2,4}, {1,3}, {3,4}), ({3,4}, {1,2}, {3,4})	({1,2}, {1,2}, {1,2}), ({1,2}, {1,3}, {1,3}), ({1,2}, {1,4}, {1,4}), ({1,2}, {2,3}, {2,3}), ({1,2}, {2,4}, {2,4}), ({1,2}, {3,4}, {3,4}), ({1,3}, {1,2}, {1,3}), ({1,3}, {1,3}, {1,4}), ({1,3}, {1,3}, {2,3}), ({1,3}, {1,4}, {2,4}), ({1,3}, {2,3}, {2,4}), ({1,3}, {2,4}, {3,4}), ({1,4}, {1,2}, {1,4}), ({1,4}, {1,3}, {2,4}), ({1,4}, {1,4}, {3,4}), ({2,3}, {1,2}, {2,3}), ({2,3}, {1,3}, {2,4}), ({2,3}, {2,3}, {3,4}), ({2,4}, {1,2}, {2,4}), ({2,4}, {1,3}, {3,4}), ({3,4}, {1,2}, {3,4})
(4, 3)	({1,2,3}, {1,2,3}, {1,2,3}), ({1,2,3}, {1,2,4}, {1,2,4}), ({1,2,3}, {1,3,4}, {1,3,4}), ({1,2,3}, {2,3,4}, {2,3,4}), ({1,2,4}, {1,2,3}, {1,2,4}), ({1,2,4}, {1,2,4}, {1,3,4}), ({1,2,4}, {1,3,4}, {2,3,4}), ({1,3,4}, {1,2,3}, {1,3,4}), ({1,3,4}, {1,2,4}, {2,3,4}), ({2,3,4}, {1,2,3}, {2,3,4})	({1,2,3}, {1,2,3}, {1,2,3}), ({1,2,3}, {1,2,4}, {1,2,4}), ({1,2,3}, {1,3,4}, {1,3,4}), ({1,2,3}, {2,3,4}, {2,3,4}), ({1,2,4}, {1,2,3}, {1,2,4}), ({1,2,4}, {1,2,4}, {1,3,4}), ({1,2,4}, {1,3,4}, {2,3,4}), ({1,3,4}, {1,2,3}, {1,3,4}), ({1,3,4}, {1,2,4}, {2,3,4}), ({2,3,4}, {1,2,3}, {2,3,4})

References

1. Manganiello, F.; Trautmann, A.L.; Rosenthal, J. On conjugacy classes of subgroups of the general linear group and cyclic orbit codes. In Proceedings of the IEEE International Symposium on Information Theory proceedings (ISIT), St. Petersburg, Russia, 31 July–5 August 2011.
2. Besana, A.; Martínez, C. Combinatorial enumeration of cyclic covers of $\mathbb{P}^1$. *Turk. J. Math.* **2018**, *42*, 2018–2034. [CrossRef]
3. Martínez, C. On the cohomology of Brill-Noether loci over Quot schemes. *J. Algebra* **2008**, *319*, 391–403. [CrossRef]
4. Climent, J.; Napp, D.; Pinto, R.; Simoes, R. Decoding of 2D convolutional codes over an erasure Channel. *Adv. Math. Commun.* **2016**, *10*, 179–193. [CrossRef]
5. Bifet, E.; Ghione, F.; Leticia, M. On the Abel-Jacobi map for divisors of higher rank on a curve. *Math. Ann.* **1994**, *299*, 641–672. [CrossRef]
6. SageMath. Available online: http://www.sagemath.org/ (accessed on 20 October 2020).
7. Fulton, W. Eigenvalues, invariant factors, highest weights and Schubert calculus. *Bull. Am. Math. Soc.* **2000**, *37*, 209–249. [CrossRef]
8. Manivel, L. On rectangular Kronecker coefficients. *J. Algebr. Comb.* **2011**, *33*, 153–162. [CrossRef]
9. Sottile, F.; Lam, T.; Lauve, A. A skew Littlewood-Richardson rule from Hopf algebras. *Int. Math. Res. Not.* **2011**, *2011*, 1205–1219.
10. Fomin, S.; Milkhalkin, G. Label floor diagrams for plane curves. *J. Eur. Math. Soc.* **2009**, *12*, 1453–1496.
11. Okounkov, A.; Pandharipande, P. Quantum cohomology of the Hilbert scheme of points in the plane. *Invent. Math.* **2010**, *179*, 523–557. [CrossRef]
12. Caporaso, L.; Harris, J. Counting plane curves of any genus. *Invent. Math.* **1998**, *131*, 345–392. [CrossRef]
13. Bezzateev, S.V.; Shekhunova, N. Class of generalized Goppa codes perfect in weighted Hamming metric. *Des. Codes Criptogr.* **2013**, *66*, 391–399. [CrossRef]
14. Bezzateev, S.V.; Shekhunova, N.A. Subclass of cyclic Goppa codes. *IEEE Trans. Inf. Theory* **2013**, *59*, 11. [CrossRef]
15. Lee, K.; O'Sullivan, M.E. List decoding of RS codes from a Gröbner basis perspective. *J. Symb. Comput.* **2008**, *43*, 645–658. [CrossRef]
16. Gmainer, J. Pascal's Triangle, Normal Rational Curves, and their Invariant Subspaces. *Eur. J. Comb.* **2001**, *22*, 37–49. [CrossRef]
17. Besana, A.; Martínez, C. A Geometrical Realisation of Quasi-Cyclic Codes. In *Combinatorics, Probability and Control*; IntechOpen: London, UK, 2020; pp. 259–271.

Article

On the Affine Image of a Rational Surface of Revolution

Juan G. Alcázar

Departamento de Física y Matemáticas, Universidad de Alcalá, E-28871 Madrid, Spain; juange.alcazar@uah.es

Received: 5 September 2020; Accepted: 16 November 2020; Published: 19 November 2020

Abstract: We study the properties of the image of a rational surface of revolution under a nonsingular affine mapping. We prove that this image has a notable property, namely that all the *affine normal lines*, a concept that appears in the context of *affine differential geometry*, created by Blaschke in the first decades of the 20th century, intersect a fixed line. Given a rational surface with this property, which can be algorithmically checked, we provide an algorithmic method to find a surface of revolution, if it exists, whose image under an affine mapping is the given surface; the algorithm also finds the affine transformation mapping one surface onto the other. Finally, we also prove that the only rational *affine surfaces of rotation*, a generalization of surfaces of revolution that arises in the context of affine differential geometry, and which includes surfaces of revolution as a subtype, affinely transforming into a surface of revolution are the surfaces of revolution, and that in that case the affine mapping must be a similarity.

Keywords: surface of revolution; affine differential geometry; affine equivalence

1. Introduction

Surfaces of revolution are classical objects in differential geometry, generated by rotating a curve around a fixed line, called the axis of revolution of the surface. These surfaces appear often in nature, in architecture, and in many common human artifacts, and are widely used in Geometric Design. Additionally, when the surface of revolution is rational, i.e., admitting a parametrization whose components are quotients of bivariate polynomials (a *rational parametrization*), the strong structure of the surface allows to perform easily certain operations like implicitizing [1], reparametrizing the surface over the real numbers [2], or analyzing the surjectivity of the parametrization [3]. We recall that every rational surface is *algebraic*, i.e., it is the zeroset of a trivariate polynomial.

In this paper we study how rational surfaces of revolution are transformed when a nonsingular affine mapping is applied. The resulting surface is certainly rational too, but in general it is not a surface of revolution. However, some properties of this image can be discovered when elements of *affine differential geometry* are used. Classical differential geometry studies objects and notions that behave well when an orthogonal transformation is applied: for instance, normal lines transform accordingly, and the Gauss curvature is preserved. Affine differential geometry [4,5], started by Blaschke in the first decades of the 20th century, however, studies objects and notions that behave well when we consider matrix transformations of the special linear group $\mathbf{SL}_3(\mathbb{R})$, i.e., the group of matrices with determinant equal to 1. Thus, in the context of affine differential geometry, for instance, normals and Gauss curvature are replaced by *affine normals* and *affine* curvature, which have good properties when transformations of the special linear group are applied.

In the context of affine differential geometry, *affine surfaces of rotation* [6,7], which generalize classical surfaces of revolution, are introduced. These surfaces can be of three different subtypes, *elliptic*, *hyperbolic* and *parabolic*, the first of them being the classical surfaces of revolution. Theoretical properties of algebraic affine surfaces of rotation are treated in some recent papers: the elliptic case is studied in [8], the hyperbolic case is addressed in [9], and the parabolic in [10]. Furthermore, an algorithm for recognizing algebraic affine surfaces of revolution is provided in [11]. In this regard, a necessary condition, although not sufficient, for a surface to be an affine surface of rotation is that all the affine normal lines of the surface intersect a fixed line, called the *affine axis* of rotation. If the affine normal lines satisfy this property, we say that the surface is ANIL (Affine Normal lines Intersecting a Line). In particular, surfaces of revolution are ANIL surfaces.

Using notions of affine differential geometry and *Plücker coordinates* (see [12]) as fundamental tools, we prove that the image of every rational ANIL surface, and therefore of every rational surface of revolution, under a nonsingular affine mapping is also ANIL. Furthermore, we also provide an algorithmic method to find, given a rational ANIL surface, a rational surface of revolution affinely transforming onto the first surface, and to compute the mapping itself. This is useful because, as we mentioned before, certain operations like implicitizing, reparametrizing over the reals or studying surjectivity can be efficiently performed on surfaces of revolution; via the affine mapping relating the surface of revolution and the given ANIL surface, the results of these operations can be carried to the original ANIL surface.

Additionally, we also explore under what conditions the image of a rational surface of revolution under a nonsingular affine mapping is an affine rotation surface. We prove that this is only possible when the affine rotation surface is another surface of revolution and the mapping is a similarity, i.e., the composition of a rigid motion and a scaling. This shows that there are in fact many ANIL surfaces which however are not affine surfaces of rotation, since the image of any surface of revolution under an affine mapping that is not a similarity is an ANIL surface, but not an affine surface of rotation. The observation is of interest since up to our knowledge, the only known examples of ANIL surfaces to this date are affine surfaces of rotation and *affine spheres*, i.e., surfaces where all the affine normals intersect at one point, called the *center* of the sphere. Affine spheres do not need to be affine surfaces of rotation [11], and their nature is preserved by affine mappings.

The structure of the paper is the following. In Section 2, we recall several notions and results on affine differential geometry, and Plücker coordinates. In Section 3, we prove that the image of a rational surface of revolution is an ANIL surface. In Section 4, we develop an algorithmic method to compute a surface of revolution affinely equivalent to a given ANIL rational surface, and to find the affine mapping between the surfaces. In Section 5, we address the conditions for the affine image of a rational surface of revolution to be an affine surface of rotation. We close in Section 6, where we present our conclusions.

2. Preliminaries

In this section we consider several preliminary notions on affine differential geometry and line geometry. Along the section, we let $S \subset \mathbb{R}^3$ be a rational surface. For certain technical reasons, which will be clear later, we assume that S is not a developable surface, i.e., isometric to the plane, so S has Gaussian curvature not identically equal to zero.

2.1. Affine Rotation Surfaces

In this subsection we recall several notions and results on affine differential geometry and a special class of surfaces, called *affine rotation surfaces*, which appear in the context of affine differential geometry and generalize surfaces of revolution. First, we recall from [13,14] some notions from affine differential geometry. The *affine co-normal vector* at each point of S is defined as

$$\nu = |K|^{-\frac{1}{4}} \cdot \mathbf{N}, \tag{1}$$

where **N** is the unit Euclidean normal vector, and K is the Gaussian curvature. The affine co-normal vector is not defined when K is zero.

The *affine normal vector* to S at a point $p \in S$ is

$$\xi(p) = [\nu(p), \nu_u(p), \nu_v(p)]^{-1} (\nu_u(p) \times \nu_v(p)), \tag{2}$$

where $\bullet_u, \bullet_v$ represent the partial derivatives of $\bullet$ with respect to the variables u, v, and $[\bullet, \bullet_u, \bullet_v]$ represents the determinant of $\bullet, \bullet_u, \bullet_v$. The *affine normal line* at $p \in S$ is the line through p, parallel to the affine normal vector. Denoting by $\mathbf{SL}_3(\mathbb{R})$ the special linear group, i.e., the group of matrices with determinant equal to 1, the affine normal lines are known to be *covariant* under affine transformations of $\mathbf{SL}_3(\mathbb{R})$ (see Prop. 3 in [13]): this means that if h represents an affine transformation of the special linear group and $\mathcal{L}_p$ represents the affine normal line at p, then $h(\mathcal{L}_p)$ coincides with $\mathcal{L}_{h(p)}$. Sometimes we will refer to this property as the *covariance property* of affine normal lines.

Also in the context of affine differential geometry, *affine rotation groups* are introduced. An *affine rotation group* is a uniparametric matrix group that is a subgroup of $\mathbf{SL}_3(\mathbb{R})$, and which leaves invariant exactly one line in 3-space, called the *affine axis of rotation*. Lee [6] shows that there are only three different types of such subgroups; in an appropriate coordinate system, these types correspond to the following uniparametric matrix groups:

$$\begin{pmatrix} \cos(\theta) & -\sin(\theta) & 0 \\ \sin(\theta) & \cos(\theta) & 0 \\ 0 & 0 & 1 \end{pmatrix}, \begin{pmatrix} \cosh(\theta) & \sinh(\theta) & 0 \\ \sinh(\theta) & \cosh(\theta) & 0 \\ 0 & 0 & 1 \end{pmatrix}, \begin{pmatrix} 1 & 0 & 0 \\ \theta & 1 & 0 \\ \frac{\theta^2}{2} & \theta & 1 \end{pmatrix}. \tag{3}$$

In the three cases of Equation (3), the invariant line is the z-axis. We name the rotations defined in each case as *elliptic* (left-most matrix, which defines a classical rotation about the z-axis), *hyperbolic* (center matrix, which defines a hyperbolic rotation about the z-axis), and *parabolic* (right-most matrix). The surfaces which, after perhaps an orthogonal change of coordinates $\mathcal{T}$, are invariant under one of the matrix groups in Equation (3) are called *affine rotation surfaces*; furthermore, in this case the preimage under $\mathcal{T}$ of the z-axis is called the *affine axis of rotation* of the surface. We say that an affine rotation surface is of elliptic, hyperbolic or parabolic type depending on the form of the matrix group. If the surface is algebraic, then we say that the surface is an *algebraic affine rotation surface*. Notice that the affine rotation surfaces of elliptic type are the classical surfaces of revolution.

Every affine rotation surface about the z-axis can be parametrized locally around a regular point using differentiable functions $f(s), g(s)$ as

$$\mathbf{x}(\theta, s) = Q_\theta \cdot [f(s), 0, g(s)]^T, \tag{4}$$

where $[f(s), 0, g(s)]^T$ parametrizes a *directrix curve* and Q_θ corresponds to one of the uniparametric matrix groups in Equation (3). We will refer to this representation as the *standard* form of the surface. Using the standard form, the curves $\mathbf{x}(\theta_0, s)$ are called *meridians*, while the curves $\mathbf{x}(\theta, s_0)$ are called *parallel curves*. In particular, the directrix is a meridian. Moreover, according to [6], the parallel curves are (a) in the elliptic case, circles centered on the z-axis, contained in planes normal to the z-axis; (b) in the hyperbolic case, equilateral hyperbolae centered on the z-axis, contained in planes normal to the z-axis, with parallel asymptotes; (c) in the parabolic case, parabolas placed in planes normal to the x-axis, with parallel axes, whose major axis is parallel to the z-axis.

Affine normals can be used to characterize affine rotation surfaces [11]. Before providing this characterization, we need to introduce two more properties.

Definition 1. *Let S be a surface which under an orthogonal change of coordinates $\mathcal{T}$, can be locally parametrized as* $\mathbf{y}(\theta, s) = A_\theta \cdot [f(s), 0, g(s)]^T$, *where A_θ is a 3×3 matrix depending on a parameter θ, and let $\mathcal{A}$ be the preimage of the z-axis under the transformation $\mathcal{T}$. We say that S has the* shadow line *property with respect to the line $\mathcal{A}$, if along every meridian $\mathbf{y}(\theta_0, s)$ the tangents to the parallel curves $\mathbf{y}(\theta, s_0)$ are parallel.*

For instance, one can see that surfaces of revolution always have the shadow line property with respect to its axis of revolution.

Definition 2. *We say that a non-developable surface S is* ANIL *(Affine Normals Intersecting a same Line), or that S has the* ANIL *property, if all the affine normal lines of S intersect a same line $\mathcal{A}$, called the* axis *of S.*

Then we have the following theorem (see [11] for a proof), which characterizes affine rotation surfaces.

Theorem 1. *The surface S is an affine rotation surface with affine axis $\mathcal{A}$ if and only if the following two conditions hold: (1) S is ANIL, with axis $\mathcal{A}$; (2) S has the shadow line property with respect to the line $\mathcal{A}$.*

From Theorem 1, it is clear that every affine rotation surface is ANIL. The converse, however, is not true: in [11] it is observed that there exist *affine spheres*, i.e., surfaces where all the affine normal lines intersect at one point (for instance, ellipsoids), called the center of the sphere, which are not affine rotation surfaces. Since all the affine normals of an affine sphere intersect at the center of the sphere, the affine normals obviously intersect every line through the center, so every affine sphere is an ANIL surface. In this paper, however, we will discover that there are many ANIL surfaces which are not affine rotation surfaces, or affine spheres: in fact, in Section 5 we will see that the images of surfaces of revolution under most nonsingular affine mappings are exactly like this.

2.2. Plücker Coordinates

Theorem 1 can be used to device an algorithm for detecting whether a given algebraic surface is an affine rotation surface, and to find the affine axis, in the affirmative case [11]. In order to do so, the key question is to efficiently exploit Condition (1) in Theorem 1, i.e., the fact that all the affine normal lines intersect the affine axis. This can be done using *Plücker coordinates* [12,15], which we recall in this subsection.

Plücker coordinates provide an alternative way to represent straight lines. A line $L \subset \mathbb{R}^3$ is completely determined when we know a point $P \in L$ and a vector w parallel to L. Therefore, we often write $L = (P, w)$. Now let $\overline{w} = P \times w$, where P here denotes the vector connecting the point P with the origin of the coordinate system. Then the *Plücker coordinates* of L are $(w, \overline{w}) \in \mathbb{R}^6$. Notice that by construction $w \cdot \overline{w} = 0$; this equation defines a quadric in $\mathbb{R}^6$ known as the *Klein quadric*.

Plücker coordinates of lines are unique up to multiplication by a constant nonzero factor. Moreover $\overline{w}$ is independent of the choice of the point $P \in L$, since if $Q \in L$, then $(Q - P) \times w = 0$. Furthermore, given the Plücker coordinates $(w, \overline{w})$ of L, we can recover a point P on L from the relationship

$$P \times w = \overline{w}, \qquad (5)$$

by writing $P = (x, y, z)$ and solving the system of linear Equations (5) for x, y, z. An alternative to solving this system of linear equations is simply to compute the pedal point $w \times \overline{w} \langle w, w \rangle^{-1}$ on the line $(w, \overline{w})$.

Let (α, β) be the Plücker coordinates of a line in $\mathbb{R}^3$, and consider all the lines $(w, \overline{w})$, written in Plücker coordinates, such that

$$\alpha \cdot \overline{w} + \beta \cdot w = 0. \qquad (6)$$

This equation (see [15,16]) expresses the condition that the lines $(w, \overline{w})$ intersect the line (α, β), so these lines span a hyperplane of $\mathbb{R}^6$. Thus, Equation (6) provides an efficient way of managing Condition (1) in Theorem 1, and therefore of detecting whether a given rational surface is ANIL: given a rational surface rationally parametrized by $\mathbf{x}(t,s)$, one can compute the affine normal line at several points $\mathbf{x}(t_i, s_i)$, where $(t_i, s_i) \in \mathbb{R}^2$. From Equation (6), each point gives a linear condition on the Plücker coordinates (α, β) of a potential line $\mathcal{A}$, intersected by all the affine normal lines of the surface. Solving the linear system of equations corresponding to all these linear conditions, the coordinates (α, β) can be efficiently computed. This, for instance, is used in [11] in order to detect whether an algebraic surface is an affine surface of rotation.

3. Affine Image of a Rational Surface of Revolution (I)

The goal of this section is to prove that the image under a nonsingular affine mapping f of a surface of revolution about an axis $\mathcal{A}$ is an ANIL surface of axis $\widehat{\mathcal{A}} = f(\mathcal{A})$. Notice that from Theorem 1, this is a necessary condition for a surface to be an affine surface of rotation. Later, in Section 5, we will explore in what cases the image of a surface of revolution under a nonsingular affine mapping is an affine surface of rotation.

In order to do this, we let $S \subset \mathbb{R}^3$ be a rational ANIL surface, rationally parametrized by $\mathbf{x}(t,s)$, where t, s are parameters, and we let $f(\mathbf{x}) = A\mathbf{x} + \mathbf{b}$, where $A \in \mathcal{M}_{3 \times 3}(\mathbb{R})$ and $\mathbf{b} \in \mathbb{R}^3$. We denote $\widehat{S} = f(S)$. By definition, if S is a surface of revolution about an axis $\mathcal{A}$ then S is an affine rotation surface of elliptic type about the affine axis $\mathcal{A}$. Hence, by Theorem 1 all the affine normals of S intersect the line $\mathcal{A}$, so S is an ANIL surface of axis $\mathcal{A}$.

For now we will assume that S is not developable; some considerations about developable surfaces will be made at the end of this section. Observe that a developable surface (see [17]) can always be, at least locally, parametrized as $\mathbf{y}(u,v) = \mathbf{a}(u) + v\mathbf{c}(u)$ where $[\mathbf{a}'(u), \mathbf{c}(u), \mathbf{c}'(u)] = 0$, so the vectors $\{\mathbf{a}'(u), \mathbf{c}(u), \mathbf{c}'(u)\}$ are coplanar. Since the images of these vectors under a nonsingular mapping $g(\mathbf{x}) = A\mathbf{x}$ are also coplanar, a surface is developable if and only if its image under a nonsingular mapping $g(\mathbf{x}) = A\mathbf{x}$ is also developable. Since translations are isometries, and therefore preserve the property of being developable, we deduce that a surface is developable if and only if the image of the surface under every nonsingular affine mapping $f(\mathbf{x}) = A\mathbf{x} + \mathbf{b}$ is also developable. In particular, and since we are assuming that S is not developable, $\widehat{S}$ is not developable either. Thus, the affine normal lines of both S and $\widehat{S}$ are well defined.

Furthermore, we will need the following technical lemma.

Lemma 1. *Let $A \in \mathcal{M}_{3\times 3}(\mathbb{R})$ be nonsingular. Then $A = kB$, where $k \in \mathbb{R}$ and $det(B) = 1$.*

Proof. Let $k = \sqrt[3]{\det(A)}$. Since $\det(A) \neq 0$, $k \neq 0$ too. Let $B = \frac{1}{k}A$. Then $\det(B) = \frac{1}{k^3} \cdot \det(A) = \frac{1}{k^3} \cdot k^3 = 1$. □

In order to show that $\widehat{S} = f(S)$ is an ANIL surface, we first consider the image $\widetilde{S}$ of S under a homothety $\widetilde{f}(\mathbf{x}) = k\mathbf{x}$ with $k \in \mathbb{R} - \{0\}$; we denote $\widetilde{S} = \widetilde{f}(S)$.

Lemma 2. *Let $S \subset \mathbb{R}^3$ be an ANIL surface of axis $\mathcal{A}$ rationally parametrized by $\mathbf{x}(t,s)$ which is not developable, and let $\widetilde{S}$ be the image of S under a homothety $\widetilde{f}(\mathbf{x}) = k\mathbf{x}$, $k \in \mathbb{R} - \{0\}$. Then $\widetilde{S} = \widetilde{f}(S)$ is an ANIL surface of axis $\widetilde{\mathcal{A}} = \widetilde{f}(\mathcal{A})$.*

Proof. First we need to consider the relationship between the affine normal lines of S and $\widetilde{S}$. In order to do this, observe that $\mathbf{y}(t,s) = k\mathbf{x}(t,s)$ parametrizes $\widetilde{S}$. Let us denote by $K_\mathbf{x}, \mathbf{N_x}, K_\mathbf{y}, \mathbf{N_y}$ the Gauss curvatures and unitary normal vectors of S and $\tilde{S}$. And let us also denote by $\mu_x, \xi_x, \mu_y, \xi_y$ the affine co-normal vectors and the affine normal vectors of S and $\tilde{S}$. One can check that

$$\mathbf{N_y} = \mathbf{N_x}, \quad K_y = \frac{1}{k^2} K_x, \quad \mu_y = \sqrt{|K_x|}\mu_x, \quad \xi_y = \frac{1}{\sqrt{|K_x|}}\xi_x. \tag{7}$$

Notice that these equalities describe the relationship between the unitary normal, co-normal and affine normal vector of $\tilde{S}$ at the point $\mathbf{y}(t,s)$, and the corresponding vector of S at the point $\mathbf{x}(t,s)$; similarly for the Gaussian curvatures. Furthermore, since by hypothesis S is not developable, the affine normal vectors of both S and $\tilde{S}$ are well defined.

Now let $P \in \mathcal{A}$, and let $\overline{w}$ be a vector parallel to $\mathcal{A}$. Then $(\alpha, \beta) = (\overline{w}, P \times \overline{w}) = (\alpha, \beta)$, where P denotes the vector connecting the point P and the origin of the coordinate system, are the Plücker coordinates of the line $\mathcal{A}$. Thus, the Plücker coordinates of the line $\tilde{\mathcal{A}} = \tilde{f}(\mathcal{A})$ are $(\overline{w}, kP \times \overline{w}) = (\alpha, k\beta)$.

Since S is an ANIL surface about the axis $\mathcal{A}$, from Equation (6) we have

$$\alpha(\mathbf{x} \times \xi_\mathbf{x}) + \beta \xi_\mathbf{x} = 0. \tag{8}$$

Taking into account that the Plücker coordinates of the line $\tilde{\mathcal{A}} = \tilde{f}(\mathcal{A})$ are $(\overline{w}, kP \times \overline{w}) = (\alpha, k\beta)$, and using Equations (7) and (8), we get

$$\alpha(\mathbf{y} \times \xi_\mathbf{y}) + k\beta \xi_\mathbf{y} = \alpha\left(k\mathbf{x} \times \frac{1}{\sqrt{|K_x|}}\xi_\mathbf{x}\right) + k\beta \frac{1}{\sqrt{|K_x|}} \xi_\mathbf{x} = \frac{K_x}{\sqrt{|K_x|}}[\alpha(\mathbf{x} \times \xi_\mathbf{x}) + \beta \xi_\mathbf{x}] = 0 \tag{9}$$

Hence, again from Equation (6) we conclude that the affine normal lines of $\tilde{S}$ all intersect the line $\tilde{\mathcal{A}}$. □

Now we consider the image $S^\star$ of S under a translation $f^\star(x) = x + b$, with $b \in \mathbb{R}^3$.

Lemma 3. *Let $S \subset \mathbb{R}^3$ be a ANIL surface of axis $\mathcal{A}$ rationally parametrized by $\mathbf{x}(t,s)$ which is not developable, and let $S^\star$ be the image of S under a translation $f^\star(x) = x + b$, $b \in \mathbb{R}^3$. Then $S^\star = f^\star(S)$ is an ANIL surface of axis $\mathcal{A}^\star = f^\star(\mathcal{A})$.*

Proof. Observing that $\mathbf{y}(t,s) = \mathbf{x}(t,s) + b$ parametrizes $S^\star$, we get that

$$\mathbf{N_y} = \mathbf{N_x}, \quad K_y = K_x, \quad \mu_y = \mu_x, \quad \xi_y = \xi_x, \tag{10}$$

where these equalities describe the relationships between the unitary normal, co-normal and affine normal vector of $S^\star$ at the point $\mathbf{y}(t,s)$, and the corresponding vector of S at the point $\mathbf{x}(t,s)$; similarly for the Gaussian curvatures. Furthermore, since by hypothesis S is not developable, the affine normal vectors of both S and $S^\star$ are well defined. Then we argue as in the proof of Lemma 1. □

Finally we can prove the main result of this section.

Theorem 2. *Let $S \subset \mathbb{R}^3$ be an ANIL surface of axis $\mathcal{A}$ rationally parametrized by $\mathbf{x}(t,s)$ which is not developable, and let $\hat{S}$ be the image of S under a nonsingular affine mapping $f(x) = Ax + b$, $A \in \mathcal{M}_{3\times 3}(\mathbb{R})$, $b \in \mathbb{R}^3$. Then $\hat{S} = f(S)$ is an ANIL surface of axis $\hat{\mathcal{A}} = f(\mathcal{A})$.*

Proof. By Lemma 1, $A = kB$, where $k \in \mathbb{R} - \{0\}$ and $\det(B) = 1$; thus, $f(x) = kBx + b$. Let $f^\dagger$ be the linear mapping defined by $f^\dagger(x) = Bx$, and let $S^\dagger$ be the image of S under $f^\dagger$, i.e., $S^\dagger = f^\dagger(S)$. By the covariance property of affine normal lines, the affine normal lines of $S^\dagger$ are the images of the affine normal lines of S under $f^\dagger$. Since by hypothesis all the affine normal lines of S intersect $\mathcal{A}$, and since linear mappings preserve incidence, all the affine normal lines of $S^\dagger$ intersect the line $\mathcal{A}^\dagger = f^\dagger(\mathcal{A})$. Then the result follows from Lemma 2 and Lemma 3. □

Corollary 1. *Let $S \subset \mathbb{R}^3$ be a rational surface of revolution about an axis $\mathcal{A}$, and assume that S is not developable. Let $f(x) = Ax + b$, $A \in \mathcal{M}_{3\times 3}(\mathbb{R})$ nonsingular, $b \in \mathbb{R}^3$. Then the image of S under the mapping f is an ANIL surface of axis $f(\mathcal{A})$.*

Furthermore, since affine mappings preserve incidence, we also have the following corollary of Theorem 2 on affine spheres.

Corollary 2. *Let $S \subset \mathbb{R}^3$ be an affine sphere of center $\mathbf{c}$ rationally parametrized by $\mathbf{x}(t,s)$ which is not developable, and let $\widehat{S}$ be the image of S under a nonsingular affine mapping $f(x) = Ax + b$, $A \in \mathcal{M}_{3\times 3}(\mathbb{R})$, $b \in \mathbb{R}^3$. Then $\widehat{S} = f(S)$ is an affine sphere of center $\widehat{\mathbf{c}} = f(\mathbf{c})$.*

The Case of Developable Surfaces

Let $S \subset \mathbb{R}^3$ be a developable surface, in which case the Gaussian curvature is zero. Since the affine normal line is not defined when the Gassian curvature is zero, the notion of an ANIL surface is not applicable to these surfaces. However, some considerations can be done in the case when S is a surface of revolution. Without loss of generality we assume that the axis of revolution of S is the z-axis. A first obvious possibility is that S is a cylinder of revolution, and therefore a quadric. If S is not a cylinder of revolution, then S admits (see Section 15.1 of [17]) an, at least local, parametrization of S as

$$\mathbf{x}(\rho, \gamma) = (\rho \cos \gamma, \rho \sin \gamma, h(\rho)).$$

Additionally, imposing that the Gaussian curvature of S is identically zero, one can see (e.g., Section 15.3 of [17]) that $h(\rho) = C_1 \rho + C_2$, with C_1, C_2 constants, C_1 nonzero, so S is a cone of revolution: indeed, eliminating ρ, γ in

$$x = \rho \cos \gamma, \; y = \rho \sin \gamma, \; z = C_1 \rho + C_2,$$

one gets $x^2 + y^2 = \left(\frac{z - C_2}{C_1}\right)^2$, which shows that S is a cone of revolution. Since affine mappings preserve incidence and parallelism, one deduces that the image of a developable surface of revolution under a nonsingular affine mapping is either cylindrical, i.e., a ruled surface whose generatrices are all of them parallel, or conical, i.e., a ruled surface whose generatrices intersect at a point, named the vertex of the surface. Furthermore, since affine mappings preserve the degree of the surface, it must also be a quadric.

4. Computing a Surface of Revolution Affinely Equivalent to an ANIL Surface

Given an ANIL surface $S_1 \subset \mathbb{R}^3$, rationally parametrized by $\mathbf{x}(t,s)$, we aim to find an algorithm to solve the following problem: find, if it exists, a rational surface of revolution $S_2 \subset \mathbb{R}^3$ which is *affinely equivalent* to S_1, i.e., such that there is a nonsingular affine mapping $f(x) = Ax + b$, where $A \in \mathcal{M}_{3\times 3}(\mathbb{R})$ and $b \in \mathbb{R}^3$, satisfying that $f(S_1) = S_2$. We say that f is an *affine equivalence* between S_1, S_2. Notice that certainly S_2 is not unique, since by composing f with any *similarity* h, the surface $(h \circ f)(S_1)$ is also a surface of revolution affinely equivalent to S_1; recall that similarities are the composition of a congruence (also called rigid motion, a mapping preserving distances) and a homothety (which preserves angles and scales the objects).

In order to solve the problem, it is useful to recall the following theorem, characterizing algebraic surfaces of revolution. In this theorem we consider classical normals, and not affine normals. We will need to apply this theorem on the surface S_2 we are seeking. Notice that by hypothesis S_1 is ANIL; since the notion of an ANIL surface is not applicable to developable surfaces, S_1 is not developable, and therefore S_2 is not developable either. In particular, S_2 is not cylindrical.

Theorem 3. *Let $S \subset \mathbb{R}^3$ be an algebraic surface which is not cylindrical. Then S is a surface of revolution about an axis $\mathcal{A}$ if and only if all the normals to the surface intersect the axis $\mathcal{A}$.*

Proof. See Theorem 4.2.1 and Lemma 4.2.2 of [12]. □

Observe that in Theorem 3 the hypothesis of S being algebraic is necessary: if the surface is not algebraic, the condition in the theorem implies that the surface is either a surface of revolution, or a *helical surface*, i.e., a surface invariant under a *helical motion* (see Section 3.1.2 of [12]). Helical motions are the mappings in $\mathbb{R}^3$ that can be written in a certain system of coordinates as $T(x) = Q_\theta x + [0, 0, p\theta]^T$, where Q_θ is the left-most matrix in Equation (3), and $p \neq 0$ (p is called the *pitch*). However, helical surfaces are not algebraic. Notice also in Theorem 3 that the condition on the shadow line property is not necessary. As it also happened with Theorem 1 and affine rotation surfaces, using Plücker coordinates one can use Theorem 3 to build an efficient algorithm for detecting surfaces of revolution (see e.g., [16]). In our case, Theorem 3 will be key in order to solve the problem we are addressing.

We still need some additional observations. First, by applying if necessary a translation followed by a rotation about a line, we can assume that the axis of S_1 is the z-axis. Furthermore, we can also assume that the axis of revolution of the surface S_2 we are looking for is the z-axis as well: since the composition of nonsingular affine mappings is a nonsingular affine mapping, if there exists a surface of revolution affinely equivalent to S_1, then there also exists a surface of revolution about the z-axis with the same property (one just needs to apply a congruence to reach this surface). Finally, since the composition of S_2 with any translation by a vector parallel to the z-axis also provides a surface of revolution about the z-axis, we can assume that the affine equivalence transforming S_1 into S_2 fixes the origin, so that $f(x) = Ax$. Our problem, then, is to find the matrix A: after computing A, the surface S_2 is immediately obtained.

Now if S_1 is parametrized by $\mathbf{x}(t,s)$ and $f(S_1) = S_2$, then $\mathbf{y}(t,s) = A\mathbf{x}(t,s)$ is a parametrization of S_2. In order to use Theorem 3, we consider the (classical) normals to S_2. Since $\mathbf{y}_t = A\mathbf{x}_t$, $\mathbf{y}_s = A\mathbf{x}_s$, and taking into account the well-known formula $M\mathbf{a} \times M\mathbf{b} = \det(M)M^{-T}(\mathbf{a} \times \mathbf{b})$ for $M \in \mathcal{M}_{3\times 3}(\mathbb{R})$, $\mathbf{a}, \mathbf{b} \in \mathbb{R}^3$, we get

$$\mathbf{y}_t \times \mathbf{y}_s = \det(A)A^{-T}(\mathbf{x}_t \times \mathbf{x}_s). \tag{11}$$

The Plücker coordinates of a generic normal line of S_1 are $(\alpha, \beta) = (\mathbf{x}_t \times \mathbf{x}_s, \mathbf{x} \times \mathbf{x}_t \times \mathbf{x}_s)$. From Equation (11), the Plücker coordinates of a generic normal line of S_2 are

$$(\mathbf{y}_t \times \mathbf{y}_s, \mathbf{y} \times \mathbf{y}_t \times \mathbf{y}_s) = (\det(A)A^{-T}(\mathbf{x}_t \times \mathbf{x}_s), A\mathbf{x} \times \det(A)A^{-T}(\mathbf{x}_t \times \mathbf{x}_s)). \tag{12}$$

Notice that since $\mathbf{y}(t,s)$ parametrizes a surface of revolution about the z-axis, $k\mathbf{y}(t,s)$ with $k \in \mathbb{R} - \{0\}$ parametrizes another surface of revolution about the z-axis too; we can prove it from Theorem 3, taking into account the relationship between the normals of the surfaces parametrized by $\mathbf{y}(t,s)$ and $k\mathbf{y}(t,s)$. This implies that we can assume $\det(A) = 1$. Therefore, and calling $\alpha = \mathbf{x}_t \times \mathbf{x}_s$, we get that the Plücker coordinates of a generic normal line of S_2 are

$$(A^{-T}\alpha, A\mathbf{x} \times A^{-T}\alpha) \tag{13}$$

Additionally, the Plücker coordinates of the z-axis, which is the axis of revolution of S_2, are $(\mathbf{0}, \mathbf{k})$, where $\mathbf{0} = (0,0,0)$ and $\mathbf{k} = (0,0,1)$. From Theorem 3, all normals to S_2 intersect the z-axis. Using Plücker coordinates, from Equation (6) this condition is translated into

$$A^{-T}\alpha \cdot \mathbf{0} + (A\mathbf{x} \times A^{-T}\alpha) \cdot \mathbf{k} = 0. \tag{14}$$

Let $\mathbf{Co}(A)$ be the cofactor matrix of A. Since we are assuming that $\det(A) = 1$, $A^{-T} = \mathbf{Co}(A)$. Then, Equation (14) is equivalent to

$$(A\mathbf{x} \times \mathbf{Co}(A)\alpha) \cdot \mathbf{k} = [A\mathbf{x}, \mathbf{Co}(A)\alpha, \mathbf{k}] = 0. \tag{15}$$

By Theorem 2, $f(\mathbf{x}) = A\mathbf{x}$ must preserve the z-axis, so $A\mathbf{k} = \lambda\mathbf{k}$ for $\lambda \neq 0$. Since additionally $\det(A) = 1$, we get that

$$A = \begin{pmatrix} a_{11} & a_{12} & 0 \\ a_{21} & a_{22} & 0 \\ a_{31} & a_{32} & a_{33} \end{pmatrix}, \tag{16}$$

where $a_{33}(a_{11}a_{22} - a_{12}a_{21}) = 1$. Since $a_{33} \neq 0$ (because otherwise A is singular), we can always assume that $a_{33} = 1$. Thus, we get

$$A = \begin{pmatrix} a_{11} & a_{12} & 0 \\ a_{21} & a_{22} & 0 \\ a_{31} & a_{32} & 1 \end{pmatrix}, \quad \mathbf{Co}(A) = \begin{pmatrix} a_{22} & -a_{21} & a_{21}a_{32} - a_{22}a_{31} \\ -a_{12} & a_{11} & -a_{11}a_{32} + a_{12}a_{31} \\ 0 & 0 & a_{11}a_{22} - a_{12}a_{21} \end{pmatrix} \tag{17}$$

Substituting the expressions for A and $\mathbf{Co}(A)$ into Equation (15), and adding the equation

$$a_{11}a_{22} - a_{12}a_{21} = 1, \tag{18}$$

we get cubic equations in $a_{11}, a_{12}, a_{21}, a_{22}, a_{31}, a_{32}$ which define an algebraic variety $\mathcal{V} \subset \mathbb{C}^6$. Any real point of $\mathcal{V}$ provides a matrix A with the desired property. So throughout the section we have proven the following result. In turn, this result provides the Algorithm 1, which solves the problem considered in this section.

Theorem 4. *Let $S_1 \subset \mathbb{R}^3$ be an ANIL surface whose axis is the z-axis. Then S_1 is affinely equivalent to a surface of revolution if and only if $\mathcal{V} \cap \mathbb{R}^6 \neq \emptyset$.*

Remark 1. *In fact, if $\mathcal{V} \cap \mathbb{R}^6 \neq \emptyset$ then $\mathcal{V}$ must contain at least a real curve, since rotating a surface of revolution S_2 with the desired properties around the z-axis also yields a surface of revolution.*

In practice, instead of deriving a system of cubic equations directly from Equation (15), it is cheaper from the computational point of view to substitute points (t_i, s_i) into Equation (15) to generate equations. The system of cubic equations derived this way can be solved by using computer algebra methods, e.g., Gröbner bases. In our case, we used the computer algebra system Maple 17, and the `Groebner` package.

Algorithm 1 Revol.

Require: A non-developable ANIL surface S_1, rationally parametrized by $\mathbf{x}(t,s)$.
Ensure: A rational surface of revolution S_2 affinely equivalent to S_1, or a certificate of its non-existence.
1: Substitute the entries of A and $\mathbf{Co}(A)$ from Equation (17) into Equation (15).
2: Solve the cubic polynomial system $\mathcal{S}$ in a_{ij}, $i \in \{1,2,3\}, j \in \{1,2\}$, consisting of the equations derived in Step 1, and Equation (18).
3: **if** the system does not have any real solution **then**
4:     return "there is no surface of revolution affinely equivalent to the surface"
5: **else**
6:     pick a real solution $a_{11}, a_{12}, a_{21}, a_{22}, a_{31}, a_{32}$ of $\mathcal{S}$.
7:     return the surface S_2 parametrized by $\mathbf{y}(t,s) = A\mathbf{x}(t,s)$, where A is the matrix in the left-hand side of Equation (17) whose entries correspond to the solution in Step 6.
8: **end if**

Example 1. *Let S_1 be the sextic surface, rationally parametrized by*

$$\mathbf{x}(t,s) = \left(-\frac{2(s^3t^2 - s^3t + s^2t^2 - s^3 + s^2 + t^2 - t - 1)}{t^2+1}, -\frac{(s^3+1)(t^2-2t-1)}{t^2+1}, -\frac{(s^3+1)(t^2-4t-1)}{t^2+1}\right).$$

Using Plücker coordinates, one can see that S_1 is an ANIL surface, and that the axis is the x-axis. Additionally, one can check that the implicit equation of the surface has the form

$$F(x,y,z) = (x - 3y + z)^6 + l.o.t.,$$

where l.o.t. stands for lower order terms. Since the form of highest order of an affine surface of rotation has a very specific structure (see Theorem 6 in [8], Theorem 6 in [10], Theorem 6 in [9]), we deduce that S_1 is not an affine surface of rotation. In order to compute cubic equations defining the variety $\mathcal{V}$, we consider Equation (15) for the points corresponding to (t_i, s_i) with t_i, s_i ranging from -3 to 3. The first of these equations is

$$912600a_{11}^2 a_{32} - 912600a_{11}a_{12}a_{31} + 638820a_{11}a_{12}a_{32} - 638820a_{12}^2 a_{31} + 912600a_{21}^2 a_{32}$$
$$-912600a_{21}a_{22}a_{31} + 638820a - 21a_{22}a_{32} - 638820a - 22^2 a - 31 + 2332200a_{11}^2$$
$$+2464020a_{11}a_{12} + 582036a_{12}^2 + 2332200a - 21^2 + 2464020a_{21}a_{22} + 582036a_{22}^2 = 0.$$

Adding also Equation (18), we get 50 cubic equations. Maple solves the polynomial system consisting of these equations in 0.265 s, and yields the following families of real solutions (there are also some complex solutions, which we do not list):

$$a_{11} = \lambda, a_{12} = -\frac{1}{2}a_{21} - \frac{3}{2}\lambda, a_{22} = -\frac{3}{2}a_{21} + \frac{1}{2}\lambda, a_{31} = 1, a_{32} = -3,$$

where λ satisfies that $\lambda^2 + a_{21}^2 - 2 = 0$, and

$$a_{11} = 0, a_{12} = -\frac{1}{2}\mu, a_{21} = \mu, a_{22} = -\frac{3}{2}\mu, a_{31} = 1, a_{32} = -3,$$

where μ satisfies that $\mu^2 - 2 = 0$. Picking $a_{21} = 1$, $\lambda = 1$ in the first family, we get

$$A = \begin{pmatrix} 1 & -2 & 0 \\ 1 & -1 & 0 \\ 1 & -3 & 1 \end{pmatrix}.$$

The affine mapping $f(x) = Ax$ maps S_1 onto the surface S_2 parametrized by

$$\mathbf{y}(t,s) = \left(\frac{(s^3+1)(t^2-1)}{t^2+1}, \frac{2(s^3+1)t}{t^2+1}, -2s^2 \right),$$

which one can recognize as the surface of revolution generated by rotating the cubic curve parametrized by $(s^3 + 1, 0, -2s^2)$ about the z-axis.

5. Affine Image of a Surface of Revolution (II)

In this section, we want to explore under what circumstances the image of a surface of revolution under a nonsingular affine mapping is an affine surface of rotation. In order to do this, we will use the preceding notations, and we will benefit from certain observations done in Section 4.

Let S_1, S_2 be two rational surfaces, none of them developable, S_1 an ANIL surface, S_2 a surface of revolution, related by a nonsingular affine mapping. Following the observations in Section 4, without loss of generality we can assume that the the affine axis of S_1 is the z-axis, the axis of revolution of S_2 is the z-axis as well, and that the nonsingular affine mapping transforming S_1 into S_2 has the form $f(x) = Ax$. Even more, we can assume that the matrix A has the form in Equation (17), and that the entries of the matrix A also satisfy Equation (18). We will separately consider the cases when S_1 is an elliptic, hyperbolic or parabolic affine surface of rotation. We begin with the parabolic and the hyperbolic cases, and we conclude with the elliptic case. In what follows, the reader is invited to review the notion of *parallel curve* of an affine rotation surface, recalled in Section 2.1.

5.1. The Parabolic Case

If S_1 is a parabolic affine rotation surface about the z-axis, we can assume (see Section 2.1) that the parallel curves are placed in planes normal to the x-axis, i.e., planes $x = x_0$, $x_0 \in \mathbb{R}$, that we denote by Π_{x_0}. Furthermore, in that case the intersection $\Pi_{x_0} \cap S_1$ is a union of parabolas lying on planes parallel to the yz-plane, and whose major axes are parallel to the z-axis. We are interested in finding the images of the planes Π_{x_0} under the mapping $f(x) = Ax$. Thus, we have

$$\begin{pmatrix} a_{11} & a_{12} & 0 \\ a_{21} & a_{22} & 0 \\ a_{31} & a_{32} & 1 \end{pmatrix} \cdot \begin{pmatrix} x_0 \\ \lambda \\ \mu \end{pmatrix} = \begin{pmatrix} a_{11}x_0 + a_{12}\lambda \\ a_{21}x_0 + a_{22}\lambda \\ a_{31}x_0 + a_{32}\lambda + \mu \end{pmatrix}. \qquad (19)$$

Elliminating the parameters λ, μ, and since $a_{11}a_{22} - a_{12}a_{21} = 1$, we get the plane $a_{22}x - a_{12}y - x_0 = 0$, that is parallel to the z-axis, and which we denote by $\widehat{\Pi}_{x_0}$. Since $f(x) = Ax$ is an affine mapping $\widehat{\Pi}_{x_0} \cap S_2$ must be a union of parabolas as well. Since S_2 is a surface of revolution about the z-axis, we deduce that S_2 is generated by rotating parabolas around the z-axis, so S_2 must be the union of several paraboloids of revolution. Because S_2 is rational and therefore irreducible, we get that S_2 must be a paraboloid of revolution, so $S_1 = f(S_2)$ must also be a paraboloid. But this is a contradiction, because from Corollary 5 in [10] the only quadrics that are affine surfaces of rotation of parabolic type are either cones (which are developable surfaces), or hyperboloids. Therefore, we have proved the following result.

Theorem 5. *The affine image of a rational surface of revolution that is not developable cannot be an affine surface of rotation of parabolic type.*

5.2. The Hyperbolic Case

Let S_1 be a hyperbolic affine surface of rotation about the z-axis. Then the parallel curves are placed in planes $z = z_0$, that we denote by Π_{z_0}. Proceeding as in Section 5.1, we can check that f maps Π_{z_0} onto the plane $\widehat{\Pi}_{z_0}$, defined by

$$A_{13}x + A_{23}y + (z - z_0) = 0, \tag{20}$$

where A_{ij} represents the cofactor of the element (i, j) of the matrix A. Since the coefficient of z in Equation (20) is nonzero, $\widehat{\Pi}_{z_0}$ is not parallel to the z-axis. Furthermore, since f is affine and $S_2 \cap \Pi_{z_0}$ is a union of equilateral hyperbolae, $f(S_1 \cap \Pi_{z_0}) = S_2 \cap \widehat{\Pi}_{z_0}$ must also be a union of hyperbolae. Additionally, since S_2 is a surface of revolution we can see S_2 as generated by rotating $S_2 \cap \widehat{\Pi}_{z_0}$ around the z-axis. We want to see that this cannot be.

In order to do that, assume first that $A_{13} = A_{23} = 0$. Then, Equation (20) corresponds to a horizontal plane, i.e., normal to the z-axis. Since S_2 is a surface of revolution about the z-axis, the horizontal sections of S_2 are unions of circles centered at the points on the z-axis. Since $S_1 \cap \widehat{\Pi}_{z_0}$ is a union of hyperbolas, this cannot happen. So let us focus on the case where A_{13}, A_{23} are not both zero, in which case the plane in Equation (20) is not horizontal. We need the following previous result.

Lemma 4. *Let S be a rational surface of revolution about the z-axis, and let $\mathcal{D}$ be a rational planar curve contained in a planar section $S \cap \Pi$ of the surface S, where Π is not normal to the z-axis. Then S is the surface obtained by rotating $\mathcal{D}$ about the z-axis.*

Proof. By rotating $\mathcal{D}$ around the z-axis we get a rational surface $S' \subset S$. Since S and S' are rational and therefore irreducible, $S = S'$. □

Now assume that $S_2 = f(S_1)$, where f is an affinity, is a surface of revolution about the z-axis, and consider two planes Π_{z_0} and Π_{z_1}, defined by $z = z_0$ and $z = z_1$, where $z_0 \neq z_1$. Let $\widehat{\Pi}_{z_0}, \widehat{\Pi}_{z_1}$ be the images of Π_{z_0}, Π_{z_1} under f. Notice that $S_1 \cap \Pi_{z_0}, S_1 \cap \Pi_{z_1}$ are unions of circles, so $\mathcal{C}_0 = f(S_1 \cap \Pi_{z_0})$, $\mathcal{C}_1 = f(S_1 \cap \Pi_{z_1})$ are unions of hyperbolas. Furthermore, since $z_0 \neq z_1, \mathcal{C}_0 \neq \mathcal{C}_1$.

From Lemma 4, and since S_2 is rational and therefore irreducible, the surface S_2 should be obtained both by rotating a rational component of $\mathcal{C}_0$ around z, and by rotating a rational component of $\mathcal{C}_1$ around z. We want to see that this is not possible, i.e., that by rotating such components we generate different surfaces, not the same surface. For simplicity, we will assume that $\mathcal{C}_0$ and $\mathcal{C}_1$ are hyperbolae, and not unions of hyperbolae; were this not the case, it suffices to consider one rational component in each case.

The situation is shown in Figure 1: in more detail, the notation in Figure 1 represents the following:

- $\mathcal{C}_0 = f(S_1 \cap \Pi_{z_0}), \mathcal{C}_1 = f(S_1 \cap \Pi_{z_1})$.
- $\mathcal{P}$ is a horizontal plane, i.e., normal to the z-axis, through one of the vertices of $\mathcal{C}_0$.

Furthermore, Figure 2 represents the plane $\mathcal{P}$ seen from above. The notation in Figure 2 represents the following:

- The point **P** is a vertex of $\mathcal{C}_0$. Furthermore, **P** is the only intersection of $\mathcal{P}$ with $\mathcal{C}_0$.
- The lines $\mathcal{L}_1$ and $\mathcal{L}_2$ are the intersections of the planes $\widehat{\Pi}_{z_0}, \widehat{\Pi}_{z_1}$ with the horizontal plane $\mathcal{P}$. These lines are also shown in blue in Figure 1. Notice that since $\widehat{\Pi}_{z_0}, \widehat{\Pi}_{z_1}$ are parallel, $\mathcal{L}_1, \mathcal{L}_2$ are parallel too.

- The points $\mathbf{Q_1}, \mathbf{Q_2}$ are the intersections of the curve $\mathcal{C}_1$ with the plane $\widehat{\Pi}_{z_1}$; it could happen that $\mathbf{Q_1} = \mathbf{Q_2}$, or even that the intersection of $\mathcal{C}_1$ with the plane $\widehat{\Pi}_{z_1}$ was empty, but in those cases we would obtain contradictions as well.
- The point $\mathbf{C}$ is the intersection of the z-axis with the horizontal plane $\mathcal{P}$.
- The circle in red, $\tilde{\mathcal{C}}$, is the circle through the point $\mathbf{P}$, centered at $\mathbf{C}$; this circle is also shown in red in Figure 1.

The following result certifies that the picture shown in Figure 2 is correct:

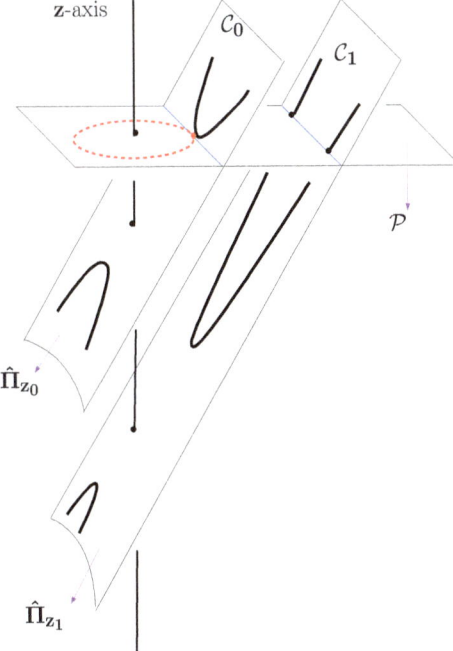

Figure 1. The case of hyperbolic affine rotation surfaces (I).

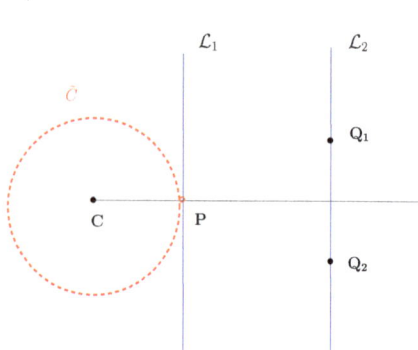

Figure 2. The case of hyperbolic affine rotation surfaces (II).

Lemma 5. *The line connecting* **C** *and* **P** *is perpendicular to the line* $\mathcal{L}_1$.

Proof. Assume that the line connecting **C**, which is the intersection of the z-axis and the plane $\mathcal{P}$, and **P**, which is the vertex of the hyperbola $\mathcal{C}_0$ contained in the plane $\mathcal{P}$, is not perpendicular to $\mathcal{L}_1$. Then the circle $\tilde{C}$ centered at **C** through **P** is not tangent to $\mathcal{L}_1$, and therefore there is another intersection point $\mathbf{P}' \neq \mathbf{P}$ of the circle $\tilde{C}$ with the line $\mathcal{L}_1$. However, since S_2 is generated by rotating the curve $\mathcal{C}_0$ around the z-axis, then $\mathbf{P}' \in S_2$. Even more, since $\mathcal{C}_0 = \widehat{\Pi}_{z_0} \cap S_2$ and $\mathbf{P}' \in \mathcal{L}_1 \subset \widehat{\Pi}_{z_0}$, we get that $\mathbf{P}' \in \mathcal{C}_0$. Furthermore, since $\mathcal{L}_1 \subset \mathcal{P}$, $\mathbf{P}' \in \mathcal{P}$, so $\mathbf{P}' \in \mathcal{C}_0 \cap \mathcal{P}$. However, since $\mathcal{P}$ is the horizontal plane through the vertex **P**, the only point of $\mathcal{C}_0 \cap \mathcal{P}$ is **P**, and therefore $\mathbf{P}' = \mathbf{P}$. □

We also need the following lemma.

Lemma 6. *Let* $d(\mathbf{C}, \mathbf{Q_i})$, *with* $i = 1, 2$, *denote the distance between* $\mathbf{C}, \mathbf{Q_i}$, *let* $d(\mathbf{C}, \mathcal{L}_j)$, *with* $j = 1, 2$, *denote the distance between the point* **C** *and the line* $\mathcal{L}_j$, *and let* $\overline{\mathbf{CP}}$ *denote the segment connecting* **C** *and* **P**. *Then* $d(\mathbf{C}, \mathbf{Q_i}) > \overline{\mathbf{CP}}$.

Proof. Since $\mathcal{L}_1, \mathcal{L}_2$ are parallel, $d(\mathbf{C}, \mathcal{L}_2) > d(\mathbf{C}, \mathcal{L}_1)$. Furthermore, $d(\mathbf{C}, \mathbf{Q_i}) \geq d(\mathbf{C}, \mathcal{L}_2)$. Thus, $d(\mathbf{C}, \mathbf{Q_i}) > d(\mathbf{C}, \mathcal{L}_1)$. But from Lemma 5, $d(\mathbf{C}, \mathcal{L}_1) = \overline{\mathbf{CP}}$. □

Corollary 3. *The circle* $\tilde{C}$ *centered at* **C** *of radius* $\overline{\mathbf{CP}}$ *is not contained in the set generated by rotating* $\mathcal{C}_1$ *around the z-axis.*

Now we can prove the following result. Here we use the preceding notation, and the help of Figures 1 and 2.

Theorem 6. *The affine image of a rational surface of revolution that is not developable cannot be an affine surface of rotation of hyperbolic type.*

Proof. Without loss of generality, we reduce to the situation analyzed before. We have already seen that A_{13}, A_{23} cannot be both zero, so we can assume that Equation (20) defines a plane which is neither horizontal, nor parallel to the z-axis, in which case we can use our last observations. In particular, if $S_2 = f(S_1)$, where f is an affinity, is a surface of revolution about the z-axis, S_2 is generated by both curves C_0 and C_1, defined before. However, the surface obtained by rotating C_0 about the z-axis contains the circle $\tilde{C}$. But from Corollary 3, $\tilde{C}$ is not contained in the surface generated by rotating C_1 about the z-axis. Thus, rotating C_0 and C_1 around the z-axis provides different surfaces (Notice that by just moving the value z_0, the union of the corresponding circles $\tilde{C}$ gives rise to another surface not contained in S_2, so one can refine the argument to show that the surfaces generated by rotating C_0 and C_1 about the z-axis differ not in one curve, but in a whole 2-dimensional subset.), which contradicts our hypothesis. □

Remark 2. *Notice that the essence of the argument in the proof of Theorem 6 is not altered if the points Q_i coincide, or if $C_1 \cap P$ is empty.*

5.3. The Elliptic Case

Assume now that S_1 is an affine surface of rotation of elliptic type, i.e., a surface of revolution. Thus, the sections Π_{z_0} of S_1 with planes $z = z_0$ are unions of circles, which are transformed by $f(x) = Ax$ into unions of ellipses contained in planes $\widehat{\Pi}_{z_0}$ like Equation (20). If A_{13} and A_{23} are not both zero, then Equation (20) defines a plane not normal to the z-axis. In this case, we can argue as in Section 5.2 to see that this cannot happen: again, we prove that by considering the affine images of different sections of S_1 normal to the axis, we get planar curves, contained in S_2, which generate different surfaces when rotating about the z-axis. So we focus on the case $A_{13} = A_{23} = 0$. Here, we observe that $\widehat{\Pi}_{z_0}$ is also the plane $z = z_0$, so f preserves the z-coordinate. Thus, the entries a_{31}, a_{32} of the matrix A are both zero, so A can be written as a block matrix

$$A = \begin{pmatrix} \mathbf{Q} & 0 \\ 0 & 1 \end{pmatrix}, \quad \mathbf{Q} = \begin{pmatrix} a_{11} & a_{12} \\ a_{21} & a_{22} \end{pmatrix}, \tag{21}$$

where $\mathbf{Q}$ defines a linear transformation $g(x) = \mathbf{Q}x$ of the plane, preserving the origin, where $\det(\mathbf{Q}) = 1$. Furthermore, since S_2 is by hypothesis a surface of revolution about the z-axis, and f preserves the z-coordinate, we deduce that f maps circles to circles, and therefore that g maps circles centered at the origin onto circles centered at the origin. Then we have the following lemma.

Lemma 7. *With the preceding notation and hypotheses, $g(x) = \mathbf{Q}x$ defines a congruence of the plane.*

Proof. Let $x = [x, y]^T$. Then the equation of a circle C_r centered at the origin is

$$x^T \cdot x = r^2, \tag{22}$$

with $r > 0$. Since $g(x) = \mathbf{Q}x$ maps circles to circles and preserves the origin, C_r is mapped onto the circle C_R of equation

$$x^T \mathbf{Q}^T \cdot \mathbf{Q}x = R^2, \tag{23}$$

where $R > 0$ and x satisfies Equation (22). Multiplying Equation (22) by an appropriate λ, we get $x^T \cdot \lambda \mathbf{I} \cdot x = R^2$, where $\mathbf{I}$ denotes the 2×2 identity matrix. Subtracting this expression from Equation (23), we get that $\mathbf{Q}^T \mathbf{Q} = \lambda \mathbf{I}$. Finally, since $\det(\mathbf{Q}) = 1$, we deduce that $\lambda = 1$, so $\mathbf{Q}$ is orthogonal. Therefore $g(x)$ is an orthogonal transformation, so $g(x)$ defines a congruence. □

Lemma 7 provides the following corollary.

Corollary 4. *The image of a rational surface of revolution that is not developable under a nonsingular affine mapping, is another surface of revolution if and only if the affine mapping corresponds to a similarity.*

From the algorithmic point of view, notice that given two surfaces of revolution about the same axis, one can check whether the surfaces are similar by intersecting both surfaces with a same plane, say, the yz-plane, and then checking whether the resulting planar curves are similar. There are efficient algorithms for doing this: if the sections are rational, one can use the algorithm in [18]; if the sections are not rational, one can use the algorithm in [19].

Finally, we summarize all the results of the section in the following theorem.

Theorem 7. *The image of a surface of revolution under a nonsingular affine mapping is an affine surface of rotation if and only if the affine mapping defines a similarity, in which case the image is also a surface of revolution.*

Corollary 5. *The image of a non-developable rational surface of revolution under a nonsingular affine mapping that is not a similarity, is an ANIL surface that is not an affine surface of rotation.*

Notice that Corollary 5 comes to show that there are many ANIL surfaces that are not affine surfaces of rotation: in fact, the image of any surface of revolution under a non-orthogonal affine mapping is that way. Taking Corollary 2 also into account, we conclude that there are many ANIL surfaces that are not either affine surfaces of rotation, or affine spheres.

6. Conclusions

Throughout the paper we have proved that the image of a non-developable rational surface of revolution under a nonsingular affine mapping is an ANIL surface which is not an affine rotation surface except for certain, well-described, cases. Furthermore, given an ANIL surface, we have provided an algorithm to determine whether it is the affine image of a surface of revolution, and to recover it, if it exists.

One can wonder whether there exist ANIL surfaces which are not the image of a surface of revolution or an affine sphere. We do not have an answer to this question. Were the answer negative, it would be nice to identify notable surfaces with this property. These are problems that we leave here as open questions.

Funding: This research received no external funding.

Acknowledgments: Juan G. Alcázar is a member of the Research Group ASYNACS (Ref. CCEE2011/R34), and is also involved in the scientific team of the project MTM2017-88796-P of the Spanish Ministerio de Economía y Competitividad, where the research in this paper is inscribed. The author is grateful to the reviewers of the paper, whose comments allowed to improve an earlier version of the paper.

Conflicts of Interest: The author declares no conflict of interest.

References

1. Shi, X.; Goldman, R. Implicitizing rational surfaces of revolucion using μ-bases. *Comput. Aided Geom. Des.* **2012**, *29*, 348–362. [CrossRef]
2. Andradas, C.; Recio, T.; Sendra, J.R.; Tabera, L.F.; Villarino, C. Reparametrizing swung surfaces over the reals. *Appl. Algebra Eng. Commun. Comput.* **2014**, *25*, 39–65. [CrossRef]
3. Sendra, J.R.; Sevilla, D.; Villarino, C. Missing sets in rational parametrizations of surfaces of revolution. *Comput. Aided Des.* **2015**, *66*, 55–61. [CrossRef]
4. Nomizu, K.; Sasaki, T. *Affine Differential Geometry*; Cambridge University Press: Cambridge, UK, 1994.
5. Schirokow, P.; Schirokow, A. *Affine Differential Geometrie*; B.G. Teubner: Leipzig, Germany, 1962.
6. Lee, I.C. On Generalized Affine Rotation Surfaces. *Results Math.* **1995**, *27*, 63–76. [CrossRef]
7. Su, B. Affine moulding surfaces and affine surfaces of revolution. *Tohoku Math. J.* **1928**, *5*, 185–210.
8. Alcázar, J.G.; Goldman, R. Finding the axis of revolution of an algebraic surface of revolution. *IEEE Trans. Vis. Comput. Graph.* **2016**, *22*, 2082–2093. [CrossRef] [PubMed]

9. Alcázar, J.G.; Goldman, R.; Hermoso, C. Algebraic surfaces invariant under scissor shears. *Graph. Model.* **2016**, *87*, 23–34. [CrossRef]
10. Alcázar, J.G.; Goldman, R. Algebraic affine rotations of parabolic type. *J. Geom.* **2019**, *110*, 46. [CrossRef]
11. Alcázar, J.G.; Goldman, R. Recognizing algebraic affine rotation surfaces. *Comput. Aided Geom. Des.* **2020**, *81*, 101905. [CrossRef]
12. Pottmann, H.; Wallner, J. *Computational Line Geometry*; Springer: Berlin/Heidelberg, Germany, 2001.
13. Freitas, N.; Andrade, M.; Martínez, D. Estimating affine normal vectors in discrete surfaces. In Proceedings of the 28th SIBGRAPI Conference on Graphics, Patterns and Images, SIBGRAPI 2015, Salvador, Brazil, 26–29 August 2015.
14. Andrade, M.; Lewiner, T. Affine-Invariant Estimators for Implicit Surfaces. *Comput. Aided Geom. Des.* **2012**, *29*, 162–173. [CrossRef]
15. Pottmann, H.; Rrup, T. Rotational and helical surface approximation for reverse engineering. *Computing* **1998**, *60*, 307–322. [CrossRef]
16. Vršek, J.; Làvička, M. Determining surfaces of revolution from their implicit equations. *J. Comput. Appl. Math.* **2014**, *290*, 125–135. [CrossRef]
17. Gray, M.; Abbena, E.; Salamon, S. *Modern Differential Geometry of Curves and Surfaces with Mathematica*; Chapman & Hall/CRC: Boca Raton, FL, USA, 2006.
18. Alcazár, J.G.; Hermoso, C.; Muntingh, G. Symmetry detection of rational space curves from their curvature and torsion. *Comput. Aided Geom. Des.* **2015**, *33*, 51–65. [CrossRef]
19. Alcázar, J.G.; Làvička, M.; Vršek, J. Symmetries and similarities of planar algebraic curves using harmonic polynomials. *J. Comput. Appl. Math.* **2019**, *357*, 302–318. [CrossRef]

Publisher's Note: MDPI stays neutral with regard to jurisdictional claims in published maps and institutional affiliations.

© 2020 by the authors. Licensee MDPI, Basel, Switzerland. This article is an open access article distributed under the terms and conditions of the Creative Commons Attribution (CC BY) license (http://creativecommons.org/licenses/by/4.0/).

Article

Geometric Algebra Framework Applied to Symmetrical Balanced Three-Phase Systems for Sinusoidal and Non-Sinusoidal Voltage Supply

Francisco G. Montoya [1,*], Raúl Baños [1], Alfredo Alcayde [1], Francisco Manuel Arrabal-Campos [1] and Javier Roldán Pérez [2]

1. Department of Engineering, University of Almeria, 04120 Almeria, Spain; rbanos@ual.es (R.B.); aalcayde@ual.es (A.A.); fmarrabal@ual.es (F.M.A.-C.)
2. IMDEA Energy Institute, Electrical Systems Unit, 28935 Madrid, Spain; javier.roldan@imdea.org
* Correspondence: pagilm@ual.es; Tel.: +34-950-214501

Abstract: This paper presents a new framework based on geometric algebra (GA) to solve and analyse three-phase balanced electrical circuits under sinusoidal and non-sinusoidal conditions. The proposed approach is an exploratory application of the geometric algebra power theory (GAPoT) to multiple-phase systems. A definition of geometric apparent power for three-phase systems, that complies with the energy conservation principle, is also introduced. Power calculations are performed in a multi-dimensional Euclidean space where cross effects between voltage and current harmonics are taken into consideration. By using the proposed framework, the current can be easily geometrically decomposed into active- and non-active components for current compensation purposes. The paper includes detailed examples in which electrical circuits are solved and the results are analysed. This work is a first step towards a more advanced polyphase proposal that can be applied to systems under real operation conditions, where unbalance and asymmetry is considered.

Keywords: geometric algebra; non-sinusoidal power; clifford algebra; power theory

1. Introduction

For more than a century, the steady-state operation of AC electrical circuits has been analysed in the frequency domain using complex numbers. The foundations of this well-established technique were initially developed by Steinmetz [1] and later refined by other authors such as Kennelly [2] or Heaviside [3]. In its basic form, an AC signal is transformed from the time to the frequency domain, where algebraic equations can be easily manipulated. This transformation is commonly referred to as phasor transformation. For example, the phasor representation of a voltage waveform such as $v(t) = \sqrt{2}V\cos(\omega t + \varphi)$ is

$$\mathcal{F}[v(t)] = \vec{V} = Ve^{j\varphi} \tag{1}$$

while the inverse transformation is given by

$$\mathcal{F}^{-1}[\vec{V}] = \Re\left[\sqrt{2}\vec{V}e^{j\omega t}\right] = v(t) \tag{2}$$

This methodology is widely applied to solve single- and three-phase electrical circuits that operate in steady state under sinusoidal conditions.

It can also be applied to circuits operating under non-sinusoidal conditions by using the superposition theorem. In this case, the voltage and current components of each harmonic frequency are calculated separately, one by one, and then added in the time domain so that the voltage and current waveforms are obtained. Nevertheless, this property can be seen as both an advantage and a disadvantage. The main reason is that bilinear operations, such as products between voltages and currents of different frequencies, are not meaningful in the algebra of complex numbers when applied to power systems. However, this procedure is strictly required to calculate power flows under distorted conditions. For example, consider a voltage $v(t) = 100\sqrt{2}\cos \omega t$ and a current $i(t) = 100\sqrt{2}\cos 2\omega t$. Their phasor representations are $\vec{V} = 100\angle 0$ and $\vec{I} = 100\angle 0$, respectively. Even though these signals are completely different, their representation in the complex domain is the same. From a mathematical perspective, the product $\vec{V}\vec{I}^*$ cannot be performed since only rotating vectors of the same frequency (and, thus, phasor quantities) can be axiomatically multiplied in the complex domain. Due to the aforementioned limitations, the principle of energy (power) conservation cannot be applied to apparent power in the complex domain in a general sense [4]. These drawbacks have given rise to a great number of proposals for the resolution of electrical circuits and the analysis of power flows [5–7]. This topic is of a paramount relevance because of the increasing energy losses in transmission systems as well as the negative effects on electrical-drives, power transformers and electronic devices.

Recently, geometric algebra (GA) has been proposed and applied to solve physical and engineering problems [8,9]. It has also been proposed for analysing electrical circuits [10,11]. The use of GA has shed some light on a number of important shortcomings of complex numbers, mainly due to the following properties:

1. It is possible to perform calculations between voltages and currents of different frequencies that generate cross-coupling power terms. Therefore, power under non-sinusoidal conditions can be adequately calculated;
2. Foundations of GA circuit analysis is defined in a multi-dimensional geometric domain ($\mathcal{G}_n$), where a definition of geometric apparent power that fulfils the principle of energy conservation can be obtained [12]. This power (M) has been named geometric apparent power in the literature. Compared to the traditional definition of apparent power (S), it considers the contribution of cross effects between voltages and currents of different frequencies and is a signed quantity.

The aforementioned statements are strongly supported by the very basic foundations of electromagnetic power theory: the Poynting Theorem. It is well-known that the density power S delivered to a load can be calculated through the Poynting vector

$$S = E \times H \qquad (3)$$

where E and H are the electric and magnetic field vector, respectively. Note the cross product in (3). If both the electric field and the magnetic field are transferred to the frequency domain [13], it is evident that the product of the harmonics content of different frequencies leads to a density power with a clear physical existence.

The concept of geometric apparent power was first introduced by Menti in 2007 [14]. It was demonstrated that the traditional apparent power is a particular case of the geometric apparent power for systems that operate under sinusoidal conditions. Later, in 2010, Castro-Núñez presented a new mathematical framework based on the use of k-blades in GA for solving and analysing electrical circuits under sinusoidal conditions [15]. The concept of geometric impedance was introduced and applied to single-phase RLC circuits. The theory was extended by the same author for non-linear circuits in the presence of harmonics [16,17]. The improvements compared to traditional theories were demonstrated through examples. However, some drawbacks and inconveniences were found in this particular formulation [18]. Castilla and Bravo [19,20] made improvements to former theories and presented an alternative formulation, called *generalized complex geometric algebra*. This theory can be used to perform power calculations, but cannot solve electrical circuits in the GA domain. Recently, Montoya et al. [10,11] have studied power flows under non-sinusoidal conditions using GA. These developments were applied in different applications such as power factor correction and non-active current compensation, but only for single-phase systems. Moreover, new GA developments have redefined the geometric apparent power so that it can be fully applied to single-phase electrical circuits operating under any type of voltage and current distortion [12]. In addition, recent publications have presented a formulation of a GA power theory in the time domain that establishes the basis for both instantaneous and averaged current decomposition [21].

GA has already been applied in a number of cases to single-phase systems, but the application to three-phase systems has been seldom addressed in the literature to date. To the best of the author's knowledge, the only attempt was undertaken by Lev-Ari [22] in 2009. However, this work only presented preliminary concepts and the effectiveness of the theory was not validated. No further results have been published to date.

In this paper, GA is applied to analyse and solve three-phase electrical circuits under sinusoidal and non-sinusoidal conditions. This can be seen as a relevant improvement compared to previous theories based on GA that only addressed single-phase electrical systems. Note that this is an initial effort towards a more complete polyphase framework based on GA, where asymmetries and unbalanced effects should be taken into account. In order to substantiate the validity of the proposed theory, several examples are presented and solved in detail. Finally, the conclusions and suggestions for further research are drawn. The main benefits of the application of geometric algebra to power systems are:

- It is possible to define a new power concept based on geometrical principles that take the interaction of voltage and current harmonics of different frequency into account. This is not possible using phasors based on complex algebra;
- Unified criteria and methods are established for the study of electrical circuits based on a single tool that makes it possible to tackle multidimensional problems, such as those existing in polyphase circuits;
- It establishes basic principles for the compensation of non-active current that allow for the optimisation of energy losses in power transmission lines.

2. GA for Electrical Applications: Overview

The proposed theory requires some basic knowledge of GA. References [23–26] provide introductory material. However, a basic overview of GA has been included in order to make the paper self-contained. For detailed information about GA and its applications to electrical systems, see [11,12].

A relevant concept in GA is the geometric product. It can be applied to voltage and current vectors to calculate the so-called geometric apparent power, M. For example, for a single-phase sinusoidal supply and a linear load, an Euclidean vector basis $\sigma = \{\sigma_1, \sigma_2\}$ can be chosen so that the voltage and the current can be represented as a vector, i.e., $u = \alpha_1 \sigma_1 + \alpha_2 \sigma_2$ and $i = \beta_1 \sigma_1 + \beta_2 \sigma_2$. The geometric product is defined as the inner plus the exterior product:

$$M = ui = u \cdot i + u \wedge i = \underbrace{(\alpha_1 \beta_1 + \alpha_2 \beta_2)}_{P} + \underbrace{(\alpha_1 \beta_2 - \alpha_2 \beta_1)}_{Q} \sigma_{12} \quad (4)$$

where σ_{12} is commonly known as a bivector. Note that it is an element that is not present in traditional linear algebra.

In order to apply the GA power theory to poly-phase systems, voltages and currents are arranged as multi-dimensional vector arrays. These will be referred to as arrays, while the term vector will be used to refer to voltages and currents of a given phase. For example, the current waveforms $[i_R(t)\ i_S(t)\ i_T(t)]$ and voltage waveforms $[u_{RN}(t)\ u_{SN}(t)\ u_{TN}(t)]$ for the three-phase system depicted in Figure 1 can be represented in the geometric domain as:

$$u = \begin{bmatrix} u_{RN} & u_{SN} & u_{TN} \end{bmatrix}, \quad i = \begin{bmatrix} i_R & i_S & i_T \end{bmatrix}^T \quad (5)$$

The transformation is based on the principle of isomorphism between vector spaces. In this case, the time domain periodic Fourier functions and Euclidean vector space. Thus, the basis used for the geometric transformation can be chosen as in single-phase systems [12]:

$$\begin{aligned}
\varphi_{DC} &= 1 & &\longleftrightarrow & \sigma_0 \\
\varphi_{c1}(t) &= \sqrt{2}\cos\omega t & &\longleftrightarrow & \sigma_1 \\
\varphi_{s1}(t) &= \sqrt{2}\sin\omega t & &\longleftrightarrow & \sigma_2 \\
&\vdots & & & \\
\varphi_{cn}(t) &= \sqrt{2}\cos n\omega t & &\longleftrightarrow & \sigma_{2n-1} \\
\varphi_{sn}(t) &= \sqrt{2}\sin n\omega t & &\longleftrightarrow & \sigma_{2n}
\end{aligned} \quad (6)$$

Any current or voltage variable $x(t)$ in the time domain (including the DC component) can be expressed as a vector x in the geometric domain by using $2n + 1$ dimensions, where n is the number of harmonics in $x(t)$

$$x = x_0\sigma_0 + \sum_{k=1}^{n}(x_{1k}\sigma_{2k-1} + x_{2k}\sigma_{2k}) \quad (7)$$

while x_{1k} and x_{2k} are the Fourier coefficients of the harmonic k and x_0 is the DC component. From now on, inter-harmonics and the DC component will not be considered for the sake of simplicity, but they can be seamlessly taken into account [11,18]. This representation cannot be obtained in the complex domain since it involves rotating vectors at different frequencies. Once the voltage and current vectors are defined, it is possible to introduce the geometric apparent power for three-phase systems as:

$$M = ui = [u_{RN}\ u_{SN}\ u_{TN}]\begin{bmatrix} i_R \\ i_S \\ i_T \end{bmatrix} = u_{RN}i_R + u_{SN}i_S + u_{TN}i_T =$$

$$= \underbrace{u_{RN}\cdot i_R + u_{SN}\cdot i_S + u_{TN}\cdot i_T}_{M_a = P} + \underbrace{u_{RN}\wedge i_R + u_{SN}\wedge i_S + u_{TN}\wedge i_T}_{M_N} \quad (8)$$

In (8), the sum of scalar products $u_{kN}\cdot i_k$, with $k = \{R, S, T\}$, leads to the geometric active power M_a, which is similar to the traditional definition of P. Meanwhile, the sum of the exterior products $u_{kN}\wedge i_k$ leads to the geometric non-active power M_N, which is similar to the traditional reactive power (Q) for a symmetric and sinusoidal voltage supply feeding a balanced load.

Other apparent power definitions based on euclidean or geometric principles can be found in the literature. For example, the RMS values of voltage and current vectors are used in [27], i.e., $S = \|U\|\|I\|$. Unfortunately, they exclusively rely on the concept of a norm. Therefore, they cannot fulfil the principle of energy conservation [12].

The norm (RMS value) of any geometric array can be calculated by using the norm definition [25]:

$$\|x\| = \sqrt{x\cdot x^T} = \sqrt{\langle x^\dagger x\rangle_0} = \sqrt{\sum_i x_i^2} \quad (9)$$

For a voltage waveform, the result is

$$\|u\|^2 = u\cdot u^T = [u_{RN}\ u_{SN}\ u_{TN}]\begin{bmatrix} u_{RN} \\ u_{SN} \\ u_{TN} \end{bmatrix} = u_{RN}u_{RN} + u_{SN}u_{SN} + u_{TN}u_{TN} \quad (10)$$

$$= \|u_{RN}\|^2 + \|u_{SN}\|^2 + \|u_{TN}\|^2 \quad (11)$$

A similar result can be obtained for the current i. It can be proved that $\|M\| = \|u\|\|i\|$ [18], i.e., the product of vector norms equals the norm of the geometric power

$$\|M\| = \sqrt{\langle MM^\dagger\rangle_0} = \sqrt{\langle ui(ui)^\dagger\rangle_0} = \sqrt{\langle uii^T u^T\rangle_0} = \sqrt{\langle \|u\|^2\|i\|^2\rangle_0} = \|u\|\|i\| \quad (12)$$

where the reverse of a general geometric array a is defined as $a^\dagger = a^T$.

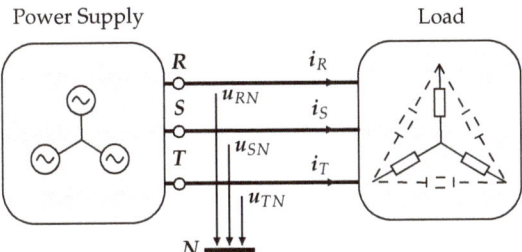

Figure 1. Three-phase three-wire electrical circuit.

3. Case I: Balanced, Symmetric and Sinusoidal

In this section, the proposed theory is applied to a three-phase circuit that operates under balanced, symmetric and sinusoidal conditions. Although it is a well-known case, already solved in the literature, we believe it is a good example to understand the GA-based methodology. For clarification, the traditional complex algebra solution is also presented. The computations were carried out using a Matlab library kown as GAPoTNumLib developed by the some of the authors for GA in electrical engineering [28].

3.1. Current, Voltage and Impedance Calculations

A three-phase three-wire electrical circuit that consists of an ideal voltage source that feeds a balanced star-connected load is shown in Figure 2. The phase voltages are $u_{RN}(t)$, $u_{SN}(t)$ and $u_{TN}(t)$ while the line currents are $i_R(t)$, $i_S(t)$ and $i_T(t)$. These waveforms are defined in the time domain. They can be transformed to the geometric domain $\mathcal{G}_n$ by using the transformation shown in (6). Since the system is balanced and sinusoidal, there is only a fundamental harmonic component, i.e., $n = 1$. Thus, the dimension of the geometric domain is two ($\mathcal{G}_2$). Under these assumptions, the chosen basis σ includes one scalar, two vectors and one bivector, i.e., $\sigma = \{1, \sigma_1, \sigma_2, \sigma_{12}\}$.

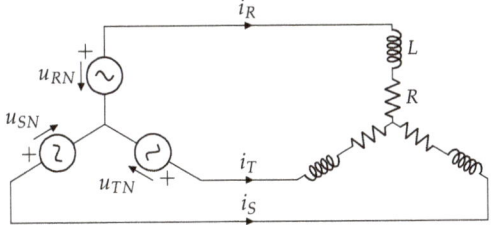

Figure 2. Symmetrical three-phase, three-wire circuit.

The following voltage waveforms are considered:

$$\begin{aligned} u_{RN}(t) &= \sqrt{2}U\cos(\omega t + \varphi) \\ u_{SN}(t) &= \sqrt{2}U\cos(\omega t + \varphi - 2\pi/3) \\ u_{TN}(t) &= \sqrt{2}U\cos(\omega t + \varphi + 2\pi/3) \end{aligned} \quad (13)$$

Without loss of generality and for simplicity reasons, it is assumed that $U = 1$ V, $\varphi = 0°$ and $\omega = 1$ rad/s. In this case, by virtue of (6), the geometric voltages become

$$u_{RN} = \sigma_1, \quad u_{SN} = -\frac{1}{2}\sigma_1 + \frac{\sqrt{3}}{2}\sigma_2, \quad u_{TN} = -\frac{1}{2}\sigma_1 - \frac{\sqrt{3}}{2}\sigma_2 \quad (14)$$

which resembles the standard phasor representation in the complex domain, being σ_1 the real part, and σ_2 the imaginary part, as in

$$\vec{u}_{RN} = 1, \quad \vec{u}_{SN} = -\frac{1}{2} + \frac{\sqrt{3}}{2}j, \quad \vec{u}_{TN} = -\frac{1}{2} - \frac{\sqrt{3}}{2}j \tag{15}$$

The Euler formula widely used in complex algebra $e^{\varphi j} = \cos\varphi + \sin\varphi j$, can also be used in GA [23]:

$$e^{\varphi\sigma_{12}} = \cos\varphi + \sin\varphi\sigma_{12} \tag{16}$$

This exponential entity is commonly known in GA as spinor, i.e., a multivector made up of a scalar plus a bivector [29]. Any vector multiplied by a spinor undergoes a rotation of φ degrees in the plane defined by the bivector (in this case σ_{12}) and a scaling. Therefore, spinors are commonly used to rotate elements in GA (unitary spinors are also known as rotors [24]). Note that the impedance and admittance of a passive element can also be represented as a spinor, as explained later. Hence, the voltage in (14) can be expressed in polar form as:

$$\begin{aligned} u_{RN} &= e^{0\sigma_{12}}\sigma_1 = & 1\angle 0 \\ u_{SN} &= e^{-120\sigma_{12}}\sigma_1 = & 1\angle -120 \\ u_{TN} &= e^{120\sigma_{12}}\sigma_1 = & 1\angle 120 \end{aligned} \tag{17}$$

The reader should keep in mind that right- and left-multiplication between vectors and rotors produce rotations in opposite directions. In the rest of the paper, rotors will left-multiply vectors. In (17), it is easy to identify $1\angle 0$, $1\angle -120$ and $1\angle 120$ as geometric vectors in the plane σ_1-σ_2. They resemble the complex phasors e^{0j}, e^{-120j} and e^{120j} in the complex plane, respectively. Line voltages can be calculated as follows:

$$\begin{aligned} u_{RS} &= u_{RN} - u_{SN} = \sqrt{3}e^{30\sigma_{12}}\sigma_1 = \sqrt{3}\angle 30 \\ u_{ST} &= u_{SN} - u_{TN} = \sqrt{3}e^{-90\sigma_{12}}\sigma_1 = \sqrt{3}\angle -90 \\ u_{TR} &= u_{TN} - u_{RN} = \sqrt{3}e^{150\sigma_{12}}\sigma_1 = \sqrt{3}\angle 150 \end{aligned} \tag{18}$$

Assuming an RL load with $R = 1/\sqrt{2}$ Ω and $L = 1/\sqrt{2}$ H, the geometric impedance becomes:

$$Z = R + X_L = Ze^{\varphi\sigma_{12}} = R + L\omega\sigma_{12} = \frac{1}{\sqrt{2}} + \frac{1}{\sqrt{2}}\sigma_{12} = e^{45\sigma_{12}}$$

Note that the traditional form in complex notation is $\frac{1}{\sqrt{2}} + \frac{1}{\sqrt{2}}j = e^{45j}$. A relevant property of vectors and multivectors in GA is that the existence of an inverse is always guaranteed, provided that they are not null. For example, for a multivector in $\mathcal{G}_2$ given by $X = X_0 + X_1\sigma_1 + X_2\sigma_2 + X_3\sigma_{12}$ and a vector $x = x_1\sigma_1 + x_2\sigma_2$, their inverses are:

$$X^{-1} = X^\dagger/(XX^\dagger), \quad x^{-1} = 1/x = x/\|x\|^2 \tag{19}$$

where

$$X^\dagger = \sum_{k=0}^{n}\langle X^\dagger\rangle_k = \sum_{k=0}^{n}(-1)^{k(k-1)/2}\langle X\rangle_k$$

is the reverse of X and $\langle X\rangle_k$ is an operator that extracts the k-th grade element of X. The admittance can be calculated as follows:

$$Y = Z^{-1} = \frac{Z^\dagger}{ZZ^\dagger} = G + B_L = \frac{1}{Z}e^{-\varphi\sigma_{12}} = \frac{1}{\sqrt{2}} - \frac{1}{\sqrt{2}}\sigma_{12} = e^{-45\sigma_{12}} \tag{20}$$

The currents can be found by applying Kirchhoff and Ohm's laws to the former expressions, yielding:

$$i_R = Yu_{RN} = \frac{1}{\sqrt{2}}(\sigma_1 + \sigma_2) = e^{-45\sigma_{12}}\sigma_1$$

$$i_S = Yu_{SN} = \frac{-1-\sqrt{3}}{2\sqrt{2}}\sigma_1 + \frac{-1+\sqrt{3}}{2\sqrt{2}}\sigma_2 = e^{-165\sigma_{12}}\sigma_1 \quad (21)$$

$$i_T = Yu_{TN} = \frac{-1+\sqrt{3}}{2\sqrt{2}}\sigma_1 + \frac{-1-\sqrt{3}}{2\sqrt{2}}\sigma_2 = e^{75\sigma_{12}}\sigma_1$$

Compare the results of (22) with that of complex notation e^{-45j}, e^{-165j} and e^{75j}. It may look like the complex notation is lighter than that of GA, but it comes at a cost: only two dimensions can be handled at a time, i.e., only one harmonic component can be solved. Figure 3 shows a graphical representation of geometric vectors that resembles the traditional Argand diagram for complex numbers. However, the concept of phase shift now leads to a negative angle, represented as a rotation in counter-clockwise direction. Meanwhile, phase lead is represented by a positive angle and a rotation in clockwise direction. This interesting fact can be explained by using the trigonometric identity $\sin\theta = \cos(\theta - \pi/2)$. Therefore, we conclude that $\sin\omega t$ lags $\cos\omega t$ (σ_2 lags σ_1).

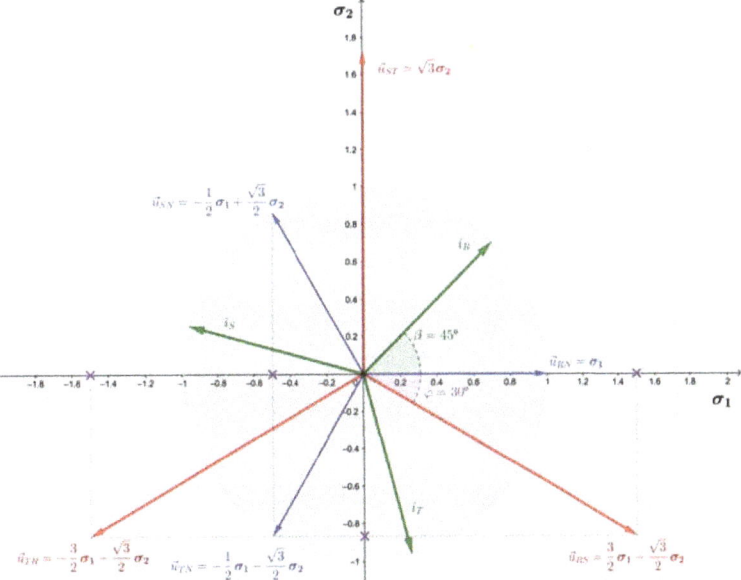

Figure 3. Representation of voltage and current geometric vectors in the plane $\sigma_1\sigma_2$.

Compared to traditional complex algebra, the graphical representation of powers and impedances/admittances in GA is slightly different. These elements are spinors, i.e., entities that consist of a scalar and a bivector part. Therefore, they should not be depicted in the plane σ_1-σ_2 but in the *scalar-bivector* one. It is a subtle difference, but it is worth to highlight this aspect. Although both GA objetcs and complex numbers can be depicted in Argand diagrams, completely different representations are used for GA objects, according to their nature (impedances/powers or voltages/currents). Figure 4 shows an example of a graphical representation, where the x axis now represents scalars and the y axis represents bivectors. This interpretation is a novel contribution of this paper.

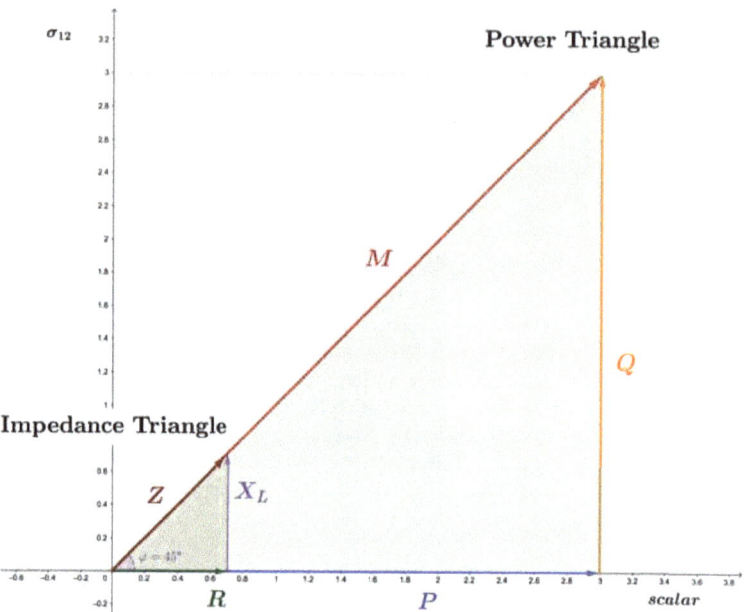

Figure 4. Power and impedance triangle in GA plane $scalar$-σ_{12}.

3.2. Power Calculations

The norm of the voltage and current arrays can be calculated by using (5) and (11):

$$\|u\|^2 = u \cdot u = uu^T = [u_{RN}\ u_{SN}\ u_{TN}] \begin{bmatrix} u_{RN} \\ u_{SN} \\ u_{TN} \end{bmatrix} = \sigma_1 \sigma_1 + \left(-\frac{1}{2}\sigma_1 + \frac{\sqrt{3}}{2}\sigma_2\right)\left(-\frac{1}{2}\sigma_1 + \frac{\sqrt{3}}{2}\sigma_2\right) +$$

$$+ \left(-\frac{1}{2}\sigma_1 - \frac{\sqrt{3}}{2}\sigma_2\right)\left(-\frac{1}{2}\sigma_1 - \frac{\sqrt{3}}{2}\sigma_2\right) = 3 \tag{22}$$

$$\|i\|^2 = i \cdot i = i^T i = [i_R\ i_S\ i_T] \begin{bmatrix} i_R \\ i_S \\ i_T \end{bmatrix} = 3 \tag{23}$$

Therefore, the concept of three-phase geometric apparent power can be used, yielding:

$$M = ui = u \cdot i + u \wedge i = M_a + M_N \tag{24}$$

where

$$M_a = \frac{1}{\sqrt{2}} + \frac{1+\sqrt{3}}{4\sqrt{2}} + \frac{\sqrt{3}(-1+\sqrt{3})}{4\sqrt{2}} + \frac{1-\sqrt{3}}{4\sqrt{2}} + \frac{\sqrt{3}(1+\sqrt{3})}{4\sqrt{2}}$$

$$M_N = \left[\frac{1}{\sqrt{2}} + \frac{1-\sqrt{3}}{4\sqrt{2}} + \frac{\sqrt{3}(1+\sqrt{3})}{4\sqrt{2}} + \frac{1+\sqrt{3}}{4\sqrt{2}} + \frac{\sqrt{3}(-1+\sqrt{3})}{4\sqrt{2}}\right]\sigma_{12}$$

After some algebraic manipulations, the geometric apparent power can be written as:

$$M = M_a + M_N = P + Q\sigma_{12} = \frac{3}{\sqrt{2}} + \frac{3}{\sqrt{2}}\sigma_{12} \tag{25}$$

It can be seen that the geometric power consists of two terms of different nature. On the one hand, the active geometric power M_a, which is a scalar number that is equal to the active power P for the sinusoidal case. Therefore, for this example, $P = R\|i\|^2 = 3/\sqrt{2}$. On the other hand, we get the non-active power M_N, which is a bivector. For the ideal case, the non-active power is equal to Q. Therefore, in this example, $Q = X\|i\|^2 = 3/\sqrt{2}$. Furthermore, it can be verified that $\|M\|$ yields to the same result as the traditional apparent power S by using (12):

$$\|M\| = \sqrt{(3/\sqrt{2})^2 + (3/\sqrt{2})^2} = \|u\|\|i\| = 3 \tag{26}$$

A relevant difference between M and S is that the result is a multivector and not a scalar nor a complex number, as in the traditional apparent power, $\bar{S}$. For this example (balanced and sinusoidal), the result is a spinor, where the scalar part is the active power P, while the bivector part is the well-known reactive power Q.

3.3. Current Decomposition

GA can be used to decompose currents in components that are relevant for engineering purposes (e.g., filter design), as in other power theories [6,30]. For sinusoidal single-phase circuits, the current can be decomposed into two terms: active and reactive, i.e., current in phase and in quadrature with respect to the voltage, respectively. However, if the source voltage is not sinusoidal, an additional term appears. This term is commonly known as *scattered current* in the CPC theory [27].

In GA, it is possible to decompose currents by applying Kirchhoff laws [31]. In the case under study, the current decomposition yields:

$$\begin{aligned} i_R &= Y_R u_{RN} = (G_R + B_R \sigma_{12}) u_{RN} \\ i_S &= Y_S u_{SN} = (G_S + B_S \sigma_{12}) u_{SN} \\ i_T &= Y_T u_{TN} = (G_T + B_T \sigma_{12}) u_{TN} \end{aligned} \tag{27}$$

which can be expressed in array form as:

$$\begin{bmatrix} Y_R u_{RN} \\ Y_S u_{SN} \\ Y_T u_{TN} \end{bmatrix} = \underbrace{\begin{bmatrix} G_R u_{RN} \\ G_S u_{SN} \\ G_T u_{TN} \end{bmatrix}}_{i_p} + \underbrace{\begin{bmatrix} B_R \sigma_{12} u_{RN} \\ B_S \sigma_{12} u_{SN} \\ B_T \sigma_{12} u_{TN} \end{bmatrix}}_{i_q} \tag{28}$$

Therefore, the current can be decomposed into a parallel current array i_p (proportional to the voltage) and a quadrature current array i_q (in quadrature with the voltage). This finding is inline with Shepherd and Zakikhany theory [4]. The squared norm of i_p can be calculated using (9):

$$\|i_p\|^2 = i_p \cdot i_p = G_R^2 \|u_{RN}\|^2 + G_S^2 \|u_{SN}\|^2 + G_T^2 \|u_{TN}\|^2 \tag{29}$$

Note that for a balanced load, $G_R = G_S = G_T = G$. Therefore, the active power can be written as:

$$P = R\|i_p\|^2 \tag{30}$$

Since the voltages of the system under study are balanced, then $\|u\|^2 = 3\|u_{RN}\|^2$. Therefore:

$$G = \frac{P}{3\|u_{RN}\|^2} \tag{31}$$

Figure 5 shows a three-phase balanced circuit equivalent to that depicted in Figure 2. When these circuits are supplied with the same voltage u, both demand the same active power P. In this case, the active power can be easily calculated:

$$P = G_e\|u_{RN}\|^2 + G_e\|u_{SN}\|^2 + G_e\|u_{TN}\|^2 \tag{32}$$

Therefore, the equivalent conductance G_e is:

$$G_e = \frac{P}{\|u_{RN}\|^2 + \|u_{SN}\|^2 + \|u_{TN}\|^2} = G \quad (33)$$

The same analysis can be carried out for the susceptance B_e by using an equivalent load written in terms of reactances. Therefore, it is possible to derive a general expression for the equivalent admittance:

$$Y_e = (Y_R + Y_S + Y_T)/3 = G_e + B_e \sigma_{12} \quad (34)$$

This expression simplifies the decomposition of currents into components that are significant for the engineering practice. As shown in (29), G_e (the scalar part of Y_e) is related to the parallel current, which, for sinusoidal systems, matches the active current, i.e., the minimum current that produces the same active power P. B_e (the bivector part of Y_e) is the equivalent susceptance, and leads to the quadrature current. This current does not produce net power transfer and increases the total current, thereby increasing losses. For this example, the geometric power associated to the current components can be obtained as in (8):

$$P = M_a = M_p = u i_p$$
$$Q = M_N = M_q = u i_q \quad (35)$$

Compared to the traditional apparent power, the three-phase geometric apparent power defined in this work fulfills the Tellegen's Theorem and is conservative (see references [10,11,20]) since:

$$M = M_a + M_N = M_p + M_q \quad (36)$$

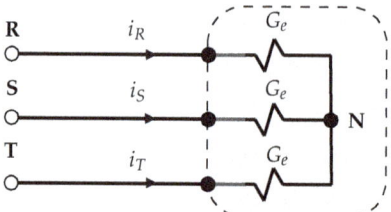

Figure 5. Equivalent three-phase balanced resistor.

3.4. Voltage Transformation Using Geometric Rotors

One of the most interesting features of GA is its ability to spatially manipulate geometric objects. For example, it is widely used in computer graphics to perform translations, reflections, or, more interestingly, rotations. In electrical engineering, a transformer can be considered as an element that causes a phase shift and a scaling of voltage or current signals between its primary and secondary terminals. This translates into a scaling and rotation in the geometrical domain, so that a voltage or current vector applied to the primary will be seen as a rotated and scaled vector in the secondary.

On the basis of the circuit in Figure 2, a three-phase transformer can be placed between the source and the load according to Figure 6. Let us assume that the transformer has a connection group Dy11. Therefore, the voltages of the secondary will be shifted by $\frac{11\pi}{6}$

and scaled by $\frac{1}{\sqrt{3}}\frac{N_1}{N_2}$ with respect to the primary. For the symmetric case, represented by Equations (13) and (14), the time domain voltage in the secondary is

$$u'_{RN}(t) = \frac{\sqrt{2} N_1}{\sqrt{3} N_2} \cos\left(\omega t - \frac{11\pi}{6}\right)$$
$$u'_{SN}(t) = \frac{\sqrt{2} N_1}{\sqrt{3} N_2} \cos\left(\omega t - \frac{3\pi}{6}\right) \quad (37)$$
$$u'_{TN}(t) = \frac{\sqrt{2} N_1}{\sqrt{3} N_2} \cos\left(\omega t - \frac{7\pi}{6}\right)$$

which translates to the geometric domain as

$$u'_{RN} = \frac{1}{\sqrt{3}} \frac{N_1}{N_2}\left(\frac{\sqrt{3}}{2}\sigma_1 + \frac{1}{2}\sigma_2\right), \quad u'_{SN} = -\frac{1}{\sqrt{3}}\frac{N_1}{N_2}\sigma_2, \quad u'_{TN} = \frac{1}{\sqrt{3}}\frac{N_1}{N_2}\left(-\frac{\sqrt{3}}{2}\sigma_1 + \frac{1}{2}\sigma_2\right) \quad (38)$$

It is easy to prove that this operation corresponds to a rotation plus a scaling in the geometric domain. For this purpose, it is enough to establish the geometric object associated with the rotation, as well as the scale factor. In this way, the result can be expressed compactly in geometrical terms as

$$u' = \frac{1}{\sqrt{3}}\frac{N_1}{N_2} R u R^\dagger \quad (39)$$

where

$$R = e^{\frac{11\pi}{12}\sigma_{12}} = \cos\frac{11\pi}{12} + \sin\frac{11\pi}{12}\sigma_{12} \quad (40)$$

Note that the rotor angle is just half of the full rotation angle, because the rotation is a *sandwich* operation that operates half on each side of the vector. The detail of the development of (39) is as follows

$$u' = \frac{1}{\sqrt{3}}\frac{N_1}{N_2} R u R^\dagger = \frac{1}{\sqrt{3}}\frac{N_1}{N_2}\begin{bmatrix} R u_{RN} R^\dagger \\ R u_{SN} R^\dagger \\ R u_{TN} R^\dagger \end{bmatrix} = \frac{1}{\sqrt{3}}\frac{N_1}{N_2}\begin{bmatrix} e^{\frac{11\pi}{12}\sigma_{12}}\sigma_1 e^{-\frac{11\pi}{12}\sigma_{12}} \\ e^{\frac{11\pi}{12}\sigma_{12}}\left(-\frac{1}{2}\sigma_1 + \frac{\sqrt{3}}{2}\sigma_2\right)e^{-\frac{11\pi}{12}\sigma_{12}} \\ e^{\frac{11\pi}{12}\sigma_{12}}\left(-\frac{1}{2}\sigma_1 - \frac{\sqrt{3}}{2}\sigma_2\right)e^{-\frac{11\pi}{12}\sigma_{12}} \end{bmatrix}$$

$$= \frac{1}{\sqrt{3}}\frac{N_1}{N_2}\begin{bmatrix} \frac{\sqrt{3}}{2}\sigma_1 + \frac{1}{2}\sigma_2 \\ -\sigma_2 \\ -\frac{\sqrt{3}}{2}\sigma_1 + \frac{1}{2}\sigma_2 \end{bmatrix} \quad (41)$$

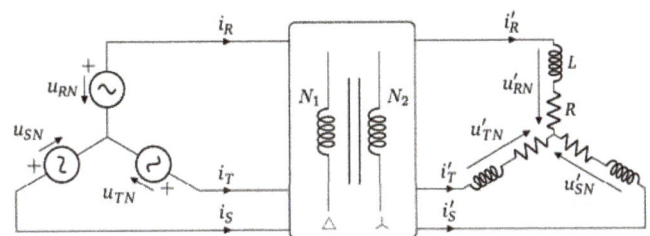

Figure 6. Three phase circuit with a Dy11 transformer with N_1 and N_2 windings for the primary and secondary, respectively.

4. Case II: Balanced, Symmetric and Non-Sinusoidal

In this section, the proposed theory is applied to a three-phase circuit that operates in balanced, symmetric and non-sinusoidal conditions.

4.1. Current, Voltage and Impedance Calculations

Consider again the example in Figure 2, but supplied with a non-sinusoidal, symmetric and positive sequence voltage:

$$\begin{aligned} u_{RN}(t) &= \sum_{k=1}^{n} \sqrt{2} U_k \cos(k\omega t + \varphi_k) \\ u_{SN}(t) &= \sum_{k=1}^{n} \sqrt{2} U_k \cos(k\omega t - k\frac{2\pi}{3} + \varphi_k) \\ u_{TN}(t) &= \sum_{k=1}^{n} \sqrt{2} U_k \cos(k\omega t + k\frac{2\pi}{3} + \varphi_k) \end{aligned} \quad (42)$$

Table 1 shows the well-known mapping for frequency and symmetrical sequence component depending on the harmonic order. Only three-wire systems will be considered in this work. Therefore, zero-sequence voltage and current will not be considered since it is guaranteed that they will not affect power calculations. The addition of the fourth wire is a relevant topic for further research.

Based on (6), the voltage in the geometric domain is:

$$\begin{aligned} u_{RN} &= \sum_{k=1}^{n} (u_{k1}\sigma_{2k-1} + u_{k2}\sigma_{2k}) \\ u_{SN} &= \sum_{k=1}^{n} \left(-\frac{1}{2} u_{k1}\sigma_{2k-1} + \frac{\sqrt{3}}{2} u_{k2}\sigma_{2k} \right) \\ u_{TN} &= \sum_{k=1}^{n} \left(-\frac{1}{2} u_{k1}\sigma_{2k-1} - \frac{\sqrt{3}}{2} u_{k2}\sigma_{2k} \right) \end{aligned} \quad (43)$$

where $u_{k1} = U_k \cos \varphi_k$ and $u_{k2} = U_k \sin \varphi_k$. The current array can be calculated as in (27), yielding:

$$i = \begin{bmatrix} i_R \\ i_S \\ i_T \end{bmatrix} = i_p + i_q = \begin{bmatrix} i_{R_p} \\ i_{S_p} \\ i_{T_p} \end{bmatrix} + \begin{bmatrix} i_{R_q} \\ i_{S_q} \\ i_{T_q} \end{bmatrix} = \underbrace{\begin{bmatrix} \sum_{k=1}^{n} G_{R_k}(u_{k1}\sigma_{2k-1} + u_{k2}\sigma_{2k}) \\ \sum_{k=1}^{n} G_{S_k}(-\frac{1}{2} u_{k1}\sigma_{2k-1} + \frac{\sqrt{3}}{2} u_{k2}\sigma_{2k}) \\ \sum_{k=1}^{n} G_{T_k}(-\frac{1}{2} u_{k1}\sigma_{2k-1} - \frac{\sqrt{3}}{2} u_{k2}\sigma_{2k}) \end{bmatrix}}_{i_p} + \\ + \underbrace{\begin{bmatrix} \sum_{k=1}^{n} B_{R_k}\sigma_{(2k-1)(2k)}(u_{k1}\sigma_{2k-1} + u_{k2}\sigma_{2k}) \\ \sum_{k=1}^{n} B_{S_k}\sigma_{(2k-1)(2k)}(-\frac{1}{2} u_{k1}\sigma_{2k-1} + \frac{\sqrt{3}}{2} u_{k2}\sigma_{2k}) \\ \sum_{k=1}^{n} B_{T_k}\sigma_{(2k-1)(2k)}(-\frac{1}{2} u_{k1}\sigma_{2k-1} - \frac{\sqrt{3}}{2} u_{k2}\sigma_{2k}) \end{bmatrix}}_{i_q} \quad (44)$$

As in the sinusoidal case, it is possible to find an equivalent admittance for each of the harmonics present in the system:

$$Y_{ek} = G_{ek} + B_{ek}\sigma_{(2k-1)(2k)} = \frac{1}{3}(Y_{R_k} + Y_{S_k} + Y_{T_k}) \quad (45)$$

Using the same rationale of (31), and similar to the approach of CPC theory by Prof. Czarnecki, the equivalent conductance and susceptance for each harmonic is:

$$G_{ek} = \frac{P_k}{\|u\|^2} = G_k, \quad B_{ek} = \frac{Q_k}{\|u\|^2} = B_k \quad (46)$$

Equation (33) is still valid even if u is non-sinusoidal, and it can be simplified as follows

$$G_e = \frac{P}{\|u\|^2} \qquad (47)$$

Table 1. Sequences for the different harmonic orders in a balanced three-phase system for $h = 1, 2, \ldots, \infty$.

Harmonic Order	Radian Frequency	Sequence
$3h - 2$	$(3h - 2)\,\omega_1$	positive (+)
$3h - 1$	$(3h - 1)\,\omega_1$	negative (−)
$3h$	$3h\,\omega_1$	zero (0)

4.2. Current Decomposition

The current consumed by loads is commonly decomposed for engineering purposes. The main idea is to split the current into virtual components that can be used, for example, to design and control active compensators. Current decomposition is based on Fryze's ideas, where active and non-active currents are defined [6]. In this work, the equivalent conductance is used to calculate the active current [12]:

$$i_a = G_e u = \frac{u}{\|u\|^2} M_a \qquad (48)$$

The current i_a is part of i_p, as already shown in the literature [32,33]. Therefore, the scattered current becomes:

$$i_s = i_p - i_a \qquad (49)$$

The total current can be decomposed as follows:

$$i = i_p + i_q = i_a + i_s + i_q \qquad (50)$$

The three components in the left-hand side part of (50) are in quadrature. Therefore:

$$i_a \cdot i_s = 0, \quad i_a \cdot i_q = 0, \quad i_s \cdot i_q = 0$$

4.3. Numerical Example

The circuit in Figure 2 will be analysed for the case of a non-sinusoidal symmetric voltage source such as:

$$\begin{aligned} u_{RN}(t) &= \sqrt{2}[230 \cos \omega t + 110 \cos 2\omega t] \\ u_{SN}(t) &= \sqrt{2}[230 \cos(\omega t - 2\pi/3) + 110 \cos(2\omega t + 2\pi/3)] \\ u_{TN}(t) &= \sqrt{2}[230 \cos(\omega t + 2\pi/3) + 110 \cos(2\omega t - 2\pi/3)] \end{aligned} \qquad (51)$$

where the harmonic sequences presented in Table 1 have been taken into account. The transformation to the geometric domain follows the rules presented in (6), thus

$$\begin{aligned} u_{RN} &= 230\sigma_1 + 110\sigma_3 \\ u_{SN} &= 230 e^{-2\pi/3 \sigma_{12}} \sigma_1 + 110 e^{2\pi/3 \sigma_{34}} \sigma_3 \\ u_{TN} &= 230 e^{2\pi/3 \sigma_{12}} \sigma_1 + 110 e^{-2\pi/3 \sigma_{34}} \sigma_3 \end{aligned} \qquad (52)$$

The voltage array u can be expressed as:

$$u = u_1 + u_2 \qquad (53)$$

where u_1 and u_2 are the voltages of the fundamental and the second harmonic, respectively.

The impedance for each frequency should be obtained in order to calculate the current. Since the load is balanced, then $Z_{R_k} = Z_{S_k} = Z_{T_k} = Z_k$. The impedances are:

$$Z_1 = 0.7071 + 0.7071\,\sigma_{12},\ Y_1 = 0.7071 - 0.7071\,\sigma_{12}$$
$$Z_2 = 0.7071 + 1.4142\,\sigma_{34},\ Y_2 = 0.2828 - 0.5657\,\sigma_{34}$$

Now, the current array can be calculated as in (22):

$$i = \sum_{k=1}^{2} i_k = i_1 + i_2 = Y_1 u_1 + Y_2 u_2 \tag{54}$$

By substituting numerical values in (54):

$$i = \begin{bmatrix} +162.63\sigma_1 + 162.63\sigma_2 + 31.11\sigma_3 + 62.22\sigma_4 \\ -222.16\sigma_1 + 59.53\sigma_2 + 38.33\sigma_3 - 58.06\sigma_4 \\ 59.53\sigma_1 - 222.16\sigma_2 - 69.44\sigma_3 - 4.16\sigma_4 \end{bmatrix} = \underbrace{\begin{bmatrix} 162.63\sigma_1 + 31.11\sigma_3 \\ -81.32\sigma_1 + 140.84\sigma_2 - 15.55\sigma_3 - 26.94\sigma_4 \\ -81.32\sigma_1 - 140.84\sigma_2 - 15.55\sigma_3 + 26.94\sigma_4 \end{bmatrix}}_{i_p}$$

$$+ \underbrace{\begin{bmatrix} +162.63\sigma_2 + 62.22\sigma_4 \\ -140.85\sigma_1 - 81.32\sigma_2 + 53.88\sigma_3 - 31.11\sigma_4 \\ 140.85\sigma_1 - 81.32\sigma_2 - 53.88\sigma_3 - 31.11\sigma_4 \end{bmatrix}}_{i_q} \tag{55}$$

It can be verified that i_p and i_q are orthogonal since $i_p \cdot i_q = 0$.

As the voltage and current of the load are known, the apparent geometric power can be calculated:

$$M = ui = M_a + M_N \tag{56}$$

where

$$M_a = P = 122,485$$
$$M_N = \underbrace{112,217\sigma_{12} + 20,534\sigma_{34}}_{Q_B} - 16,100\sigma_{13} - 5,366\sigma_{14} - 5,366\sigma_{23} + 16,100\sigma_{24}$$

The units of active power are Watts [W] and every bivector in M_N has units of VoltAmperes [VA]. The terms $\sigma_{(2k-1)(2k)}$ refer to the reactive power of the harmonic k, in the Budeanu's sense [34] (voltage and current components of the same frequency that are in quadrature). The non-active power M_N includes all the components that do not produce active power P. The active current can be obtained by using (48):

$$i_a = G_e u = \frac{P}{\|u\|^2} u = \frac{122,485}{195,300} \begin{bmatrix} +230\sigma_1 + 110\sigma_3 \\ -115\sigma_1 + 199.19\sigma_2 - 55\sigma_3 - 95.26\sigma_4 \\ -115\sigma_1 - 199.19\sigma_2 - 55\sigma_3 + 95.26\sigma_4 \end{bmatrix} \tag{57}$$

$$= \begin{bmatrix} +144.25\sigma_1 + 68.99\sigma_3 \\ -72.12\sigma_1 + 124.92\sigma_2 - 34.49\sigma_3 - 59.74\sigma_4 \\ -72.12\sigma_1 - 124.92\sigma_2 - 34.49\sigma_3 + 59.74\sigma_4 \end{bmatrix}$$

The norm of the active current is $\|i_a\| = 159.9$ A and the norm of the total current is $\|i\| = 240.29$ A. The scattered current can be calculalated by using (49):

$$i_s = i_p - i_a \tag{58}$$

$$i_s = \underbrace{\begin{bmatrix} +162.63\sigma_1 & +31.11\sigma_3 \\ -81.32\sigma_1 - 140.84\sigma_2 - 26.94\sigma_3 - 15.56\sigma_4 \\ -81.32\sigma_1 + 140.84\sigma_2 + 26.94\sigma_3 - 15.56\sigma_4 \end{bmatrix}}_{i_p} - \underbrace{\begin{bmatrix} +144.25\sigma_1 & +68.99\sigma_3 \\ -72.12\sigma_1 + 124.92\sigma_2 - 34.49\sigma_3 - 59.74\sigma_4 \\ -72.12\sigma_1 - 124.92\sigma_2 - 34.49\sigma_3 + 59.74\sigma_4 \end{bmatrix}}_{i_a} =$$

$$= \begin{bmatrix} +18.39\sigma_1 & -37.88\sigma_3 \\ -9.19\sigma_1 + 15.92\sigma_2 + 18.94\sigma_3 + 32.80\sigma_4 \\ -9.19\sigma_1 - 15.92\sigma_2 + 18.94\sigma_3 - 32.80\sigma_4 \end{bmatrix} \tag{59}$$

The norm of the scattered current is $\|i_s\| = 42.10$ A. The above results are in line with those obtained by using complex numbers (which are omitted for the sake of brevity).

Figure 7 shows the waveforms of the source voltage and several current components. The active current is proportional to the sum of the fundamental and second harmonic components of the voltage waveform. This current is part of the parallel current, along with the scattered current. It can observed that all the currents are balanced, as expected.

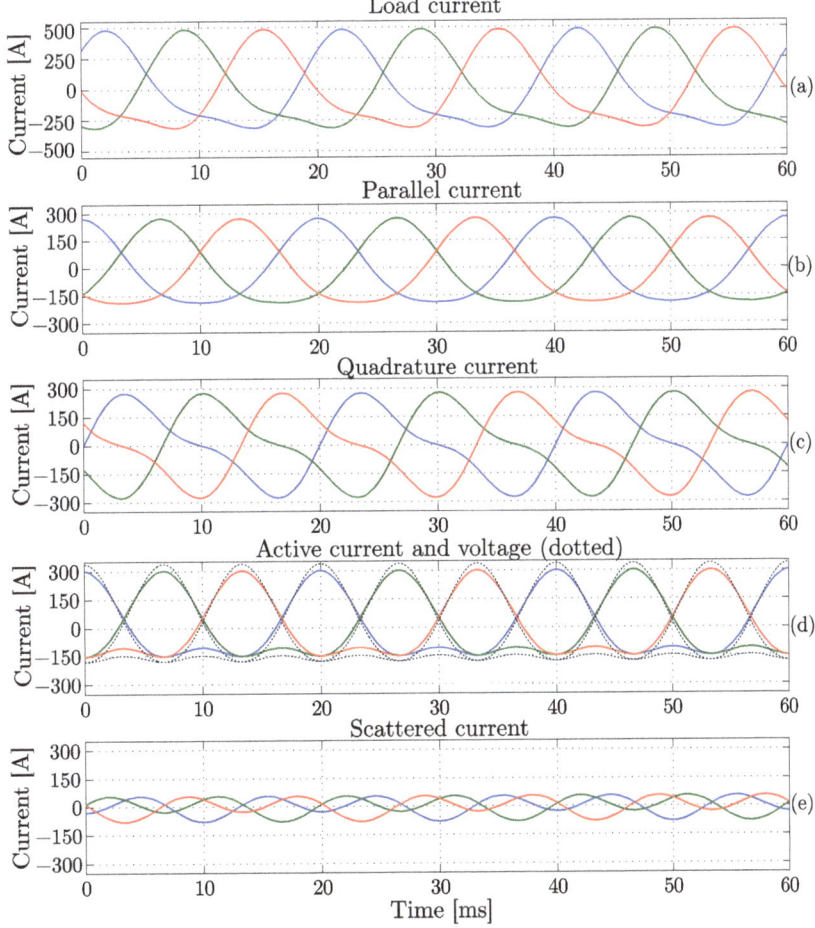

Figure 7. Voltage and current waveforms for Example 2. (**a**) Load current, (**b**) parallel current, (**c**) quadrature current, (**d**) active current and voltage, and (**e**) scattered current.

5. Conclusions

In this paper, GA was applied in order to analyse and solve symmetric and balanced three-phase electrical circuits that operate under sinusoidal and non-sinusoidal conditions. The concept of geometric vector was presented in polar coordinates so that operations such as voltage and current vector rotations can easily be performed. The Argand diagram σ_1-σ_2 was used to depict vectors, while the *scalar-bivector* one was introduced in order to depict impedances/admittances and power components. This is a clear difference compared to traditional representations based on complex numbers. It has been shown that the proposed theory can be applied directly over three-phase electrical circuits using Kirchhoff and Ohm's law. The use of the geometric apparent power M and the current decomposition with relevant engineering meaning provide additional features compared to traditional power theories. The examples presented in the paper verify the validity of the proposed theory. Further developments will include the addition of a fourth wire and unbalanced loads under asymmetrical and distorted voltage conditions. This fact requires the use of orthogonal transformations, such as the one derived from the application of the symmetrical components. It can be addressed through the addition a higher number of dimensions.

Author Contributions: Data curation, F.M.A.-C.; Formal analysis, F.G.M. and R.B.; Investigation, F.M.A.-C.; Methodology, F.G.M. and J.R.P.; Resources, A.A.; Software, A.A.; Supervision, R.B.; Writing—original draft, F.G.M. and J.R.P. All authors contributed equally to this work. All authors have read and agreed to the published version of the manuscript.

Funding: This research was funded by Ministry of Science, Innovation and Universities grant number PGC2018-098813-B-C33.

Institutional Review Board Statement: Not applicable.

Informed Consent Statement: Not applicable.

Data Availability Statement: Not applicable.

Acknowledgments: This research has been supported by the Ministry of Science, Innovation and Universities at the University of Almeria under the programme "Proyectos de I+D de Generacion de Conocimiento" of the national programme for the generation of scientific and technological knowledge and strengthening of the R+D+I system with grant number PGC2018-098813-B-C33.

Conflicts of Interest: The authors declare no conflict of interest.

References

1. Steinmetz, C.P. Complex quantities and their use in electrical engineering. In Proceedings of the International Electrical Congress, Chicago, IL, USA, 21–25 August 1893; pp. 33–74.
2. Kennelly, A. Impedance. *Trans. Am. Inst. Electr. Eng.* **1893**, *10*, 172–232. [CrossRef]
3. Heaviside, O. *Electrical Papers*; Macmillan and Company: New York, NY, USA, 1892; Volume 2.
4. Shepherd, W.; Zand, P. Energy flow and power factor in nonsinusoidal circuits. *Electron. Power* **1980**, *26*, 263.
5. Emanuel, A.E. Powers in nonsinusoidal situations—A review of definitions and physical meaning. *IEEE Trans. Power Deliv.* **1990**, *5*, 1377–1389. [CrossRef]
6. Staudt, V. Fryze-Buchholz-Depenbrock: A time-domain power theory. In Proceedings of the IEEE International School on Nonsinusoidal Currents and Compensation (ISNCC 2008), Lagow, Poland, 10–13 June 2008; pp. 1–12.
7. Czarnecki, L.S. Currents' physical components (CPC) concept: A fundamental of power theory. In Proceedings of the IEEE International School on Nonsinusoidal Currents and Compensation (ISNCC 2008), Lagow, Poland, 10–13 June 2008; pp. 1–11.
8. Yu, Z.; Li, D.; Zhu, S.; Luo, W.; Hu, Y.; Yuan, L. Multisource multisink optimal evacuation routing with dynamic network changes: A geometric algebra approach. *Math. Methods Appl. Sci.* **2018**, *41*, 4179–4194. [CrossRef]
9. Papaefthymiou, M.; Papagiannakis, G. Real-time rendering under distant illumination with conformal geometric algebra. *Math. Methods Appl. Sci.* **2018**, *41*, 4131–4147. [CrossRef]
10. Montoya, F.G.; Alcayde, A.; Arrabal-Campos, F.M.; Baños, R. Quadrature Current Compensation in Non-Sinusoidal Circuits Using Geometric Algebra and Evolutionary Algorithms. *Energies* **2019**, *12*, 692. [CrossRef]
11. Montoya, F.G.; nos, R.B.; Alcayde, A.; Arrabal-Campos, F.M. Analysis of power flow under non-sinusoidal conditions in the presence of harmonics and interharmonics using geometric algebra. *Int. J. Electr. Power Energy Syst.* **2019**, *111*, 486–492. [CrossRef]

12. Montoya, F.G.; Alcayde, A.; Arrabal-Campos, F.M.; Baños, R.; Roldán-Pérez, J. Geometric Algebra Power Theory (GAPoT): Revisiting Apparent Power under Non-Sinusoidal Conditions. *arXiv* **2020**, arXiv:2002.10011.
13. Castilla, M.; Bravo, J.C.; Ordonez, M.; Montano, J.C. An approach to the multivectorial apparent power in terms of a generalized poynting multivector. *Prog. Electromagn. Res.* **2009**, *15*, 401–422. [CrossRef]
14. Menti, A.; Zacharias, T.; Milias-Argitis, J. Geometric algebra: A powerful tool for representing power under nonsinusoidal conditions. *IEEE Trans. Circuits Syst. I Regul. Pap.* **2007**, *54*, 601–609. [CrossRef]
15. Castro-Nuñez, M.; Castro-Puche, R.; Nowicki, E. The use of geometric algebra in circuit analysis and its impact on the definition of power. In Proceedings of the IEEE 2010 International School on Nonsinusoidal Currents and Compensation (ISNCC), Lagow, Poland, 15–18 June 2010; pp. 89–95.
16. Castro-Nuñez, M.; Castro-Puche, R. Advantages of geometric algebra over complex numbers in the analysis of networks with nonsinusoidal sources and linear loads. *IEEE Trans. Circuits Syst. I Regul. Pap.* **2012**, *59*, 2056–2064. [CrossRef]
17. Castro-Núñez, M.; Castro-Puche, R. The IEEE Standard 1459, the CPC power theory, and geometric algebra in circuits with nonsinusoidal sources and linear loads. *IEEE Trans. Circuits Syst. I Regul. Pap.* **2012**, *59*, 2980–2990. [CrossRef]
18. Montoya, F.G.; Baños, R.; Alcayde, A.; Arrabal-Campos, F.M. A new approach to single-phase systems under sinusoidal and non-sinusoidal supply using geometric algebra. *Electr. Power Syst. Res.* **2020**, *189*, 106605. [CrossRef]
19. Castilla, M.; Bravo, J.C.; Ordoñez, M. Geometric algebra: A multivectorial proof of Tellegen's theorem in multiterminal networks. *IET Circuits Devices Syst.* **2008**, *2*, 383–390. [CrossRef]
20. Castilla, M.; Bravo, J.C.; Ordonez, M.; Montaño, J.C. Clifford theory: A geometrical interpretation of multivectorial apparent power. *IEEE Trans. Circuits Syst. I Regul. Pap.* **2008**, *55*, 3358–3367. [CrossRef]
21. Montoya, F.G.; Roldán-Pérez, J.; Alcayde, A.; Arrabal-Campos, F.M.; Banos, R. Geometric Algebra Power Theory in Time Domain. *arXiv* **2020**, arXiv:2002.05458.
22. Lev-Ari, H.; Stanković, A.M. A geometric algebra approach to decomposition of apparent power in general polyphase networks. In Proceedings of the 41st North American Power Symposium, Starkville, MS, USA, 4–6 October 2009; pp. 1–6.
23. Jancewicz, B. *Multivectors and Clifford Algebra in Electrodynamics*; World Scientific: Singapore, 1989.
24. Dorst, L.; Fontijne, D.; Mann, S. *Geometric Algebra for Computer Science: An Object-Oriented Approach to Geometry*; Elsevier: Amsterdam, The Netherlands, 2010.
25. Hestenes, D.; Sobczyk, G. *Clifford Algebra to Geometric Calculus: A Unified Language for Mathematics and Physics*; Springer Science & Business Media: Berlin/Heidelberg, Germany, 2012; Volume 5.
26. Chappell, J.M.; Drake, S.P.; Seidel, C.L.; Gunn, L.J.; Iqbal, A.; Allison, A.; Abbott, D. Geometric algebra for electrical and electronic engineers. *Proc. IEEE* **2014**, *102*, 1340–1363. [CrossRef]
27. Czarnecki, L.S. Currents' Physical Components (CPC) in Circuits with Nonsinusoidal Voltages and Currents. Part 2: Three-Phase Three-Wire Linear Circuits. *Electr. Power Qual. Util. J.* **2006**, *12*, 3–13.
28. Eid, A.H.; Montoya, F.G. Geometric Algebra Power Theory Numerical Library. Available online: https://github.com/ga-explorer/GAPoTNumLib (accessed on 16 March 2021).
29. Hestenes, D. *New Foundations for Classical Mechanics*; Springer Science & Business Media: Berlin/Heidelberg, Germany, 2012; Volume 15.
30. Czarnecki, L.S. Considerations on the Reactive Power in Nonsinusoidal Situations. *IEEE Trans. Instrum. Meas.* **1985**, *IM-34*, 399–404. [CrossRef]
31. Montoya, F.; Baños, R.; Alcayde, A.; Montoya, M.; Manzano-Agugliaro, F. Power Quality: Scientific Collaboration Networks and Research Trends. *Energies* **2018**, *11*, 2067. [CrossRef]
32. Sharon, D. Reactive-power definitions and power-factor improvement in nonlinear systems. *Proc. Inst. Electr. Eng.* **1973**, *120*, 704–706. [CrossRef]
33. Czarnecki, L.S. Currents' physical components (CPC) in circuits with nonsinusoidal voltages and currents. Part 1, Single-phase linear circuits. *Electr. Power Qual. Util. J.* **2005**, *11*, 3–14.
34. Willems, J.L. Budeanu's reactive power and related concepts revisited. *IEEE Trans. Instrum. Meas.* **2011**, *60*, 1182–1186. [CrossRef]

Article

Lower-Estimates on the Hochschild (Co)Homological Dimension of Commutative Algebras and Applications to Smooth Affine Schemes and Quasi-Free Algebras

Anastasis Kratsios

Department of Mathematics, ETH Zürich, Rämistrasse 101, 8092 Zurich, Switzerland; anastasis.kratsios@math.ethz.ch; Tel.: +41-078-933-99-19

Abstract: The Hochschild cohomological dimension of any commutative k-algebra is lower-bounded by the least-upper bound of the flat-dimension difference and its global dimension. Our result is used to show that for a smooth affine scheme X satisfying Pointcaré duality, there must exist a vector bundle with section M and suitable n which the module of algebraic differential n-forms $\Omega^n(X, M)$. Further restricting the notion of smoothness, we use our result to show that most k-algebras fail to be smooth in the quasi-free sense. This consequence, extends the currently known results, which are restricted to the case where $k = \mathbb{C}$.

Keywords: hochschild cohomology; homological dimension theory; non-commutative geometry; quasi-free algebras; pointcaré duality; higher differential forms

Citation: Kratsios, A. Lower-Estimates on the Hochschild (Co)Homological Dimension of Commutative Algebras and Applications to Smooth Affine Schemes and Quasi-Free Algebras. *Mathematics* **2021**, *9*, 251. https://doi.org/10.3390/math9030251

Academic Editor: Sonia Pérez-Díaz
Received: 12 January 2021
Accepted: 25 January 2021
Published: 27 January 2021

Publisher's Note: MDPI stays neutral with regard to jurisdictional claims in published maps and institutional affiliations.

Copyright: © 2021 by the authors. Licensee MDPI, Basel, Switzerland. This article is an open access article distributed under the terms and conditions of the Creative Commons Attribution (CC BY) license (https://creativecommons.org/licenses/by/4.0/).

1. Introduction

Non-commutative geometry is a rapidly developing area of contemporary mathematical research that studies non-commutative algebras using formal geometric tools. The field traces its most evident origins back to the results of [1], which show that any compact Hausdorff space can be fully reconstructed, and largely understood, from its associated C*-algebra of functions $C(X)$. However, the trend of understanding geometric properties via algebraic dual theories is echoed throughout mathematics; with notable examples coming from the duality between finitely generated algebras and affine schemes (see [2]), the description of any smooth manifold M through its commutative algebra $C^\infty(M)$, and ultimately culminating with the result of [3,4] describing the duality relationship between algebra and geometry in full generality.

Though a large portion of the interest in non-commutative geometry stems from its connections with physics, see [5–7]. A. Connes largely made these connections through the cyclic cohomology theory of [8], a generalized de Rham cohomology theory for non-commutative spaces, which closely tied through the Connes complex to one of the central tools of non-commutative geometry and the central object of study of this paper, namely Hochschild (co)homology.

Hochschild (co)homology, originally introduced in [9], is a cohomology theory for non-commutative k-algebras. Since its introduction, it has become a key tool and object of study in non-commutative geometry since the results of [10] (and more recently generalized in [11] to characteristic p fields); which identifies the Hochschild homology of commutative k-algebras over a characteristic 0 field k, to the module of Khäler differentials over their associated affine scheme. Likewise, the result identifies Hochschild's cohomology theory with the modules of derivations and, therefore, with the tangential structure over the commutative algebra's associated affine scheme. Likewise, in these cases, Pointcaré duality-like results can also be entirely formulated between these structures and the Hochschild (co)homology theories as shown in [12].

This article focuses on a fundamental non-commutative geometric invariant derived from the Hochschild (co)homology, namely its (co)homological *dimension*. We focus on the interplay between this (co)homological invariant of commutative *k*-algebras over general commutative rings *k*, and its implications on various notions of smoothness of its associated dual non-commutative space; such as the quasi-freeness (or formal smoothness) of [13,14], or more generally, the vanishing of their higher modules of differential forms as seen in [12].

The relationship between the Hochschild (co)homology theory and smoothness has seen study in the case where *k* is a field in [15,16]. However, the general case is still far from understood and this is likely due to it requiring a more subtle treatment offered by the less-standard tools of relative homological-algebraic (see [17,18] for example). Indeed, this paper proposes a set of lower-estimates of this invariant, which can be easily computed from local data of any commutative *k*-algebra over a commutative ring *k* with unity.

The paper's main results are used to show that for any smooth affine scheme *X* there must exist a vector bundle on *X* with section *M* and a suitably small natural number *n* for which the module of algebraic differential *n*-forms with values in *M*, denoted by $\Omega^n(X, M)$ is non-trivial. Our results are also used to derive simple tests for a *k*-algebras' quasi-freeness. This latter application extends known results of [14] in the special case where $k = \mathbb{C}$. Using this result, we conclude that typical *k*-algebras are not quasi-free. Concrete applications are considered within the scope of arithmetic geometry.

Organization of the Paper

The paper is organized as follows. Section 2 contains the paper's main theorems as well as its non-commutative geometric questions consequences. Each result is followed by examples which unpack the general implications in the context of algebraic geometry. Appendix A contains detailed background material in the relative homological algebraic tools required for the paper's proofs is included after the paper's conclusion. Likewise, the paper's proofs and any auxiliary technical lemma is also relegated to Appendices B–D.

2. Main Result

From here on out, *A* will always be a commutative *k*-algebra. The remainder of this paper will focus on establishing the following result. An analogous statement was made in [14] that all affine algebraic varieties over $\mathbb{C}$ of dimension at greater than 1 fail to have a quasi-free $\mathbb{C}$-algebra of functions. Once, the assumption that $k = \mathbb{C}$ is relaxed, we find an analogous claim is true; however, the analysis is more delicate. Our principle result is the following.

Theorem 1 (Lower-Bound on Hochschild Cohomological Dimension). *Let A be a commutative k-algebra and* $\mathfrak{m}$ *be a non-zero maximal ideal in A such that* $A_\mathfrak{m}$ *is has finite* $k_{i-1[\mathfrak{m}]}$*-flat dimension and* $D(k_{i-1[\mathfrak{m}]})$ *is finite. Then:*

$$fd_{A_\mathfrak{m}}(M_\mathfrak{m}) - D(k_{i-1[\mathfrak{m}]}) - fd_{k_\mathfrak{m}}(A_\mathfrak{m}) \leq HCdim(A|k)$$

Theorem 1 allows for an easily computable lower-bound on the Hochschild cohomological dimension of nearly any commutative *k*-algebra *A*, granted that it is smooth in the classical sense at-least at one point. The next result, obtains an even simpler criterion under the additional assumption that *A* is *k*-flat.

Theorem 2. *Let k be of finite global dimension, A be a k-algebra which is flat as a k-module. Then M:*

$$fd_A(M) - D(k) \leq HCdim(A|k). \tag{1}$$

Example 1. *Let A be a commutative k-algebra and* $\mathfrak{m}$ *be a non-zero maximal ideal in A such that* $A_\mathfrak{m}$ *is has finite* $k_{i-1[\mathfrak{m}]}$*-flat dimension,* $D(k_{i-1[\mathfrak{m}]})$*, and A is Cohen-Macaulay at some maximal ideal* $\mathfrak{m}$*. Then*

$$Krull(A_\mathfrak{m}) - D(k_{i-1[\mathfrak{m}]}) - fd_\mathfrak{m}(A_\mathfrak{m}) \leq HCdim(A|k).$$

Example 2. *Let k be of finite global dimension, A be a k-algebra which is flat as a k-module. Then, for every A-module M, if $x_1, .., x_n$ is a regular sequence in A then:*

$$n - D(k) \leq HCdim(A|k). \tag{2}$$

Furthermore if A is commutative and Cohen-Macaulay at a maximal ideal $\mathfrak{m}$ then:

$$Krull(A_\mathfrak{m}) - D(k) \leq HCdim(A|k). \tag{3}$$

Next, we consider the implications of our dimension-theoretic formulas within the scope of algebraic geometry from the non-commutative geometric vantage-point.

2.1. Non-Triviality of Higher Differential Forms

The paper's provides a homological argument showing that a smooth affine scheme must have some non-trivial module of higher-differential forms. These begin with the non-triviality of the Hochschild homology modules.

To show our result, we begin by recalling the terminology introduced in [12]. Recall that a k-algebra is satisfies *Pointcaré duality in dimension d* if the *dualising module* $\omega_A \triangleq Ext^d_{\mathscr{E}^k_A}(A, A)$ satisfies $Ext^i_{\mathscr{E}^k_A}(A, k) = 0$ for every $i \neq d$ and if in addition $pd_{\mathscr{E}^k_A}(\omega_A) < \infty$. We also recall that an A-bimodule M is invertible if and only if there exits another A-bimodule, which we denote by M^{-1}, for which $M \otimes_A M^{-1} \cong M^{-1} \otimes_A M \cong A$ in $_A Mod_A$.

Corollary 1 (Non-Triviality of Hochschild Homology Modules). *Let k be a commutative ring and X be a d-dimensional smooth affine scheme over k whose coordinate ring satisfies Pointcaré duality in dimension d and is invertible. Then, there is an A-bimodule M and some $0 \leq n \leq d - fd_A(M) + D(k)$ satisfying*

$$HH_n(A, M) \not\cong 0.$$

On applying the Hochschild-Kostant-Rosenberg Theorem to Corollary 1, we immediately obtain the claimed result. Recall that $\Omega^n(X, M)$ denotes the algebraic differential n-forms on the affine scheme X with coefficients in the vector bundle whose section is the $k[A]$-bimodule M.

Corollary 2. *Let k be a commutative ring and X be a d-dimensional smooth affine scheme over k whose coordinate ring satisfies Pointcaré duality in dimension d and is invertible. Then, there exists a some $0 \leq n \leq d - fd_A(M) + D(k)$ and a vector bundle whose section is the $k[A]$-module M for which the algebraic differential n-forms for which*

$$\Omega^n(X, M) \not\cong 0.$$

Next, we use Theorem 1 to demonstrate the rarity of commutative quasi-free k-algebras.

2.2. Quasi-Free Algebras are Uncommon

Corollary 3 (Krull Dimension-Theoretic Criterion for Quasi-Freeness). *Let A be a commutative k-algebra and $\mathfrak{m}$ be a non-zero maximal ideal in A such that $A_\mathfrak{m}$ is has finite $k_{i-1[\mathfrak{m}]}$-flat dimension, $D(k_{i-1[\mathfrak{m}]})$, and A is Cohen-Macaulay at some maximal ideal $\mathfrak{m}$. Then, A is not Quasi-free if*

$$Krull(A_\mathfrak{m}) \leq 2 + D(k_{i-1[\mathfrak{m}]}) - fd_\mathfrak{m}(A_\mathfrak{m}).$$

Let us also consider the simpler form implied by Theorem 2.

Corollary 4. *If k is of finite global dimension, A is a k-algebra which is flat as a k-module, and if $A_\mathfrak{m}$'s Krull dimension is at least $2 + D(k)$ then A is not Quasi-free.*

We unpack Theorem 2 in the context of classical algebraic and arithmetic geometry.

Examples

To build intuition before proceeding, we consider a counter-intuitive consequence. Namely, that most examples of smooth commutative algebras fail to be quasi-free, even when $k \neq \mathbb{C}$. This makes smoothness, in the sense of [14], very rare in the non-commutative category. The following example from arithmetic geometry is of interest.

Let be an affine algebraic $\mathbb{C}$-variety $V(A)$. For any point x in $V(A)$ the ideal generated by the collection of regular functions on $V(A)$ vanishing at the point x is denoted by $\mathscr{I}(x)$; in fact $\mathscr{I}(x)$ is a maximal ideal in A [19]. Moreover, for any affine-algebraic variety $V(A)$ there exists a point x such that $A_{\mathscr{I}(x)}$ is regular. Since every regular local $\mathbb{C}$-algebra is Cohen Macaulay at its maximal ideal, then A is Cohen-Macaulay at $\mathscr{I}(x)$. Since $\mathbb{C}$ is a field it is a regular local ring of Krull dimension 0; the Auslander-Buchsbaum-Serre theorem thus implies $D(k) = Krull(k) = 0$, moreover $A_{\mathscr{I}(x)}$ is a $\mathbb{C}$-vector space whence it is a $\mathbb{C}$-free and so is a $\mathbb{C}$-flat module. Therefore Theorem 2 applies if $Krull(A) \geq 2$. We summarize this finding as follows.

Corollary 5. *If X is an affine $\mathbb{C}$-variety and $k[A]$'s Krull dimension is greater than 1 then the $\mathbb{C}$-algebra A is not quasi-free*

Remark 1. *Corollary 5 implies that any affine algebraic $\mathbb{C}$-variety which is not a disjoint union of curves or points has a coordinate ring which fails to be quasi-free over $\mathbb{C}$.*

Example 3. *The $\mathbb{C}$-algebra $\mathbb{C}[x_{1,1}, x_{1,2}, x_{2,1}, x_{2,2}]_{(det)}$ is not quasi-free.*

Proof. $\mathbb{C}[x_{1,1}, x_{1,2}, x_{2,1}, x_{2,2}]_{(det)}$ is of Krull dimension $4 > 1$ [20] therefore Theorem 2 applies. □

Corollary 6 (Arithmetic Polynomial-Algebras). *The $\mathbb{Z}$-algebra $\mathbb{Z}[x_1, .., x_n]$ fails to be quasi-free for values of $n > 1$.*

Proof. Since $\mathbb{Z}[x_1, ...x_n]$ is Cohen-Macaulay at the maximal ideal $(x_1, ...x_n, p)$ and is of Krull dimension $n + 1 = Krull(\mathbb{Z}[x_1, ...x_n])$. Moreover, one computes that $D(\mathbb{Z}) = 1$. Whence by point 2 of Theorem 2: $\mathbb{Z}[x_1, .., x_n]$ fails to be Quasi-free if $2 \leq Krull(\mathbb{Z}[x_1, ...x_n]) - D(\mathbb{Z}) = (n+1) - 1 = n$. □

The contributions of the paper are now summarized.

3. Conclusions

This paper's main result derived a general lower bound on the Hochschild cohomological dimension of an arbitrary commutative k-algebra A over a general commutative ring k. Theorem 1 derived, the lower-bound for this (co)homological invariant was expressed in terms of other (co)homological dimension-theoretic invariants, namely the flat dimension over A, the global dimension of A, and the flat dimension of A over k; where each quantity was appropriately localized. Examples 1 and 2, built on these results to lower-bound the Hochschild cohomological dimension purely in terms of easily computable quantities, such as the Krull dimension, when A was Cohen-Macaulay. Theorem 2 then expresses a non-localized analog of Theorem 1 wherein no commutativity of A was required.

The paper's results have then been applied the results to purely geometric questions. First, the dimension-theoretic formula was used in Corollary 2 to show infer the non-triviality of certain higher algebraic differential forms of any smooth affine scheme with values in a vector bundle with a non-trivial section. The dual result was also considered in Corollary 1 where dimension-theoretic conditions were obtained for the non-vanishing of some of the Hochschild homology modules under Pointcaré duality in the sense of [12].

Next, using the general (co)homological dimension-theoretic estimates, a result of [14], which showed that most commutative affine k-algebras fail to be smooth in the non-commutative sense formalized by quasi-freeness, was extended from the simple case where

k was a field to the general case where k is simply a commutative ring. Specifically, in Corollaries 3 and 4, easily applicable dimension-theoretic tests for the non-quasi-freeness (non-formal smoothness) of a commutative k-algebra over a general ring k were derived. The tools are simple and only require a simple computation involving the Krull dimension of A, the flat-dimension of k at one point, and the base ring's global dimension to identify if A's associated non-commutative space is quasi-free or not.

Author Contributions: All relevant work was carried out by the author. All authors have read and agreed to the published version of the manuscript.

Funding: This research was funded by the ETH Zürich Foundation.

Institutional Review Board Statement: Not applicable.

Informed Consent Statement: Not applicable.

Data Availability Statement: Not Applicable.

Acknowledgments: I would like to thank Abraham Broer for the numerous helpful and insightful conversations.

Conflicts of Interest: The author declares no conflict of interest.

Appendix A. Background

This appendix contains the necessary background material for the formulation of this paper's main results. We refer the reader in further reading to the notes of [21].

Appendix A.1. Relative Homological Algebra

The results in this paper are formulated using the *relative homological algebra*, see [17] for example. The theory is analogous to standard homological algebra; see [22] for example, but in this case, one builds the entire theory relative to a suitable subclass of epi(resp. mono)-morphisms. In our case, these are defined as follows.

Definition A1 ($\mathscr{E}_A^k$-**Epimorphism**)**.** *For any k-algebra A, an epimorphism ϵ in $_A\mathrm{Mod}$ is an $\mathscr{E}_A^k$-epimorphism if and only if ϵ's underlying morphism of k-modules is a k-split epimorphism in $_k\mathrm{Mod}$. The class of these epimorphisms is denoted $\mathscr{E}_A^k$.*

Definition A2 ($\mathscr{E}_A^k$-**Exact sequence**)**.** *An exact sequence of A-modules:*

$$\ldots \xrightarrow{\phi_{i-1}} M_i \xrightarrow{\phi_i} M_{i+1} \xrightarrow{\phi_{i+1}} M_{i+2} \xrightarrow{\phi_{i+2}} \ldots \tag{A1}$$

is said to be $\mathscr{E}_A^k$-exact if and only if for every integer i the there exists a morphism of k-modules $\psi_i : M_{i+1} \to M_i$ such that:

$$\phi_i = \phi_i \circ \psi_i \circ \phi_i. \tag{A2}$$

In particular, a short exact sequence of A-modules which is $\mathscr{E}_A^k$-exact is called an $\mathscr{E}_A^k$-short exact sequence.

Remark A1. *Property* (A2) *is called $\mathscr{E}_{A^e}^k$-admissibility [18]. Alternatively, it is called $\mathscr{E}_{A^e}^k$-allowable [23].*

Example A1. *The augmented bar complex $\hat{CB}_\star(A)$ of a k-algebra A is $\mathscr{E}_{A^e}^k$-exact.*

Definition A3 ($\mathscr{E}_A^k$-**Projective module**)**.** *If A is a k-algebra and P is an A-module, then P is said to be $\mathscr{E}_A^k$-projective if and only if for every $\mathscr{E}_A^k$-short exact sequence:*

$$0 \to M \xrightarrow{\eta} N \xrightarrow{\epsilon} N' \to 0 \tag{A3}$$

the sequence of k-modules:

$$0 \to Hom_A(P, M) \xrightarrow{\eta^*} Hom_A(P, N) \xrightarrow{\epsilon^*} Hom_A(P, N') \to 0 \quad (A4)$$

is exact.

Remark A2. *This definition is equivalent to requiring that P verify the universal property of projective modules only on $\mathcal{E}_A^k$-epimorphisms [23].*

Example A2. *$A^{\otimes n+2}$ is $\mathcal{E}_{A^e}^k$-projective for all $n \in \mathbb{N}$.*

$\mathcal{E}_A^k$-projective A-modules have analogous properties to projective A-modules. For example, $\mathcal{E}_A^k$-projective A-modules admit the following characterization.

Proposition A1. *For any A-module P the following are equivalent:*
- *$\mathcal{E}_A^k$-**Short exact sequence preservation property** P is $\mathcal{E}_A^k$-projective.*
- *$\mathcal{E}_A^k$-**lifting property** For every $\mathcal{E}_A^k$-epimorphism $f : N \to M$ if there exists an A-module morphism $g : P \to M$ then there exists an A-module map $\tilde{f} : P \to N$ such that $f \circ \tilde{f} = g$.*
- *$\mathcal{E}_A^k$-**splitting property** Every short $\mathcal{E}_A^k$-exact sequence of the form:*

$$\mathfrak{E}_\pi : 0 \to M \to N \to P \to 0 \quad (A5)$$

is A-split-exact.
- *$\mathcal{E}_A^k$-**free direct summand property** There exists a k-module F, an A-module Q and an isomorphism of A-modules $\phi : P \oplus Q \xrightarrow{\cong} A \otimes_k F$.*

Remark A3. *If F is a free k-module, some authors call $A \otimes_k F$ an $\mathcal{E}_A^k$-free module. In fact this gives an alternative proof that $A^e \otimes_k A^{\otimes n} \cong A^{\otimes n+2}$ is $\mathcal{E}_{A^e}^k$-free for every $n \in \mathbb{N}$.)*

Proof. See [23] pages 261 for the equivalence of 1, 2 and 3 and page 277 for the equivalence of 1 and 4. □

For a homological algebraic theory to be possible, one needs enough projective (resp. injective) objects. The next result shows that there are indeed enough $\mathcal{E}_A^k$-projectives in $_A\text{Mod}$.

Proposition A2 (Enough $\mathcal{E}_A^k$-projectives). *If A is a k-algebra and M is an A-module then there exists an $\mathcal{E}_A^k$-epimorphism $\epsilon : P \to M$ where P is an $\mathcal{E}_A^k$-projective.*

Proof. By Proposition A1 $A \otimes_k M$ is $\mathcal{E}_A^k$-projective. Moreover, the A-map $\zeta : A \otimes_k M \to M$ described on elementary tensors as $(\forall a \otimes_k m \in A \otimes_k M) \zeta(a \otimes_k m) := a \cdot m$ is epi and is k-split by the section $m \mapsto 1 \otimes_k m$. □

Since there are enough projective objects, then one can build a resolution of any A-module by $\mathcal{E}_A^k$-projective modules.

Definition A4 ($\mathcal{E}_A^k$-projective resolution). *If M is an A^e-module then a resolution $P_\star$ of M is called an $\mathcal{E}_A^k$-projective resolution of M if and only if each P_i is an $\mathcal{E}_A^k$-projective module and $P_\star$ is an $\mathcal{E}_A^k$-exact sequence.*

Example A3. *The augmented bar complex $\hat{C}B_\star(A)$ of A is an $\mathcal{E}_{A^e}^k$-projective resolution of A.*

Remark A4. *A nearly completely analogous argument to Example A3 shows that for any (A, A)-bimodule M, $M \otimes_A \hat{C}B_\star(A)$ is an $\mathcal{E}_{A^e}^k$-projective resolution of M, see for details [24].*

Following [18], the $\mathscr{E}_A^k$-relative derived functors of the tensor product and the Hom_A-functors are introduced, as follows.

Definition A5. $\mathscr{E}_A^k$-*relative Tor*
If N is a right A-module, M is an A-module and $P_\star$ is an $\mathscr{E}_A^k$-projective resolution of N then the k-modules $H_\star(P_\star \otimes_A M)$ are called the $\mathscr{E}_A^k$-relative Tor k-modules of N with coefficients in the A-module M and are denoted by $Tor^n_{\mathscr{E}_A^k}(N, M)$.

Let $H_\star$ (resp. $H^\star$) denote the (co)homology functor from the category of chain (co)complexes on an A-module to the category of A-modules. The $\mathscr{E}_A^k$-relative Tor functors are defined as follows.

Example A4. *The $\mathscr{E}_A^k$-relative Tor functors may differ from the usual (or "absolute") Tor functors. For example consider all the $\mathbb{Z}$-algebra $\mathbb{Z}$, any $\mathbb{Z}$-modules N and M are $\mathscr{E}_\mathbb{Z}^\mathbb{Z}$-projective. In particular, this is true for the $\mathbb{Z}$-modules $\mathbb{Z}$ and $\mathbb{Z}/2\mathbb{Z}$. Therefore $Tor^n_{\mathscr{E}_\mathbb{Z}^\mathbb{Z}}(\mathbb{Z}, \mathbb{Z}/2\mathbb{Z})$ vanish for every positive n, however $Tor^n_\mathbb{Z}(\mathbb{Z}, \mathbb{Z}/2\mathbb{Z})$ does not. For example, $Tor^1_\mathbb{Z}(\mathbb{Z}, \mathbb{Z}/2\mathbb{Z}) \cong \mathbb{Z}/2\mathbb{Z}$ [22].*

Similarly there are $\mathscr{E}_A^k$-relative Ext functors.

Definition A6 ($\mathscr{E}_A^k$-*relative Ext*). If N is and M are A-modules and $P_\star$ is an $\mathscr{E}_A^k$-projective resolution of N then the k-modules $H^\star(Hom_A(P_\star, M))$ are called the $\mathscr{E}_A^k$-relative Ext k-modules of N with coefficients in the A-module M and are denoted by $Ext^n_{\mathscr{E}_A^k}(N, M)$.

The $\mathscr{E}_A^k$-relative homological algebra is indeed well defined, since both the definitions of $\mathscr{E}_A^k$-relative Ext and $\mathscr{E}_A^k$-relative Tor are independent of the choice of $\mathscr{E}_A^k$-projective resolution.

Theorem A1 ($\mathscr{E}_A^k$-*Comparison theorem*). *If $P_\star$ and $P'_\star$ are $\mathscr{E}_A^k$-projective resolutions of an A-module N then for any A-module M there are natural isomorphisms:*

$$H^\star(Hom_{\mathscr{E}_A^k}(P_\star, N)) \xrightarrow{\cong} H^\star(Hom_{\mathscr{E}_A^k}(P'_\star, N)) \tag{A6}$$

and if $P_\star$ and $P'_\star$ are $\mathscr{E}_A^k$-projective resolutions of a right A-module N then:

$$H_\star(P_\star \otimes_A N) \xrightarrow{\cong} H_\star(P'_\star \otimes_A N) \tag{A7}$$

Proof. Nearly identical to the usual comparison theorem, see [23]. □

Example A5. *The $Ext_\mathbb{Z}$ and $\mathscr{E}_\mathbb{Z}^\mathbb{Z}$-relative Ext may differ. For example, one easily computes $Ext^1_\mathbb{Z}(\mathbb{Z}, \mathbb{Z}/2\mathbb{Z}) \cong \mathbb{Z}/2\mathbb{Z}$. However, $Ext^1_{\mathscr{E}_\mathbb{Z}^\mathbb{Z}}(\mathbb{Z}, \mathbb{Z}/2\mathbb{Z}) \cong 0$.*

Analogous to the fact that for any A-module P, P is projective if and only if $Ext^1_A(P, N) \cong 0$ for every A-module N there is the following result, which can be found in ([18], Chapter IX).

Proposition A3. *P is an $\mathscr{E}_A^k$-projective module if and only if for every A-module N:*

$$Ext^1_{\mathscr{E}_A^k}(P, N) \cong 0 \tag{A8}$$

Using the theory of relative (co)homology, we are now in-place to review the Hochschild cohomology theory over general k-algebras.

Appendix A.2. Hochschild (Co)homological Dimension

Since $CB_\star(A)$ is an $\mathscr{E}^k_{A^e}$-projective resolution of A then Theorem A1 and the definition of the $Ext^\star_{\mathscr{E}^k_{A^e}}(A,-)$ functors imply that the Hochschild cohomology of A with coefficients in of [9], denoted by $HH^\star(A,N)$, can be expressed using the $Ext^\star_{\mathscr{E}^k_{A^e}}$. We maintain this perspective throughout this entire article.

Proposition A4. *For every A^e module N there are k-module isomorphisms, natural in N:*

$$HH^\star(A,N) \xrightarrow{\cong} Ext^\star_{\mathscr{E}^k_{A^e}}(A,N) \tag{A9}$$

Taking short $\mathscr{E}^k_{A^e}$-exact sequences to isomorphic long exact sequences.

Definition A7 (Hochschild Homology). *The Hochschild homology $HH_\star(A,N)$ of a k-algebra A with coefficient in the (A,A)-bimodule N is defined as:*

$$HH_\star(A,N) := H_\star(P_\star \otimes_A N) \tag{A10}$$

where $P_\star$ is an $\mathscr{E}^k_{A^e}$-projective resolution of A.

Following the results of [10], the Hochschild cohomology has become the central tool for obtaining non-commutative algebraic geometric analogues of classical commutative algebraic geometric notions. The one of central focus in this paper, is the Hochschild cohomological dimension,

Definition A8 (Hochschild cohomological dimension). *The Hochschild cohomological dimension of a k-algebra A is defined as:*

$$HCdim(A|k) := \sup_{M \in {}_{A^e}Mod} (sup\{n \in \mathbb{N}^\# | HH^n(A,M) \not\cong 0\}). \tag{A11}$$

where $\mathbb{N}^\#$ is the ordered set of extended natural numbers.

The Hochschild cohomological dimension may be related to the following cohomological dimension.

Definition A9 ($\mathscr{E}^k_A$-projective dimension). *If n is a natural number and M is an A-module then M is said to be of $\mathscr{E}^k_A$-projective dimension at most n if and only if there exists a deleted $\mathscr{E}^k_A$-projective resolution of M of length n. If no such $\mathscr{E}^k_A$-projective resolution of M exists then M is said to be of $\mathscr{E}^k_A$-projective dimension ∞. The $\mathscr{E}^k_A$-projective dimension of M is denoted $pd_{\mathscr{E}^k_A}(M)$.*

The following is a translation of a classical homological algebraic result into the setting of $\mathscr{E}^k_{A^e}$-projective dimension, $\Omega^n(A/k)$ and Hochschild cohomology. Here, $\Omega^n(A/k) \triangleq Ker(b'_{n-1})$ and b'_{n-1} is the $(n-1)^{th}$ differential in the augmented Bar resolution of A; see [24] for details on the augmenter Bar complex.

Theorem A2. *For every natural number n, the following are equivalent:*
- $HCdim(A|k) \leq n$
- A is of $\mathscr{E}^k_{A^e}$-projective dimension at most n
- $\Omega^n(A/k)$ is an $\mathscr{E}^k_{A^e}$-projective module.
- $HH^{n+1}(A,M)$ vanishes for every (A,A)-bimodule M.
- $Ext^{n+1}_{\mathscr{E}^k_{A^e}}(A,M)$ vanishes for every A^e-module M.

Proof. (1 ⇒ 4) By definition of the Hochschild cohomological dimension. (4 ⇔ 5) By Proposition A4. (3 ⇒ 2) Since $\Omega^n(A/k)$ is $\mathscr{E}_{A^e}^k$-projective:

$$0 \to \Omega^n(A/k) \to CB_{n-1}(A) \xrightarrow{b'_{n-1}} \dots \xrightarrow{b'_0} A \to 0$$

is a $\mathscr{E}_{A^e}^k$-projective resolution of A of length n. Therefore $pd_{\mathscr{E}_{A^e}^k}(A) \leq n$.

(3 ⇔ 4) By Proposition A9 there are isomorphism natural in M:

$$(\forall M \in {}_{A^e}\text{Mod})\ HH^{1+n}(A, M) \cong Ext^{1+n}_{\mathscr{E}_{A^e}^k}(A, M) \cong Ext^1_{\mathscr{E}_{A^e}^k}(\Omega^n(A/k), M).$$

Therefore for every A^e-module M:

$$Ext^1_{\mathscr{E}_{A^e}^k}(\Omega^n(A/k), M) \cong 0 \text{ if and only if } HH^{1+n}(A, M) \cong 0.$$

By Proposition A3 $\Omega^n(A/k)$ is $\mathscr{E}_A^k$-projective if and only if $Ext^1_{\mathscr{E}_{A^e}^k}(\Omega^n(A/k), M) \cong 0$.

(2 ⇒ 1) If A admits an $\mathscr{E}_{A^e}^k$-projective resolution $P_\star$ of length n then Theorem A1 implies there are natural isomorphisms of A^e-modules:

$$(\forall M \in {}_{A^e}\text{Mod}) Ext^\star_{\mathscr{E}_{A^e}^k}(A, M) \cong H^\star(Hom_{A^e}(P_\star, M)). \tag{A12}$$

Since $P_\star$ is of length n all the maps $p_j : P_{j+1} \to P_j$ are the zero maps therefore so are the maps $p_j^\star : Hom_{A^e}(P_j) \to Hom_{A^e}(P_{j+1})$. Whence (A12) entails that for all $j > n+1$ $Ext^\star_{\mathscr{E}_{A^e}^k}(A, M)$ vanishes. By Proposition A4 this is equivalent to $HH^j(A, M)$ vanishing for all $j > n+1$ for all $M \in {}_{A^e}\text{Mod}$. Hence A is of Hochschild cohomological dimension at most n. □

Next, the non-commutative geometric object focused on in this paper is reviewed.

Appendix A.3. Quasi-Free Algebras

Many of the properties of an algebra are summarized by its Hochschild cohomological dimension, see [10,17] for example. However, this article focuses on the following non-commutative analogue of smoothness of [13], introduced by [14].

Remark A5. *Due to their lifting property, the quasi-free k-algebras are considered a non-commutative analogue to smooth k-algebras; that is k-algebras for which $\Omega_{A|k}$ is a projective A-module.*

This notion of smoothness has played a key role in a number of places in non-commutative algebraic geometry, especially in the cyclic (co)homology of [25].

Definition A10 (Quasi-free k-algebra). *A k-algebra for which all k-Hochschild extensions of A by an (A, A)-bimodule lift is called a **quasi-free** k-algebra.*

Corollary A1. *For a k-algebra A, the following are equivalent:*
- *A is $HCdim(A|k) \leq 1$.*
- *$\Omega^1(A/k)$ is a $\mathscr{E}_{A^e}^k$-projective A^e-module.*
- *A is quasi-free.*

One typically construct quasi-free algebras using Morita equivalences. However, the next proposition, which extends a result of [14] to the case where k need not be a field, may also be used without any such restrictions on k.

Proposition A5. *If A is a quasi-free k-algebra and P is an $\mathscr{E}_{A^e}^k$-projective (A, A)-bimodule then $T_A(P)$ is a quasi-free A-algebra.*

Proof. Differed until the appendix. □

Example A6. *Let $n \in \mathbb{N}$. The $\mathbb{Z}$-algebra $T_\mathbb{Z}\left(\bigoplus_{i=0}^{n}\mathbb{Z}\right)$ is quasi-free.*

Proof. Since all free $\mathbb{Z}$-modules are projective $\mathbb{Z}$-modules and all projective $\mathbb{Z}$-modules are $\mathcal{E}_\mathbb{Z}^\mathbb{Z}$-projective modules, the free $\mathbb{Z}$-module $\bigoplus_{i=0}^{n}\mathbb{Z}$ is $\mathcal{E}_\mathbb{Z}^\mathbb{Z}$-projective. Whence Proposition A5 implies $T_\mathbb{Z}\left(\bigoplus_{i=0}^{n}\mathbb{Z}\right)$ is a quasi-free $\mathbb{Z}$-algebra. □

Example A7. *If A is a quasi-free k-algebra then $T_A(\Omega^1(A/k))$ is a quasi-free A-algebra.*

Proof. By Corollary A1 if A is quasi-free $\Omega^1(A/k)$ must be an $\mathcal{E}_{A^e}^k$-projective (A, A)-bimodule; whence Proposition A5 applies. □

Next, we overview some relevant dimension-theoretic notions and terminology.

Appendix A.4. Classical Cohomological Dimensions

We remind the reader of a few important algebraic invariants which we will require. The reader unfamiliar with certain of these notions from commutative algebra and algebraic geometry is referred to [2,26] or to [19].

Definition A11 (*A-Flat Dimension*)**.** *If A is a commutative ring then the A-flat dimension $fd_A(M)$ of an A-module M is the extended natural number n, defined as the shortest length of a resolution of M by A-flat A-modules. If no such finite n exists n is taken to be ∞.*

We will require the following result, whose proof can be found in [24].

Proposition A6. *If n is a positive integer and if there exists a regular sequence $x_1, .., x_n$ in A of length n then:*

$$n = fd_A(A/(x_1, .., x_n)). \tag{A13}$$

One more ingredient related to the flat dimension will soon be needed.

Proposition A7. *If A is a commutative ring and $\mathfrak{m}$ is a maximal ideal of A then for any A-module M $fd_{A_\mathfrak{m}}(M_\mathfrak{m})$ is a lower-bound for $fd_A(M)$.*

Definition A12. *A-Projective Dimension*

If A is a commutative ring and M is an A-module then the A-projective dimension $pd_A(M)$ of M is the extended natural number n, defined as the shortest length of a deleted A-projective resolution of M. If no such finite n exists n is taken to be ∞.

Lemma A1. *If A is a commutative ring and M is an A-module then $fd_A(M) \leq pd_A(M)$.*

Proof. Since all A-projective A-modules are A-flat, then any A-projective resolution is a A-flat resolution. □

Lemma A2. *If A is a commutative ring then for any A-module M the following are equivalent:*
- *The A-projective dimension of M is at most n.*
- *For every A-module N, the A-module $Ext^A_{n+1}(M, N)$ is trivial.*
- *For every A-module N and every integer $m \geq n+1$: $Ext^A_m(M, N) \cong 0$.*

Proof. Nearly identical to the proof of Theorem A2, see page 456 of [22] for details. □

Definition A13 (Cohen-Macaulay at an Ideal). *A commutative ring A is said to be Cohen-Macaulay at a maximal ideal $\mathfrak{m}$ if and only if either:*

- *$Krull(A_{\mathfrak{m}})$ is finite and there is an $A_{\mathfrak{m}}$-regular sequence $x_1, ..., x_d$ in $A_{\mathfrak{m}}$ of maximal length $d = Krull(A_{\mathfrak{m}})$ such that $\{x_1, .., x_d\} \subseteq \mathfrak{m}$.*
- *$Krull(A_{\mathfrak{m}})$ is infinite and for every positive integer d there is an $A_{\mathfrak{m}}$-regular sequence $x_1, .., x_d$ in $\mathfrak{m}$ on A of length d.*

Proposition A8 ([24]). *If A is a commutative ring which is Cohen Macaulay at the maximal ideal $\mathfrak{m}$ and $Krull(A_{\mathfrak{m}})$ is finite then:*

$$Krull(A_{\mathfrak{m}}) = fd_{A_{\mathfrak{m}}}(A_{\mathfrak{m}}/(x_1,..,x_n)) \leq pd_A(A_{\mathfrak{m}}/(x_1,..,x_n)) \tag{A14}$$

Definition A14. *Global Dimension*

The **global dimension** $D(A)$ of a ring A, is defined as the supremum of all the A-projective dimensions of its A-modules. That is:

$$D(A) := \sup_{M \in {}_A Mod} pd_A(M). \tag{A15}$$

The following modification of the global dimension of a k-algebra, does not ignore the influence of k on a k-algebra A, as will be observed in the next section of this paper.

Definition A15. *$\mathscr{E}^k$-Global dimension*

The *$\mathscr{E}^k$-global Dimension* $D_{\mathscr{E}^k}(A)$ of a k-algebra A is defined as the supremum of all the $\mathscr{E}^k_A$-projective dimensions of its A-modules. That is:

$$D_{\mathscr{E}^k}(A) := \sup_{M \in {}_A Mod} pd_{\mathscr{E}^k_A}(M). \tag{A16}$$

Appendix B. Proofs

This appendix contains certain technical lemmas or auxiliary results that otherwise detracted from the overall flow of the paper.

Appendix C. Technical Lemmas

We make use of the following result appearing in a technical note of Hochschild circa 1958, see [27].

Theorem A3 ([27]). *If k is of finite global dimension, A is a k-algebra which is flat as a k-module and M is an A-module then:*

$$pd_A(M) - D(k) \leq pd_{\mathscr{E}^k_A}(M) \tag{A17}$$

Proposition A9 (Dimension Shifting). *If*

$$... \xrightarrow{d_{n+1}} P_{n+j}P_n \xrightarrow{d_n} ... \xrightarrow{d_2} P_1 \xrightarrow{d_1} P_0 \to 0 \tag{A18}$$

is a deleted $\mathscr{E}^k_A$-projective resolution of an A-module M then for every A-module N and for every positive integer n there are isomorphisms natural in N:

$$Ext^1_{\mathscr{E}^k_A}(Ker(d_n), N) \cong Ext^{n+1}_{\mathscr{E}^k_A}(A, N) \tag{A19}$$

Proof. By definition the truncated sequence is exact:

$$... \xrightarrow{d_{n+j}} P_{n+j} \xrightarrow{d_{n+j-1}} ... \xrightarrow{d_{n+1}} P_{n+1} \xrightarrow{\eta} Ker(d_n) \to 0,, \tag{A20}$$

where η is the canonical map satisfying $d_n = ker(d_n) \circ \eta$ (arising from the universal property of $ker(d_n)$). Moreover, since (A30) is $\mathscr{E}_A^k$-exact, d_n is k-split; whence η must be k-split. Moreover, for every $j \geq n+1$, d_j was by assumption k-split therefore (A20) is $\mathscr{E}_A^k$-exact and since for every natural number $m > n$ P_m is by hypothesis $\mathscr{E}_A^k$-projective then (A20) is an augmented $\mathscr{E}_A^k$-projective resolution of the A-module $Ker(d_n)$.

For every natural number m, relabel:

$$Q_m := P_{m+n} \text{ and } p_m := d_{n+m}. \tag{A21}$$

By Theorem A1, for all $N \in_A \text{Mod}$ and all $m \in \mathbb{N}$, we have that:

$$Ext^m_{\mathscr{E}_A^k}(Ker(d_n), N) \cong H^m(Hom_A(Q_\star, N))$$
$$= Ker(Hom_A(p_n, N))/Im(Hom_A(p_{n+1}, N))$$
$$= Ker(Hom_A(d_{n+m}, N))/Im(Hom_A(d_{n+m+1}, N)) \tag{A22}$$
$$= H^{m+n}(Hom_A(P_\star, N))$$
$$\cong Ext^m_{\mathscr{E}_A^k}(A, N).$$

Therefore, the result follows. □

Appendix D. Auxiliary Results

Proof of Proposition A5. Let

$$0 \to M \to B \xrightarrow{\pi} T_A(P) \to 0 \tag{A23}$$

be a k-Hochschild extension of $T_A(P)$ by M. We use the universal property of $T_A(P)$ to show that there must exist a lift l of (A23).

Let $p : T_A(P) \to A$ be the projection k-algebra homomorphism of $T_A(P)$ onto A. p is k-split since the k-module inclusion $i : A \to T_A(P)$ is a section of p; therefore p is an $\mathscr{E}_{A^e}^k$-epimorphism and

$$0 \to Ker(p \circ \pi) \to B \to A \to 0 \tag{A24}$$

is a k-Hochschild extension of A by the (A, A)-bimodule $Ker(p \circ \pi)$. Since A is a quasi-free k-algebra there exists a k-algebra homomorphism $l_1 : A \to B$ lifting $p \circ \pi$. Hence B inherits the structure of an (A, A)-bimodule and π may be viewed as an (A, A)-bimodule homomorphism. Moreover, l_1 induces an A-algebra structure on B.

Let $f : P \to T_A(P)$ be the (A, A)-bimodule homomorphism satisfying the universal property of the tensor algebra on the (A, A)-bimodule P. Since $\pi : B \to A$ is an $\mathscr{E}_{A^e}^k$-epimorphism and since P is an $\mathscr{E}_{A^e}^k$-projective (A, A)-bimodule, Proposition A1 implies that that there exists an (A, A)-bimodule homomorphism $l_2 : P \to B$ satisfying $\pi \circ l_2 = f$.

Since $l_2 : P \to B$ is an (A, A)-bimodule homomorphism to a A-algebra the universal property of the tensor algebra $T_A(P)$ on the (A, A)-bimodule P, see [28], implies there is an A-algebra homomorphism $l : T_A(P) \to B$ whose underlying function satisfies: $l \circ f = l_2$.

Therefore $l \circ \pi \circ l_2 = l_2$; whence $l \circ \pi = 1_{T_A(P)}$; that is l is a A-algebra homomorphism which is a section of π, that is l lifts π. □

Appendix D.1. Proof of Theorem 1

Our first lemma is a generalization of the central theorem of [27]; which does not rely on the assumption that A is k-flat.

Lemma A3. *If k is of finite global dimension and A is a k-algebra which is of finite flat dimension as a k-module, then for every A-module M:*

$$pd_A(M) - D(k) - fd_k(A) \leq pd_{\mathscr{E}_A^k}(M) \tag{A25}$$

The proof of Lemma A3 relies on the following lemma.

Lemma A4. *If A is a k-algebra such that $fd_k(A) < \infty$ then:*

$$(\forall M \in_k Mod) \; pd_A(A \otimes_k M) - fd_k(A) \leq pd_k(M) \tag{A26}$$

Proof. For every k-module M and every A-module N there is a convergent third quadrant spectral sequence (see [22], page 667):

$$Ext_A^p(Tor_q^k(A,M), N) \underset{p}{\Rightarrow} Ext_k^{p+q}(M, Hom_A(A,N)). \tag{A27}$$

Moreover, the adjunction $- \otimes_k A \dashv Hom_A(A,-)$ extends to a natural isomorphism:

$$(\forall p, q \in \mathbb{N}) Ext_k^{p+q}(M, Hom_A(A,N)) \cong Ext_A^{p+q}(M \otimes_k A, N). \tag{A28}$$

Therefore there is a convergent third-quadrant spectral sequence:

$$Ext_A^p(Tor_q^k(A,M), N) \underset{p}{\Rightarrow} Ext_A^{p+q}(M \otimes_k A, N). \tag{A29}$$

If $pd_A(N) < \infty$, then the result is immediate. Therefore assume that: $pd_A(N) < \infty$. If $p + q > fd_k(A) + pd_A(N)$ then either $p > pd_A(N)$ or $q > fd_k(A)$. In the case of th

$$0 \cong E_2^{p,q} \cong E_\infty^{p,q} \cong Ext_A^{p+q}(M \otimes_k A, N)$$

and in the latter case

$$0 \cong E_2^{p,q} \cong E_\infty^{p,q} \cong Ext_A^{p+q}(M \otimes_k A, N)$$

also. Therefore

$$(\forall N \in {}_AMod) \; 0 \cong Ext_A^n(M \otimes_k A, N) \text{ if } n > fd_k(A) + pd_A(N);$$

hence: $pd_A(M \otimes_k A) \leq fd_k(A) + pd_A(M)$.

Finally, the result follows since $fd_k(A)$ is finite and, therefore, can be subtracted unambiguously. □

Lemma A5. *If A is a k-algebra then for any k-module M there is an $\mathscr{E}_A^k$-exact sequence:*

$$0 \to Ker(\alpha) \to A \otimes_k M \xrightarrow{\alpha} M \to 0 \tag{A30}$$

where α be the map defined on elementary tensors $(a \otimes_k m)$ in $A \otimes_k M$ as $a \otimes_k m \mapsto a \cdot m$.

Proof. α is k-split by the map $\beta : M \to A \otimes_k M$ defined on elements $m \in M$ as $m \mapsto 1 \otimes_k m$. Indeed if $m \in M$ then:

$$\alpha \circ \beta(m) = \alpha(1 \otimes_k m) = 1 \cdot m = m. \tag{A31}$$

□

Lemma A6. *If M and N are A-modules then:*

$$pd_A(M) \leq pd_A(M \oplus N). \tag{A32}$$

Proof.

$$(\forall n \in \mathbb{N})(\forall X \in_A Mod) \; Ext_A^n(M, X) \oplus Ext_A^n(N, X) \cong Ext_A^n(M \oplus N, X). \tag{A33}$$

Therefore $Ext_A^n(M \oplus N, X)$ vanishes only if both $Ext_A^n(M, X)$ and $Ext_A^n(N, X)$ vanish. Lemma A2 then implies: $pd_A(M) \leq pd(M \oplus N)$. □

Proof of Lemma A3
Proof.
Case 1: $pd_{\mathscr{E}_A^k}(M) = \infty$

By definition $pd_A(M) \leq \infty$ therefore trivially if $pd_{\mathscr{E}_A^k}(M) = \infty$ then:

$$pd_A(M) \leq pd_{\mathscr{E}_A^k}(M) + D(k). \tag{A34}$$

Since k's global dimension is finite hence (A34) implies:

$$pd_A(M) - D(k) \leq \infty = pd_{\mathscr{E}_A^k}(M). \tag{A35}$$

Case 2: $pd_{\mathscr{E}_A^k}(M) < \infty$

Let $d := pd_{\mathscr{E}_A^k}(M) + D(k) + fd_k(A)$. The proof will proceed by induction on d.

Base: $d = 0$

Suppose $pd_{\mathscr{E}_A^k}(M) = 0$.

By Theorem A2 M is $\mathscr{E}_A^k$-projective. Lemma A5 implies there is an $\mathscr{E}_A^k$-exact sequence:

$$0 \to Ker(a) \to A \otimes_k M \xrightarrow{\alpha} M \to 0. \tag{A36}$$

Proposition A1 implies that (A36) is A-split therefore M is a direct summand of the A-module $A \otimes_k M$. Hence Lemma A6 implies:

$$pd_A(M) \leq pd_A(M \otimes_k A). \tag{A37}$$

Lemma A4 together with (A37) imply:

$$pd_A(M) \leq pd_A(M \otimes_k A) \leq pd_k(M). \tag{A38}$$

Definition A15 and (A38) together with the assumption that $pd_{\mathscr{E}_A^k}(M) = 0$ imply:

$$pd_A(M) \leq pd_k(M) \leq D(k) = D(k) + 0 + 0 = D(k) + pd_{\mathscr{E}_A^k}(M) + fd_k(A). \tag{A39}$$

Since k's global dimension and $fd_k(A)$ are finite then (A39) implies:

$$pd_A(M) - D(k) - fd_k(A) \leq pd_{\mathscr{E}_A^k}(M). \tag{A40}$$

Inductive Step: $d > 0$

Suppose the result holds for all A-modules K such that $pd_{\mathscr{E}_A^k}(K) + D(k) + fd_k(A) = d$ for some integer $d > 0$. Again appealing to Lemma A5, there is an $\mathscr{E}_A^k$-exact sequence:

$$0 \to Ker(a) \to A \otimes_k M \xrightarrow{\alpha} M \to 0. \tag{A41}$$

Proposition A1 implies $A \otimes_k M$ is $\mathscr{E}_A^k$-projective; whence (A41) implies:

$$pd_{\mathscr{E}_A^k}(Ker(\alpha)) + 1 = pd_{\mathscr{E}_A^k}(M). \tag{A42}$$

Since $Ker(\alpha)$ is an A-module of strictly smaller $\mathscr{E}_A^k$-projective dimension than M the induction hypothesis applies to $Ker(\alpha)$ whence:

$$\begin{aligned}pd_A(Ker(\alpha)) + 1 &\leq pd_{\mathscr{E}_A^k}(Ker(\alpha)) + 1 + D(k) + fd_k(A) \\ &\leq pd_{\mathscr{E}_A^k}(M) + D(k) + fd_k(A).\end{aligned} \tag{A43}$$

The proof will be completed by demonstrating that: $pd_A(M) \leq pd_A(Ker(\alpha)) + 1$. For any $N \in_A Mod$ $Ext_A^*(-, N)$ applied to (A41) gives way to the long exact sequence in homology, particularly the following of its segments are exact:

$$Ext_A^{n-1}(A \otimes_k M, N) \to Ext_A^{n-1}(Ker(\alpha), N) \xrightarrow{\partial^n} Ext_A^n(M, N) \to Ext_A^n(A \otimes_k M, N) \quad (A44)$$

Since $A \otimes_k M$ is $\mathscr{E}_A^k$-projective $pd_{\mathscr{E}_A^k}(A \otimes_k M) = 0$, therefore by the base case of the induction hypothesis $pd_A(A \otimes_k M) \leq pd_{\mathscr{E}_A^k} + D(k) + fd_k(A) = D(k) + fd_k(A)$; thus for every positive integer $n \geq D(k)$ (in particular d is at least n):

$$(\forall N \in_A Mod) \; Ext_A^{n-1}(A \otimes_k M, N) \cong 0 \cong Ext_A^n(A \otimes_k M, N); \quad (A45)$$

whence ∂^n must be an isomorphism. Therefore Lemma A2 implies $pd_A(M)$ is at most equal to $pd_A(Ker(\alpha)) + 1$.

Therefore:

$$pd_A(M) \leq pd_A(Ker(\alpha)) + 1 \quad (A46)$$
$$\leq pd_{\mathscr{E}_A^k}(Ker(\alpha)) + 1 + D(k) + fd_k(A) \quad (A47)$$
$$\leq pd_{\mathscr{E}_A^k}(M) + D(k) + fd_k(A). \quad (A48)$$

Finally since k is of finite global dimension and A is of finite k-flat dimension then (A48) implies:

$$pd_A(M) - D(k) - fd_k(A) \leq pd_{\mathscr{E}_A^k}(M); \quad (A49)$$

thus concluding the induction. □

We will also require the following result.

Remark A6. *Let A be a k-algebra, $i : k \to A$ the morphism defining the k-algebra A and $\mathfrak{m}$ a maximal ideal in A. For legibility the $\mathscr{E}_{A_\mathfrak{m}}^{k_{i^{-1}[\mathfrak{m}]}}$-projective dimension of an $A_\mathfrak{m}$-module N will be abbreviated by $pd_{\mathscr{E}_{\mathfrak{m},k}}(N)$ (instead of writing $pd_{\mathscr{E}_{A_\mathfrak{m}}^{k_{i^{-1}[\mathfrak{m}]}}}(N)$).*

Lemma A7. *If A is a commutative k-algebra and $\mathfrak{m}$ is a non-zero maximal ideal in A then for every A-module M:*

$$pd_{\mathscr{E}_{\mathfrak{m},k}}(M_\mathfrak{m}) \leq pd_{\mathscr{E}_A^k}(M), \quad (A50)$$

where $i : k \to A$ is the inclusion of k into A.

Proof. Since $\mathfrak{m}$ is a prime ideal in A, $i^{-1}[\mathfrak{m}]$ is a maximal ideal in $k_{i^{-1}[\mathfrak{m}]}$, whence the localized ring $k_{i^{-1}[\mathfrak{m}]}$ is a well-defined sub-ring of $A_\mathfrak{m}$. Let

$$\ldots \xrightarrow{d_{n+1}} P_n \xrightarrow{d_n} \ldots \xrightarrow{d_2} P_1 \xrightarrow{d_1} P_0 \xrightarrow{d_0} M \to 0 \quad (A51)$$

be an $\mathscr{E}_A^k$-projective resolution of an A-module M. The exactness of localization [26] implies:

$$\ldots \xrightarrow{d_{n+1} \otimes_A A_\mathfrak{m}} P_n \otimes_A A_\mathfrak{m} \xrightarrow{d_n \otimes_A A_\mathfrak{m}} \ldots \xrightarrow{d_2 \otimes_A A_\mathfrak{m}} P_1 \otimes_A A_\mathfrak{m} \xrightarrow{d_1 \otimes_A A_\mathfrak{m}} P_0 \otimes_A A_\mathfrak{m} \xrightarrow{d_0 \otimes_A A_\mathfrak{m}} M \otimes_A A_\mathfrak{m} \to 0 \quad (A52)$$

is exact. It will now be verified that (A52) is a $\mathscr{E}_{\mathfrak{m},k}$-projective resolution of the $A_\mathfrak{m}$-module $M_\mathfrak{m}$.

The $d_n \otimes_A A_\mathfrak{m}$ are $k_{i^{-1}[\mathfrak{m}]}$-split

Since (A51) was k-split then for every $i \in \mathbb{N}$ there existed a k-module homomorphism $s_i : P_{n-1} \to P_n$ (where for convenience write $P_{-1} := M$) satisfying $d_i = d_i \circ s_i \circ d_i$.

Since $A_\mathfrak{m}$ is a $k_{i-1[\mathfrak{m}]}$-algebra $A_\mathfrak{m}$ may be viewed as a $k_{i-1[\mathfrak{m}]}$-module therefore the maps: $s_i \otimes_A 1_{A_\mathfrak{m}}$ are $k_{i-1[\mathfrak{m}]}$-module homomorphisms; moreover they must satisfy:

$$d_i \otimes_A 1_{A_\mathfrak{m}} = d_i \otimes_A 1_{A_\mathfrak{m}} \circ s_i \otimes_A 1_{A_\mathfrak{m}} \circ d_i \otimes_A 1_{A_\mathfrak{m}}. \tag{A53}$$

Therefore (A52) is $k_{i-1[\mathfrak{m}]}$-split-exact.

The $P_i \otimes_A A_\mathfrak{m}$ are $\mathscr{E}_{\mathfrak{m},k}$-projective

For each $i \in \mathbb{N}$ if P_i is $\mathscr{E}_A^k$-projective therefore Proposition A1 implies there exists some A-module Q and some k-module X satisfying:

$$P_i \oplus Q \cong A \otimes_k X. \tag{A54}$$

Therefore we have that:

$$\begin{aligned}(P_i \otimes_A A_\mathfrak{m}) \oplus (Q \otimes_A A_\mathfrak{m}) &\cong (P_i \otimes_A Q) \otimes_A A_\mathfrak{m} \\ &\cong (A \otimes_k X) \otimes_A A_\mathfrak{m} \\ &\cong (A \otimes_k X) \otimes_A (A_\mathfrak{m} \otimes_{k_{i-1[\mathfrak{m}]}} k_{i-1[\mathfrak{m}]})\end{aligned} \tag{A55}$$

Since A, k and $k_{i-1[\mathfrak{m}]}$ are commutative rings the tensor products $- \otimes_A -, - \otimes_k -$ and $- \otimes_{k_{i-1[\mathfrak{m}]}} -$ are symmetric [22], hence (A55) implies:

$$\begin{aligned}(P_i \otimes_A A_\mathfrak{m}) \oplus (Q \otimes_A A_\mathfrak{m}) &\cong (A \otimes_k X) \otimes_A (A_\mathfrak{m} \otimes_{k_{i-1[\mathfrak{m}]}} k_{i-1[\mathfrak{m}]}) \\ &\cong (A_\mathfrak{m} \otimes_A A) \otimes_{k_{i-1[\mathfrak{m}]}} (k_{i-1[\mathfrak{m}]} \otimes_k X)\end{aligned} \tag{A56}$$

Since A is a subring of $A_\mathfrak{m}$ then (A56) implies:

$$(P_i \otimes_A A_\mathfrak{m}) \oplus (Q \otimes_A A_\mathfrak{m}) \cong A_\mathfrak{m} \otimes_{k_{i-1[\mathfrak{m}]}} (k_{i-1[\mathfrak{m}]} \otimes_k X). \tag{A57}$$

$(k_{i-1[\mathfrak{m}]} \otimes_k X)$ may be viewed as a $k_{i-1[\mathfrak{m}]}$-module with action $\hat{\cdot}$ defined as:

$$(\forall c \in k)(\forall (c' \otimes_k x) \in k_{i-1[\mathfrak{m}]} \otimes_k X) \; c\hat{\cdot}(c' \otimes_k x) := c \cdot c' \otimes x. \tag{A58}$$

Since $(k_{i-1[\mathfrak{m}]} \otimes_k X)$ is a $k_{i-1[\mathfrak{m}]}$-module then for each $i \in \mathbb{N}$ $(P_i \otimes_A A_\mathfrak{m})$ is a direct summand of an $A_\mathfrak{m}$-module of the form $A_\mathfrak{m} \otimes_{k_{i-1[\mathfrak{m}]}} X'$ where X' is a $k_{i-1[\mathfrak{m}]}$-module, thus Proposition A1 implies that $P_i \otimes_A A_\mathfrak{m}$ is $A_\mathfrak{m}$-projective.

Hence (A52) is an $\mathscr{E}_{\mathfrak{m},k}$-projective resolution of $M \otimes_A A_\mathfrak{m} \cong M_\mathfrak{m}$; whence:

$$pd_{\mathscr{E}_{\mathfrak{m},k}}(M_\mathfrak{m}) \leq pd_{\mathscr{E}_A^k}(M). \tag{A59}$$

□

All the homological dimensions discussed to date are related as follows:

Proposition A10. *If A is a commutative k-algebra and $\mathfrak{m}$ be a non-zero maximal ideal in A such that $A_\mathfrak{m}$ has finite $k_{i-1[\mathfrak{m}]}$-flat dimension and $D(k_{i-1[\mathfrak{m}]})$ is finite then there is a string of inequalities:*

$$\begin{aligned}fd_{A_\mathfrak{m}}(M_\mathfrak{m}) - D(k_{i-1[\mathfrak{m}]}) - fd_k(A) &\leq pd_{A_\mathfrak{m}}(M_\mathfrak{m}) - D(k_{i-1[\mathfrak{m}]}) - fd_k(A) \\ &\leq pd_{\mathscr{E}_{\mathfrak{m},k}}(M_\mathfrak{m}) \\ &\leq pd_{\mathscr{E}_A^k}(M) \\ &\leq D_{\mathscr{E}_A^k}(A).\end{aligned}$$

Proof.

- By definition: $pd_{\mathscr{E}_A^k}(M) \leq D_{\mathscr{E}^k}(A)$.
- By Lemma A7: $pd_{\mathscr{E}_{m,k}}(M_m) \leq pd_{\mathscr{E}_A^k}(M)$.
- Since A_m is flat as a $k_{i-1[m]}$-module and $D(k_{i-1[m]})$ is finite Lemma A3 entails: $pd_{A_m}(M_m) - D(k_{i-1[m]}) - fd_k(A) \leq pd_{\mathscr{E}_{m,k}}(M_m)$
- Lemma A1 implies:
$$fd_{A_m}(M_m) \leq pd_{A_m}(M_m). \tag{A60}$$

Since the global dimension of $k_{i-1[m]}$ was assumed to be finite (A60) implies:

$$fd_{A_m}(M_m) - D(k_{i-1[m]}) \leq pd_{A_m}(M_m) - D(k_{i-1[m]}). \tag{A61}$$

□

Lemma A8. *If A is a commutative k-algebra and M and N be A-modules, then there are natural isomorphisms:*

$$Ext^n_{\mathscr{E}_A^k}(M,N) \cong HH^n(A, Hom_k(M,N)) \cong Ext^n_{\mathscr{E}_{A^e}^k}(A, Hom_k(M,N)). \tag{A62}$$

Proof.

- For any (A,A)-bimodule X, $X \otimes_A M$ is an (A,A)-bimodule [22] [Cor. 2.53].
- Moreover, there are natural isomorphisms [22]:

$$Hom_{A Mod}(X \otimes_A M, N) \xrightarrow{\cong} Hom_{A Mod_A}(X, Hom_{k Mod}(M,N)) \text{ [Thrm. 2.75]}. \tag{A63}$$

In particular (A63) implies that for every n in $\mathbb{N}$ there is an isomorphism which is natural in the first input:

$$Hom_{A Mod}(A^{\otimes n} \otimes_A M, N) \xrightarrow{\psi_n} Hom_{A Mod_A}(A^{\otimes n}, Hom_{k Mod}(M,N)). \tag{A64}$$

whence if $b'_{n+1} : A^{\otimes n+3} \to A^{\otimes n+2}$ is the nth map in the Bar complex (recall Example A3) and for legibility denote $Hom_{A Mod_A}(b'_n, Hom_k(M,N))$ by β_n. The naturality of the maps ψ_n imply the following diagram of k-modules commutes:

$$\begin{array}{ccc} Hom_{A Mod}(A^{\otimes n+2} \otimes_A M, N) & \xrightarrow{\psi_n} & Hom_{A Mod_A}(A^{\otimes n+2}, Hom_{k Mod}(M,N)) \\ {\scriptstyle \psi_{n+1}^{-1} \circ \beta_n \circ \psi_n} \downarrow & & \downarrow {\scriptstyle \beta_n} \\ Hom_{A Mod}(A^{\otimes n+3} \otimes_A M, N) & \xrightarrow{\psi_{n+1}} & Hom_{A Mod_A}(A^{\otimes n+3}, Hom_{k Mod}(M,N)) \end{array} \tag{A65}$$

- Therefore for every n in $\mathbb{N}$:

$$(\psi_{n+2}^{-1} \circ \beta_{n+1} \circ \psi_{n+1}) \circ (\psi_{n+1}^{-1} \circ \beta_n \circ \psi_n) = \beta_{n+1} \circ \beta_n \tag{A66}$$
$$=0.$$

Whence $< Hom_{A Mod}(A^{\otimes \star+2} \otimes_A M, N), (\psi_{\star+1}^{-1} \circ \beta_\star \circ \psi_\star) >$ is a chain complex. Moreover, the commutativity of (A65) implies that:

$$(\forall n \in \mathbb{N})\ H^n(Hom_{A Mod}(A^{\otimes \star+2} \otimes_A M, N)) = Ker(\psi_{\star+1}^{-1} \circ \beta_\star \circ \psi_\star)/Im(\psi_{n+2}^{-1} \circ \beta_{n+1} \circ \psi_{n+1})$$
$$\cong Ker(\beta_n)/Im(\beta_{n+1}) \tag{A67}$$
$$= H^n(Hom_{A Mod_A}(A^{\otimes \star+2}, Hom_{k Mod}(M,N)))$$
$$= HH^n(A, Hom_k(M,N)).$$

Furthermore Proposition A4 implies there are natural isomorphisms:

$$HH^n(A, Hom_k(M,N)) \cong Ext^n_{\mathscr{E}_{A^e}^k}(A, Hom_k(M,N)); \tag{A68}$$

Whence for all n in $\mathbb{N}$ there are natural isomorphisms:

$$H^n(Hom_{AMod}(A^{\otimes \star+2} \otimes_A M, N)) \cong HH^n(A, Hom_k(M,N)) \cong Ext^n_{\mathscr{E}^k_{A^e}}(A, Hom_k(M,N)). \tag{A69}$$

- Finally if M is an A-module then $< Hom_{AMod}(A^{\otimes \star+2} \otimes_A M, N), (\psi_{\star+1}^{-1} \circ \beta_\star \circ \psi_\star) >$ calculates the $\mathscr{E}^k_A$-relative Ext groups of M with coefficients in N; therefore, by ([24], pg. 289), there are natural isomorphisms:

$$H^n(Hom_{AMod}(A^{\otimes \star+2} \otimes_A M, N)) \cong Ext^n_{\mathscr{E}^k_A}(M,N). \tag{A70}$$

- Putting it all together, for every n in $\mathbb{N}$ there are natural isomorphisms:

$$Ext^n_{\mathscr{E}^k_{A^e}}(A, Hom_k(M,N)) \cong HH^n(A, Hom_k(M,N)) \cong Ext^n_{\mathscr{E}^k_{A^e}}(A, Hom_k(M,N)). \tag{A71}$$

□

We may now prove Theorem 1.

Proof of Theorem 1.

- For any A-modules M and N Lemma A8 implied:

$$Ext^\star_{\mathscr{E}^k_A}(N,M) \cong HH^\star(A, Hom_k(N,M)). \tag{A72}$$

Therefore taking supremums over all the A-modules M, N, of the integers n for which (A85) is non-trivial implies:

$$D_{\mathscr{E}^k}(A) = \sup_{M,N \in _A Mod} (sup(\{n \in \mathbb{N}^\# | Ext^n(M,N) \neq 0\})) \tag{A73}$$

$$= \sup_{M,N \in _A Mod} (sup(\{n \in \mathbb{N}^\# | HH^n(A, Hom_k(N,M)) \neq 0\})). \tag{A74}$$

$Hom_k(N,M)$ is only a particular case of an A^e-module; therefore taking supremums over *all* A-modules bounds (A87) above as follows:

$$D_{\mathscr{E}^k}(A) = \sup_{M,N \in _A Mod} (sup(\{n \in \mathbb{N}^\# | HH^\star(A, Hom_k(N,M)) \neq 0\})) \tag{A75}$$

$$\leq \sup_{\tilde{M} \in _{A^e} Mod} (sup(\{n \in \mathbb{N}^\# | HH^n(A, \tilde{M}) \neq 0\})). \tag{A76}$$

The right hand side of (A89) is precisely the definition of the Hochschild cohomological dimension. Therefore

$$D_{\mathscr{E}^k}(A) \leq HCdim(A|k) \tag{A77}$$

Proposition A10 applied to (A90), which draws out the conclusion.
- **Case 1:** $Krull(A_\mathfrak{m})$ **is finite**
  Since A is Cohen-Macaulay at $\mathfrak{m}$ there is an $A_\mathfrak{m}$-regular sequence $x_1, .., x_d$ in $\mathfrak{m}$ of length $d := Krull(A_\mathfrak{m})$ in $A_\mathfrak{m}$. Therefore Proposition A6 implies:

$$Krull(A_\mathfrak{m}) = fd_{A_\mathfrak{m}}(A_\mathfrak{m}/(x_1,..,x_n)). \tag{A78}$$

Part 1 of Theorem 1 applied to (A78) implies:

$$Krull(A_\mathfrak{m}) - D(k_{i-1[\mathfrak{m}]}) - fd_\mathfrak{m}(A_\mathfrak{m}) = fd_{A_\mathfrak{m}}(A_\mathfrak{m}) - D(k_{i-1[\mathfrak{m}]}) - fd_\mathfrak{m}(A_\mathfrak{m}) \leq HCdim(A|k). \tag{A79}$$

Moreover, the *characterization of quasi-freeness* given in Corollary A1 implies that A cannot be quasi-free if:

$$2 + D(k_{i-1[\mathfrak{m}]}) - fd_\mathfrak{m}(A_\mathfrak{m}) \leq Krull(A_\mathfrak{m}). \tag{A80}$$

- **Case 2:** $Krull(A_\mathfrak{m})$ **is infinite**

  For every positive integer d there exists an $A_\mathfrak{m}$-regular sequence $x_1^d, .., x_d^d$ in $\mathfrak{m}$ of length d. Therefore Proposition A6 implies:

  $$(\forall d \in \mathbb{Z}^+)\ d = fd_{A_\mathfrak{m}}(A_\mathfrak{m}/(x_1^d, .., x_d^d)). \tag{A81}$$

  Therefore part one of Theorem 1 implies:

$$(\forall d \in \mathbb{Z}^+)\ d - D(k_{i-1[\mathfrak{m}]}) - fd_\mathfrak{m}(A_\mathfrak{m}) = fd_{A_\mathfrak{m}}(A_\mathfrak{m}/(x_1^d, .., x_d^d)) - D(k_{i-1[\mathfrak{m}]}) - fd_\mathfrak{m}(A_\mathfrak{m}) \le HCdim(A|k). \tag{A82}$$

  Since $D(k)$ and $fd_\mathfrak{m}(A_\mathfrak{m})$ are finite:

  $$\infty - D(k_{i-1[\mathfrak{m}]}) - fd_\mathfrak{m}(A_\mathfrak{m}) = \infty \le HCdim(A|k). \tag{A83}$$

  Since $Krull(A_\mathfrak{m})$ is infinite (A83) implies:

  $$Krull(A_\mathfrak{m}) - D(k_{i-1[\mathfrak{m}]}) - fd_\mathfrak{m}(A_\mathfrak{m}) = \infty = HCdim(A|k). \tag{A84}$$

  In this case Corollary A1 implies that A is not quasi-free.

  □

Appendix D.2. Proof of Theorem 2

Proof of Theorem 2. For any A-modules M and N Lemma A8 implied:

$$Ext^\star_{\mathscr{E}_A^k}(N, M) \cong HH^\star(A, Hom_k(N, M)). \tag{A85}$$

Therefore taking supremums over all the A-modules M, N, of the integers n for which (A85) is non-trivial implies:

$$D_{\mathscr{E}^k}(A) = \sup_{M,N \in {_A}Mod} (\sup(\{n \in \mathbb{N}^\# | Ext^n(M,N) \ne 0\})) \tag{A86}$$

$$= \sup_{M,N \in {_A}Mod} (\sup(\{n \in \mathbb{N}^\# | HH^n(A, Hom_k(N,M)) \ne 0\})). \tag{A87}$$

$Hom_k(N, M)$ is only a particular case of an A^e-module; therefore taking supremums over all A-modules bounds (A87) above as follows:

$$D_{\mathscr{E}^k}(A) = \sup_{M,N \in {_A}Mod} (\sup(\{n \in \mathbb{N}^\# | HH^\star(A, Hom_k(N,M)) \ne 0\})) \tag{A88}$$

$$\le \sup_{\tilde{M} \in {_{A^e}}Mod} (\sup(\{n \in \mathbb{N}^\# | HH^n(A, \tilde{M}) \ne 0\})). \tag{A89}$$

The right hand side of (A89) is precisely the definition of the Hochschild cohomological dimension. Therefore

$$D_{\mathscr{E}^k}(A) \le HCdim(A|k) \tag{A90}$$

Proposition A10 applied to (A90) then draws out the conclusion.
Proposition A6 implies that:

$$n = fd_A(A/(x_1, .., x_n)). \tag{A91}$$

Therefore (1) applied to the A-module $A/(x_1, .., x_n)$ together with (A91) imply:

$$n - D(k) = fd_A(A/(x_1, .., x_n)) \le D_{\mathscr{E}^k} \le HCDim(A/k). \tag{A92}$$

If $\Omega^1(A/k)$ is generated by a regular sequence $x_1, .., x_n$ then Proposition A6 implies:

$$n = fd_{A^e}(A \otimes_k A/\Omega^1(A/k)) \tag{A93}$$

However by definition of $\Omega^1(A/k)$ as the kernel of $\mu_A \colon A \otimes_k A/\Omega^1(A/k) \cong A$. Therefore:
$$n = fd_{A^e}(A). \tag{A94}$$

Lemma A1 together with Lemma A3 imply:
$$n = fd_{A^e}(A) \leq pd_{A^e}(A) \leq pd_{\mathscr{E}_{A^e}^k}(A) + D(k). \tag{A95}$$

Since $D(k)$ is finite then (A95) entails:
$$n - D(k) \leq pd_{\mathscr{E}_{A^e}^k}(A). \tag{A96}$$

By Theorem A2 (A96) is equivalent to:
$$n - D(k) \leq HCDim(A). \tag{A97}$$

If A is Cohen-Macaulay at one of its maximal ideals $\mathfrak{m}$ then there exists a maximal regular $x_1, .., x_d$ in $A_\mathfrak{m}$ with $d = Krull(A_\mathfrak{m})$. Therefore (2) implies:
$$Krull(A_\mathfrak{m}) - D(k) = d - D(k) \leq D(A_\mathfrak{m}) - D(k). \tag{A98}$$

Since $D(A_\mathfrak{m}) \leq D(A)$, then
$$Krull(A_\mathfrak{m}) - D(k) \leq D(A_\mathfrak{m}) - D(k) \leq D(A) - D(k). \tag{A99}$$

Finally (1) applied to (A99) implies:
$$Krull(A_\mathfrak{m}) - D(k) \leq D(A) - D(k) \leq HCDim(A). \tag{A100}$$

□

Appendix D.3. Proofs of Consequences

Proof of Corollary 1. Since X is a smooth affine scheme its coordinate ring satisfies Poincaré duality in dimension d then Van den Bergh's Theorem ([12]) applied. Hence, we have that for every $M \in_{A^e} Mod$

$$HH^n(A, M) \cong HH_{d-n}(A, H_{d-n}(A, \omega_A^{-1} \otimes_A M)). \tag{A101}$$

Since $k[X]$ is flat as a k-module then we may apply Theorem 2 to the left-hand side of (A101) to conclude that

$$0 \not\cong HH^n(A, M) \cong HH_{d-n}(A, \omega_A^{-1} \otimes_A M), \tag{A102}$$

for some A-bimodule M and some $n \geq fd_A(M) - D(k)$. Again by Van den Bergh's theorem we conclude that $HH_m(A, \omega_A^{-1} \otimes_A M) \cong 0$ for any $m > d$. Hence, (A102) must hold for some A-bimodule M and some

$$fd_A(M) - D(k) \leq n \leq d.$$

Thus, there exists an A-bimodule M' and some non-negative integer n' for which

$$0 \leq n' \leq d - fd_A(M) + D(k),$$

and $HH_{n'}(A, M') \not\cong 0$; where $M' \triangleq \omega_A^{-1} \otimes_A M$. Relabeling the index we obtain the conclusion. □

Proof of Corollary 2. By the Hochschild-Kostant-Rosenberg ([10]) there are isomorphisms of A-bimodules
$$HH_n(A, M) \cong \Omega^n(A, M) \triangleq \Omega^n(A) \otimes_A M, \tag{A103}$$
for every A-bimodule M. In particular, (A103) holds for the A-bimodule M of Corollary 1. □

References

1. Gel'and, I.; Neumark, M. On the imbedding of normed rings into the ring of operators in Hilbert space. *C*-algebras 1943–1993 (San Antonio TX 1993)* **1994**, *167*, 2–19. Corrected reprint of the 1943 original [MR 5, 147]. [CrossRef]
2. Hartshorne, R. *Algebraic Geometry*; Graduate Texts in Mathematics, No. 52; Springer: New York, NY, USA, 1977; p. xvi+496.
3. Isbell, J.R. Structure of categories. *Bull. Am. Math. Soc.* **1966**, *72*, 619–655. [CrossRef]
4. Isbell, J.R. Normal completions of categories. In *Reports of the Midwest Category Seminar*; Springer: Berlin, Germany, 1967; pp. 110–155.
5. Seiberg, N.; Witten, E. String theory and noncommutative geometry. *J. High Energy Phys.* **1999**, *1999*, 032. [CrossRef]
6. Lurie, J. On the classification of topological field theories. In *Current Developments in Mathematics, 2008*; International Press: Somerville, MA, USA, 2009; pp. 129–280.
7. Hawkins, E. A cohomological perspective on algebraic quantum field theory. *Commun. Math. Phys.* **2018**, *360*, 439–479. [CrossRef]
8. Connes, A. Noncommutative differential geometry. *Inst. Hautes Études Sci. Publ. Math.* **1985**, *62*, 141–144. [CrossRef]
9. Hochschild, G. On the cohomology groups of an associative algebra. *Ann. Math.* **1945**, *46*, 58–67. [CrossRef]
10. Hochschild, G.; Kostant, B.; Rosenberg, A. Differential forms on regular affine algebras. *Trans. Am. Math. Soc.* **1962**, *102*, 383–408. [CrossRef]
11. Antieau, B.; Vezzosi, G. A remark on the Hochschild-Kostant-Rosenberg theorem in characteristic p. *Ann. Sc. Norm. Super. Pisa Cl. Sci.* **2020**, *20*, 1135–1145. [CrossRef]
12. Van den Bergh, M. A relation between Hochschild homology and cohomology for Gorenstein rings. *Proc. Am. Math. Soc.* **1998**, *126*, 1345–1348. [CrossRef]
13. Grothendieck, A.; Raynaud, M. Revêtements étales et groupe fondamental (SGA 1). *arXiv* **2002**, arXiv:math/0206203
14. Cuntz, J.; Quillen, D. Algebra extensions and nonsingularity. *J. Am. Math. Soc.* **1995**, *8*, 251–289. [CrossRef]
15. Avramov, L.L.; Vigué-Poirrier, M. Hochschild homology criteria for smoothness. *Int. Math. Res. Not.* **1992**, *1992*, 17–25. [CrossRef]
16. Avramov, L.L.; Iyengar, S. Gaps in Hochschild cohomology imply smoothness for commutative algebras. *Math. Res. Lett.* **2005**, *12*, 789–804. [CrossRef]
17. Beck, J.M. Triples, Algebras and Cohomology. Reprints in Theory and Applications of Categories. 2003, pp. 1–59. Available online: http://www.tac.mta.ca/tac/reprints/articles/2/tr2abs.html (accessed on 20 November 2020)
18. Hilton, P.J.; Stammbach, U. A Course in Homological Algebra. In *Graduate Texts in Mathematics*, 2nd ed.; Springer: New York, NY, USA, 1997; Volume 4, p. xii+364.
19. De Jong, A.J. The Stacks Project. Algebr. Stacks. 2013. Available online: https://stacks.math.columbia.edu/ (accessed on 20 November 2020).
20. Springer, T.A. *Linear Algebraic Groups*, 2nd ed.; Modern Birkhäuser Classics; Birkhäuser Boston, Inc.: Boston, MA, USA, 2009; p. xvi+334.
21. Keller, B. Introduction to Kontsevich's Quantization Theorem. Preprint. 2004. Available online: https://webusers.imj-prg.fr/bernhard.keller/publ/emalca.pdf (accessed on 20 November 2020)
22. Rotman, J.J. *An Introduction to Homological Algebra*, 2nd ed.; Universitext; Springer: New York, NY, USA, 2009; p. xiv+709.
23. MacLane, S. *Homology*, 1st ed.; Springer: Berlin, Germany; New York, NY, USA, 1967; p. x+422.
24. Weibel, C.A. *An Introduction to Homological Algebra*; Cambridge Studies in Advanced Mathematics; Cambridge University Press: Cambridge, UK, 1994; Volume 38, p. xiv+450.
25. Connes, A.; Cuntz, J.; Guentner, E.; Higson, N.; Kaminker, J.; Roberts, J.E. Noncommutative Geometry. In *Lectures Given at the C.I.M.E. Summer School Held in Martina Franca, 3–9 September 2000*; Doplicher, S., Longo, R., Eds.; Fondazione CIME/CIME Foundation Subseries; Springer: Berlin, Germamy, 2004; Volume 1831, p. xiv+343. [CrossRef]
26. Eisenbud, D. *Commutative Algebra*; Graduate Texts in Mathematics; Springer: New York, NY, USA, 1995; Volume 150, p. xvi+785.
27. Hochschild, G. Note on relative homological dimension. *Nagoya Math. J.* **1958**, *13*, 89–94. [CrossRef]
28. Bourbaki, N. *Commutative Algebra*; Elements of Mathematics (Berlin); Springer: Berlin, Germany, 1998; Chapters 1–7. p. xxiv+625.

Article

Polarized Rigid Del Pezzo Surfaces in Low Codimension

Muhammad Imran Qureshi

Department of Mathematics, King Fahd University of Petroleum & Minerals (KFUPM), Dhahran 31261, Saudi Arabia; imran.qureshi@kfupm.edu.sa

Received: 11 August 2020; Accepted: 8 September 2020; Published: 11 September 2020

Abstract: We provide explicit graded constructions of orbifold del Pezzo surfaces with rigid orbifold points of type $\left\{k_i \times \frac{1}{r_i}(1, a_i) : 3 \le r_i \le 10, k_i \in \mathbb{Z}_{\ge 0}\right\}$ as well-formed and quasismooth varieties embedded in some weighted projective space. In particular, we present a collection of 147 such surfaces such that their image under their anti-canonical embeddings can be described by using one of the following sets of equations: a single equation, two linearly independent equations, five maximal Pfaffians of 5×5 skew symmetric matrix, and nine 2×2 minors of size 3 square matrix. This is a complete classification of such surfaces under certain carefully chosen bounds on the weights of ambient weighted projective spaces and it is largely based on detailed computer-assisted searches by using the computer algebra system MAGMA.

Keywords: orbifold del pezzo surfaces; hypersurfaces; complete intersections; pfaffians; graded ring constructions

1. Introduction

A del Pezzo surface is a two dimensional algebraic variety with an ample anti-canonical divisor class. The classification of nonsingular del Pezzo surfaces is well known and there are 10 deformation families of such surfaces: $\mathbb{P}^1 \times \mathbb{P}^1$, $\mathbb{P}^2$ and the blow up of $\mathbb{P}^2$ in d general points for $1 \le d \le 8$. An orbifold del Pezzo surface X is a del Pezzo surface with at worst isolated orbifold points, classically known as a log del Pezzo surface with cyclic quotient singularities. We describe X to be locally qGorenstein(qG)-rigid if it contains only rigid isolated orbifold points, i.e., the orbifold points are rigid under qG-deformations. If it admits a qG-degeneration to a normal toric del Pezzo surface then it is called a del Pezzo surface of class TG. The Fano index of X is the largest integer I such that $K_X = ID$ for an element D in the class group of X.

The classification of orbifold del Pezzo surfaces has been an interesting area of research from various points of view, such as the existence of Kahler–Einstein metric [1,2]. Recently, the classification of orbifold del Pezzo surfaces has received much attention, primarily due to the mirror symmetry program for Fano varieties by Coates, Corti et al. [3]. The mirror symmetry for orbifold del Pezzo surface has been formulated in [4] in the form of a conjecture expecting a one to one correspondence between mutation equivalence classes of Fano polygons with the (qG)-deformation equivalence classes of locally qG-rigid del Pezzo surfaces of class TG. Therefore the construction of rigid orbifold del Pezzo surfaces has important links with the mirror symmetry due to this conjecture. The conjecture has been proved for smooth del Pezzo surfaces by Kasprzyk, Nill and Prince in [5]. Corti and Heuberger [6] gave the classification of locally qG-rigid del Pezzo surfaces with $\frac{1}{3}(1,1)$ singular points. The del Pezzo surfaces with a single orbifold point of type $\frac{1}{r}(1,1)$ have been classified by Cavey and Prince [7]. The mutation equivalence classes of Fano polygons with rigid singularities of type

$$\left\{k_1 \times \frac{1}{3}(1,1), k_2 \times \frac{1}{6}(1,1) : k_1 > 0, k_2 \ge 0\right\} \text{ and } \left\{k \times \frac{1}{5}(1,1) : k > 0\right\} \tag{1}$$

have been computed in [8]. This is equivalent to the classification of del Pezzo surfaces of class TG with the above given baskets; though it may be missing surfaces which do not admit a toric degeneration and having one of the above type of baskets of singularities. By using birational techniques, the classification of orbifold del Pezzo surfaces with basket consisting of a combination of $\frac{1}{3}(1,1)$ and $\frac{1}{4}(1,1)$ orbifold points was given by Miura [9].

In [6] the classification gave a total of 29 deformation families of del Pezzo surfaces with $\frac{1}{3}(1,1)$ orbifold points which were divided into 6 different cascades; one of the cascades was first studied by Reid and Suzuki in [10]. Moreover, good model constructions for all 29 surfaces were presented as complete intersections inside the so called rep-quotient varieties (mainly simplicial toric varieties): A geometric quotient $V//G$ of a representation V of a complex Lie group G. Among those, six of them can be described as a hypersurface in $\mathbb{P}^3(a_i)$ or as a complete intersection in $\mathbb{P}^4(a_i)$ or as complete intersection in weighted Grassmannian wGr$(2,5)$ [11]. This motivated us to classify rigid del Pezzo surfaces with certain basket of singularities which can be described by relatively small sets of equations.

1.1. Summary of Results

We classify polarized rigid del Pezzo surfaces, under the bounds chosen in Section 3.2, which contain baskets of orbifold points

$$\left\{ k_i \times \frac{1}{r_i}(1, a_i) : 3 \leq r_i \leq 10, k_i \geq 0 \right\};$$

such that their images under their anti-canonical embedding can be described by one of the following ways.

(i) as a hypersurface, i.e., by a single weighted homogenous equation; $X_d \hookrightarrow \mathbb{P}^3(a_i)$.
(ii) as a codimension 2 weighted complete intersection, i.e., by 2 weighted homogeneous equations; $X_{d_1, d_2} \hookrightarrow \mathbb{P}^4(a_i)$.
(iii) as a codimension 3 variety described by using five maximal Pfaffians of a 5×5 skew symmetric matrix;

$$X_{d_1,\ldots,d_5} \hookrightarrow \mathbb{P}^5(a_i).$$

In other words they are weighted complete intersections in weighted Grassmannian wGr$(2,5)$ or (weighted) projective cone(s) over it [11–13].

(iv) as a codimension 4 variety described by using nine 2×2 minors of a size 3 square matrix

$$X_{d_1,\ldots,d_9} \hookrightarrow \mathbb{P}^6(a_i).$$

Equivalently, they are weighted complete intersections in some weighted $\mathbb{P}^2 \times \mathbb{P}^2$ variety or (weighted) projective cone(s) over it [14].

We summarize the classification in form of the following theorem.

Theorem 1. *Let X be an orbifold del Pezzo surface having at worst a basket*

$$\mathcal{B} = \left\{ k_i \times \frac{1}{r}(1, a) : 3 \leq r \leq 10, k_i \geq 0 \right\}$$

of rigid orbifold points and their image $X \hookrightarrow \mathbb{P}(a_i)$ under their anti-canonical embedding can be described as a hypersurface or as a codimension 2 complete intersection or as a weighted complete intersection in wGr$(2,5)$ or as a weighted complete intersection of weighted $\mathbb{P}^2 \times \mathbb{P}^2$ variety. Then, subject to Section 3.2, X is one of the del Pezzo surfaces listed in Tables A1–A4. In total there are 147 families of such del Pezzo surfaces, divided as follows in each codimension.

Hypersurface	Complete intersection	4×4 Pfaffians	2×2 Minors
81	25	21	20

We construct these examples by first computing all possible candidate varieties with required basket of orbifold points using an algorithmic approach developed in [15,16], under the bounds given in Section 3.2. In case of codimension 1 and 2, the equations of these varieties are generic weighted homogeneous polynomials of given degrees. In cases of codimension 3 and 4 they are induced from the equations of the corresponding ambient weighted projective variety. We perform a detailed singularity analysis of equations of these candidate varieties to prove the existence or non-existence of given candidate surface. We calculate the qG-deformation invariants like the anti-canonical degree $-K_X^2$ and first plurigenus $h^0(-K_X)$ in all cases. We calculate their Euler number and Picard rank in hypersurface case. In complete intersection case, we were able to calculate their Euler number and identify the non-prime examples, i.e., those with the Picard rank greater than 1 by computing their orbifold Euler number.

The computer search used to find these surfaces, based on the algorithm approach of [15,16], is an infinite search. The search is usually performed in the order of increasing sum of the weights ($W = \sum a_i$) of the ambient weighted projective spaces. In each codimension and for each Fano index I, we provide complete classification of rigid del Pezzo surfaces $X \subset \mathbb{P}(a_i)$ satisfying $W - I \leq N$ where $N \geq 50$. If the last candidate example for computer search appears for $W - I = q$ then we search for all cases with $N = \text{maximum}(50, 2q)$, to minimize the possibility of any further examples. This indeed does not rule out a possibility of further other examples for larger value of W and I. It is evident that for larger values of W most weights of $\mathbb{P}(a_i)$ will be larger than 10, the highest local index of allowed orbifold points in our classification, consequently the basket of orbifold points will very likely contain orbifold points of local index $r \geq 11$. In cases of hypersurfaces and complete intersections, the classifications of tuples $(d_j; \underline{a_i})$ which give rise to a quasismooth del Pezzo surfaces can be found in [17,18] where d_j denote the degrees of the defining equations and $\underline{a_i}$ are weights of the ambient weighted projective space. These classifications of tuples can perhaps be analyzed to give the bound free proof of completeness of our results in codimension 1 and 2. However, their classification neither contains computation of any of the invariants like $h^0(-K_X)$, $-K_X^2$ and $e(X)$ and nor do they compute the basket of orbifold points lying on those surfaces.

1.2. Links with Existing Literature

A part of our search results recovers some existing examples in the literature, though a significant subset of them have not been previoudly described in terms of equations. For example, the classification of Fano polygons (equivalently of rigid del Pezzo surfaces of class TG) with basket of orbifold points (1) is given in [8]. We give descriptions in terms of equations for six of their examples; listed as 14, 16, 23, 85, 109 and 130 in our tables. We also recover the classical smooth del Pezzo surfaces of degrees 1, 2, 3, 4, 5, 6 and 8; listed as 3, 2, 1, 82, 107, 128 and 12 respectively in Tables A1–A4. Moreover, 7 of the 29 examples from [6] also appear in our list with one of them seemingly having a new description as a complete intersection in a $w(\mathbb{P}^2 \times \mathbb{P}^2)$ variety, listed as 129 in Table A4. Some examples of Fano index 1 and 2 in codimension 3 and 4 given in Tables A3 and A4 can be found in [19], primarily appearing implicitly as a part of some infinite series of orbifold del Pezzo surfaces.

2. Background and Notational Conventions

2.1. Notation and Conventions

- We work over the field of complex numbers $\mathbb{C}$.
- All of our varieties are projectively Gorenstein.
- For two orbifold points where $\frac{1}{r}(1, a) = \frac{1}{r}(1, b)$ we choose a presentation $\frac{1}{r}(1, \min(a, b))$.

- In all the tables, integers appearing as subscripts of X denote the degree of the defining equations of the given variety, where d^m means that there are m equations of degree d. Similarly, $\mathbb{P}(\cdots, a_i^m, \cdots)$ means that there are m weights of degree a_i.
- We use the same notation for canonical divisor class K_X and canonical sheaf ω_X, if no confusion can arise. We usually write $K_X = \mathcal{O}(k)$ to represent $K_X = kD$.

2.2. Graded Rings and Polarized Varieties

We call a pair (X, D) a *polarized variety* if X is a normal projective algebraic variety and D a $\mathbb{Q}$-ample Weil divisor on X, i.e., some integer multiple of D is a Cartier divisor. One gets an associated finitely generated graded ring

$$R(X, D) = \bigoplus_{n \geq 0} H^0\left(X, \mathcal{O}_X(nD)\right).$$

It is called a *projectively Gorenstein* if the ring $R(X, D)$ is a Gorenstein ring. A surjective morphism from a free graded ring $k[x_0, ..., x_n]$ to $R(X, D)$ gives the embedding

$$i : X = \operatorname{Proj} R(X, D) \hookrightarrow \mathbb{P}(a_0, \cdots, a_n)$$

where $a_i = \deg(x_i)$ and with the divisorial sheaf $\mathcal{O}_X(D)$ being isomorphic to $\mathcal{O}_X(1) = i^* \mathcal{O}_\mathbb{P}(1)$. The Hilbert series of a polarized projective variety (X, D) is given by

$$P_{(X,D)}(t) = \sum_{m \geq 0} h^0(X, mD)\, t^m, \tag{2}$$

where $h^0(X, mD) = \dim H^0(X, \mathcal{O}_X(mD))$. We usually write $P_X(t)$ for the Hilbert series and by the standard Hilbert–Serre theorem [20] (Theorem 11.1), $P_X(t)$ has the following compact form

$$P_X(t) = \frac{N(t)}{\prod_{i=0}^{a}(1 - t^{a_i})}, \tag{3}$$

where $N(t)$ is a palindromic polynomial of degree q, as X is projectively Gorenstein.

2.3. Rigid Del Pezzo Surfaces

Definition 1. *An isolated orbifold point Q of type $\frac{1}{r}(a_1, \ldots, a_n)$ is the quotient of $\mathbb{A}^n$ by the cyclic group μ_r,*

$$\epsilon : (x_1, \ldots, x_n) \mapsto (\epsilon^{a_1} x_1, \ldots, \epsilon^{a_n} x_n)$$

such that $\operatorname{GCD}(r, a_i) = 1$ *for* $1 \leq i \leq n$, $0 < a_i < r$, *and ϵ is a primitive generator of μ_r.*

A *del Pezzo surface* X is a two dimensional algebraic variety with an ample anti-canonical divisor class $-K_X$. If, at worst, X contains isolated orbifold points then we call it an *orbifold or a log del Pezzo surface*. The *Fano index* I of X is the largest positive integer I such that $-K_X = I \cdot D$ for some divisor D in the divisor class group of X. An orbifold del Pezzo surfaces $X \subset \mathbb{P}(a_i)$ of codimension c is well-formed if the singular locus of X consists of at most isolated points. It is quasismooth if the affine cone $\widetilde{X} = \operatorname{Spec} R(X, D) \subset \mathbb{A}^{n+1}$ is smooth outside its vertex $\underline{0}$.

A singularity admitting a $\mathbb{Q}$-Gorenstein smoothing is called *a T-singularity* [21]. A singularity which is rigid under $\mathbb{Q}$-Gorenstein smoothing is called a rigid or *R-singularity* [22]. The following characterization of a T-singularity and R-singularity are useful in our context [7].

Definition 2. *Let $Q = \frac{1}{r}(a, b)$ be an orbifold point and take $m = \operatorname{GCD}(a + b, r)$, $s = (a + b)/m$ and $k = r/m$ then Q has a form $\frac{1}{mk}(1, ms - 1)$. Moreover Q is called a T-singularity if $k \mid m$ [21] and an R-singularity if $m < k$ [22].*

In the two dimensional case, any orbifold point $\frac{1}{r}(a,b)$ can be represented as $\frac{1}{r}(1,a')$ by choosing a different primitive generator of the cyclic group μ_r and the following Lemma follows from it.

Lemma 1. *Let $Q_1 = \frac{1}{r}(1,a)$ and $Q_2 = \frac{1}{r}(1,b)$ be isolated orbifold points. Then $Q_1 = Q_2$ if and only if $a = b$ or $ab \equiv 1 \mod r$.*

By using the fact that each orbifold point on a surface can be written as $\frac{1}{r}(1,a)$ and by applying Lemma 1 on the all possible isolated rigid orbifold points of type $\frac{1}{r}(1,a)$; $3 \leq r \leq 10$, we get to the following Lemma.

Lemma 2. *Let $3 \leq r \leq 10$ then any isolated rigid orbifold point $\frac{1}{r}(a,b)$ is equivalent to one of the following.*

$$\left\{ \begin{array}{l} \frac{1}{3}(1,1), \ \frac{1}{5}(1,1), \ \frac{1}{5}(1,2), \ \frac{1}{6}(1,1), \ \frac{1}{6}(1,5), \ \frac{1}{7}(1,1), \ \frac{1}{7}(1,2), \\ \frac{1}{7}(1,3), \ \frac{1}{8}(1,1), \ \frac{1}{8}(1,5), \ \frac{1}{9}(1,1), \ \frac{1}{9}(1,4), \ \frac{1}{10}(1,1), \ \frac{1}{10}(1,3) \end{array} \right\}$$

2.4. Ambient Varieties

In this section we briefly recall the definition of weighted Grassmannian wGr(2,5) and w($\mathbb{P}^2 \times \mathbb{P}^2$) which we use, apart from weighted projective spaces, as rep-quotient varieties for the construction of our rigid orbifold del Pezzo surfaces; following the notion introduced in [6].

2.4.1. Weighted Grassmannian wGr(2,5)

This part is wholly based on material from ([11], Section 2). Let $w := (w_1, \cdots, w_5)$ be a tuple of all integers or all half integers such that

$$w_i + w_j > 0, \ 1 \leq i < j \leq 5,$$

Then the quotient of the affine cone over Grassmannian minus the origin $\widetilde{Gr(2,5)} \setminus \{\underline{0}\}$ by $\mathbb{C}^\times$ given by:

$$\epsilon : x_{ij} \mapsto \epsilon^{w_i + w_j} x_{ij}$$

is called weighted Grassmannians wGr(2,5) where x_{ij} are Plücker coordinates of the embedding $Gr(2,5) \hookrightarrow \mathbb{P}\left(\wedge^2 \mathbb{C}^5\right)$. Therefore we get the embedding

$$wGr(2,5) \hookrightarrow \mathbb{P}\left(a_{ij} : 1 \leq i < j \leq 5, a_{ij} = w_i + w_j\right).$$

The image of Gr(2,5) and wGr(2,5) under the Plücker embedding is defined by five 4×4 Pfaffians of the 5×5 skew symmetric matrix

$$\begin{pmatrix} x_{12} & x_{13} & x_{14} & x_{15} \\ & x_{23} & x_{24} & x_{25} \\ & & x_{34} & x_{35} \\ & & & x_{45} \end{pmatrix}, \tag{4}$$

where we only write down the upper triangular part. Explicitly, the defining equations are:

$$Pf_i = x_{jk}x_{lm} - x_{jl}x_{km} + x_{jm}x_{lm},$$

where $1 \leq j < k < l < m \leq 5$ are four integers and i makes up the fifth one in $\{1,2,3,4,5\}$. In examples we usually write down the corresponding matrix of weights, replacing x_{ij} with a_{ij} to represent the given wGr(2,5).

If wGr(2,5) is wellformed then the orbifold canonical divisor class is

$$K_{\text{wGr}(2,5)} = \left(-\frac{1}{2}\sum_{1\leq i<j\leq 5} a_{ij}\right) D, \tag{5}$$

for a divisor D in the class group of wGr(2,5).

2.4.2. Weighted $\mathbb{P}^2 \times \mathbb{P}^2$

This section recalls the definition of weighted $\mathbb{P}^2 \times \mathbb{P}^2$ from [14,23]. Let $b = (b_1, b_2, b_3)$ and $c = (c_1, c_2, c_3)$ be two integer or half integer vectors satisfying

$$b_1 + c_1 > 0, \quad b_i \leq b_j \text{ and } \quad c_i \leq c_j \text{ for } 1 \leq i \leq j \leq 3,$$

and Σ_P denotes the Segre embedding $\mathbb{P}^2 \times \mathbb{P}^2 \hookrightarrow \mathbb{P}^8(x_{ij})$. If $\widetilde{\Sigma_P}$ is the affine of this Segre embedding, then the weighted $\mathbb{P}^2 \times \mathbb{P}^2$ variety $w\Sigma_P$ is the quotient of the punctured affine cone $\widehat{\Sigma_P}\setminus\{\underline{0}\}$ by $\mathbb{C}^\times$:

$$\epsilon : x_{ij} \mapsto \epsilon^{b_i+c_j} x_{ij}, \ 1 \leq i,j \leq 3.$$

Thus for a choice of b,c, written together as a single input parameter $p = (b_1, b_2, b_3; c_1, c_2, c_3)$, we get the embedding

$$w\Sigma_P \hookrightarrow \mathbb{P}^8(a_{ij} : a_{ij} = b_i + c_j; 1 \leq i,j \leq 3).$$

The equations are defined by 2×2 minors of a size 3 square matrix which we usually refer to as the weight matrix and write it as

$$\begin{pmatrix} a_{11} & a_{12} & a_{13} \\ a_{21} & a_{22} & a_{23} \\ a_{31} & a_{32} & a_{33} \end{pmatrix} \text{ where } a_{ij} = b_i + c_j; 1 \leq i,j \leq 3. \tag{6}$$

If $w\Sigma_P$ is wellformed then the canonical divisor class is given by

$$K_{w\Sigma_P} = \left(-\sum_{i=j} a_{ij}\right) D, \tag{7}$$

for a divisor D in the class group of $w\Sigma_P$.

3. Computational Steps of The Proof

In this section we provide details of various steps of our calculations which together provide the proof of Theorem 1. In summary, for each codimension and Fano index, we first search for the list of candidate varieties using the algorithmic approach of [15,16]. The candidate lists comes with a suggestive basket(s) of orbifold points and invariants. Then we perform theoretical analysis of each candidate to establish the existence or non-existence of candidate surfaces with given basket and invariants.

3.1. Algorithm

We briefly recall the algorithm from [16] which we used to compute the candidate lists of examples. The key part of it is based on the orbifold Riemann–Roch formula of Bukcley, Reid and Zhou [24] which provides a decomposition of the Hilbert series of X into a smooth part and a singular part. It roughly states that if X is an algebraic variety with basket $\mathcal{B} = \{k_i \times Q_i : m_i \in \mathbb{Z}_{>0}\}$ of isolated orbifold points then its Hilbert series has a decomposition into a smooth part $P_{\text{sm}}(t)$ and orbifold part $\sum k_i P_{Q_i}(t)$;

$$P_X(t) = P_{\text{sm}}(t) + \sum k_i P_{Q_i}(t). \tag{8}$$

The algorithm searches for all orbifolds of fixed dimension n having fixed orbifold canonical class $K_X = \mathcal{O}(k)$ in a given ambient rep-quotient variety. Indeed, if X is a Fano variety of index I then $k = -I$. The algorithm has the following steps.

(i) Compute the Hilbert series and orbifold canonical class of ambient rep-quotient variety.
(ii) Find all possible embeddings of n-folds X with $\omega_X = \mathcal{O}(k)$ by applying the adjunction formula.
(iii) For each possible n-fold embedding of X, compute the Hilbert series $P_X(t)$ and the smooth term $P_{sm}(t)$.
(iv) Compute the list of all possible n-fold isolated orbifold points from the ambient weighted projective space containing X.
(v) For each subset of the list of possible orbifold points determine the multiplicities k_i given in Equation (8) of the orbifold terms $P_{Q_i}(t)$.
(vi) If $k_i \geq 0$ then X is a candidate n-fold with suggested basket of isolated orbifold points.

3.2. Bounds on Search Parameters

We perform our search in the order of increasing sum of the weights on the ambient weight projective space $\mathbb{P}(a_0, \ldots, a_n)$ containing X. The search is theoretically unbounded in each codimension in two directions: there is no bound on the sum of weights $W = \sum a_i$ of the ambient weighted projective space containing X and the Fano index I is also unbounded.

In each codimension, we at least search for polarized rigid del Pezzo surfaces $X \hookrightarrow \mathbb{P}(a_i)$ such that

$$W - I \leq 50, \text{ for } 1 \leq I \leq 10.$$

If the last candidate example is found for the adjunction number $q = W - I$ of the Hilbert numerator $N(t)$, then we further search for all possible cases such that

$$W - I \leq N \text{ where } N = \text{maximum}(2q, 50),$$

to absolutely minimize the possibility of any missing examples. Similarly, in each codimension if we find the last example in search domain $W - I \leq 50$ for index $I > 5$ then we search for examples up to index $2I$. For example, in the hypersurface case the maximum value of I across all candidates was 8, so we searched until index 16 in this case. Similarly, for index 2 hypersurfaces we got the last candidate when $W - 2 = 36$ so we searched for all cases with $W - 2 \leq 72$. Further details in each case can be found in Table 1.

Table 1. The following table summarises the number of surfaces we obtained for each Fano index I in each codimension and exact search domain in each case. First column contains the codimension of each surface and the rest of the columns contain a pair of numbers. First number is the number of examples of given index and the second one gives the maximum value of $q = W_{max} - I$ for which the last candidate surface was found; the classification is complete until $N = \text{maximum}(50, 2q)$. The entries with no second number means that no examples were found for $q \leq 50$.

Codimension	Fano Index, (q)								
	1	2	3	4	5	6	7	8	9–16
1	11 (28)	44 (36)	6 (15)	6 (21)	6 (21)	2 (16)	2 (17)	4 (15)	0
2	15 (22)	8 (29)	1 (26)	1 (22)	0	0	0	0	
3	12 (33)	7 (43)	1 (19)	1 (26)	0	0	0	0	
4	12 (42)	6 (48)	0	0	1 (30)	0	0	1 (42)	0

3.3. Computing Invariants

We describe how we calculate each of the following qG-deformation invariants appearing in Tables A1–A4.

(i) **First plurigenus $h^0(-K_X)$**: If it is equal to zero then we can easily conclude that X does not admit a qG-deformation to a toric variety and such surfaces are not of class TG. We compute it as the coefficient of t^I in the Hilbert series (2) where I is the Fano index of X.

(ii) **Intersection number $-K_X^2$**: It can be defined as an anti-canonical degree of X which we calculate from the Hilbert series $P_X(t)$ of X. In a surface case

$$P_X(t) = \frac{H(t)}{(1-t)^3},$$

where $H(t)$ is a rational function with only positive coefficients. Then for a generic divisor D in the class group, we have $D^2 = H(1)$. Consequently for an orbifold del Pezzo surface of index I, we have $-K_X^2 = I^2 D^2$.

(iii) **Euler Characteristics $e(X)$**: We were able to compute the Euler characteristics of X in hypersurface and complete intersection cases by using Blache's formula ([25], 2.11-14);

$$e(X) = e_{\text{orb}}(X) + \sum_{r(Q) \in \mathcal{B}} \frac{r-1}{r} \tag{9}$$

where r is the local index of each orbifold point. It was applied in the Appendix of [26] to illustrate the computation for a hypersurface. The formula has natural generalization to the cases of complete intersections

$$X_{d_1,\ldots,d_k} \subset \mathbb{P}(a_0,\ldots,a_n)$$

in higher codimension. We can computer $e_{\text{orb}}(X)$ as:

$$e_{\text{orb}}(X) = \text{coefficient of } t^{n-k} \text{ in the series expansion of } \left(\frac{\prod(1+a_i t)}{\prod(1+d_i t)} \deg(X) \right). \tag{10}$$

(iv) **Picard rank $\rho(x)$**: We were able to calculate it explicitly when X is a hypersurface in $\mathbb{P}^3(a_i)$ by using ([27], Sec. 4.4.1). Given a hypersurface

$$X_d \hookrightarrow \mathbb{P}(a_0, a_1, a_2, a_3),$$

let

$$l = \text{coefficient of } t^{2d - \sum a_i} \text{ in the series expansion of } \left(\prod \frac{t^{d-a_i} - 1}{t^{a_i} - 1} \right),$$

then $\rho(X) = l + 1$. In cases of complete intersection examples we were able to identify those examples which are not prime, i.e., the Picard rank greater than 1. From [28], we know that if the Picard rank of a log del Pezzo surface is 1 then $0 < e_{\text{orb}}(X) \leq 3$. Therefore, for each codimension 2, we complete the intersection in Table A2, we list $e_{\text{orb}}(X)$ and those with $e_{\text{orb}}(X) > 3$ have Picard rank greater than 1.

3.4. Theoretical Singularity Analysis

The last step of the calculation is the theoretical singularity analysis of each candidate orbifold. We prove that the general member X in each family is wellformed and quasismooth. We first compute the dimensions of intersection of all orbifold strata with X to establish that X is wellformed. This should be less than or equal to zero for a surface to be wellformed, i.e., it does not contain any singular lines.

The next step is to show that X is quasismooth. This is not so difficult when X is a hypersurface or complete intersection: one can use the criteria given in ([29], Sec. 8). In cases of codimension 3 and 4 examples, we consider X as complete intersections in $\text{wGr}(2,5)$ or in the Segre embedding of weighted $\mathbb{P}^2 \times \mathbb{P}^2$ or in some projective cone(s) over either of those ambient varieties. So X may not only have singularities from the ambient weighted projective but it may also contain singularities on the base loci of linear systems of the intersecting weighted homogeneous forms. In such cases we mostly prove the quasismoothness on the base locus by using computer algebra system MAGMA [30]. We write down explicit equations for X over the rational numbers and show that it is smooth, see ([19], Sec. 2.3) for more details. To prove quasismoothness on an orbifold point Q of type $\frac{1}{r}(a,b)$, which is mostly a coordinate point corresponding to some variables x_i with $\deg(x_i) = r$, we proceed as follows. If c is the codimension of X then we find c tangent variables x_m [31], i.e., we find c polynomials having a monomial of type $x_i^l x_m$. We can locally remove these variables by using the implicit function theorem. Moreover, if two other variables have weights a and b modulo r then Q is a quasismooth point of type $\frac{1}{r}(a,b)$.

4. Sample Calculations

In this section we provide sample calculations of examples given in Tables A1–A4.

Example 1. #81 *Consider the weighted projective space* $\mathbb{P}(1,5,7,10)$ *with variables* x, y, z *and* w *respectively, then the canonical class* $K_\mathbb{P} = \mathcal{O}(-23)$. *The generic weighted homogenous polynomial of degree 15,*

$$f_{15} = k_1 x^{15} + k_2 y^3 + k_3 yw + k_4 xz^2 + \cdots, \quad k_i \in \mathbb{C};$$

defines a del Pezzo surface $X_{15} \hookrightarrow \mathbb{P}(x,y,z,w)$ *of Fano index 8, i.e.,* $K_X = \mathcal{O}(-8)$. *The polynomial* f_{15} *does not contain monomials of pure power in* w *and* z *so* X *contains the orbifold points* $p_1 = (0,0,0,1)$ *and* $p_2 = (0,0,1,0)$. *By applying the implicit function theorem we can remove the variable* y *near the point* p_1 *by using the monomial* yw *and* x, z *are local variables near this point. Therefore* X *contains an orbifold point of type* $\frac{1}{10}(1,7) = \frac{1}{10}(1,3)$(Lemma 1). *Similarly, near* p_2 *the local variables are* y *and* w, *so we get an orbifold point of type*

$$\frac{1}{7}(5,10) = \frac{1}{7}(3,5) = \frac{1}{7}(1,4) = \frac{1}{7}(1,2).$$

The coordinate point of weight 5 does not lie on X *but one dimensional singular stratum* $\mathbb{P}^1(y,w)$ *intersects with* X *non-trivially and by ([29], Lemma 9.4) the intersection is in two points. One of them is* p_1 *and the other can be taken as* $p_3 = (0,1,0,0)$ *which corresponds to weight 5 variable. By using the above arguments we can show that it is a singular point of type* $\frac{1}{5}(1,2)$. *Thus* X *contains exactly the same basket of singularities as given by the computer search and it is a wellformed and quasismooth rigid del Pezzo surface of Fano index 8. Moreover, the vector space*

$$H^0(X, -K_X) = H^0(X, 8D) = <x^8, x^3y, xz>,$$

so $h^0(-K_X) = 3$.

Example 2. #126 *Consider the weighted Grassmannian* $\operatorname{wGr}(2,5)$

$$\operatorname{wGr}(2,5) \hookrightarrow \mathbb{P}(1^2, 3^3, 5^4, 7) \text{ with weight matrix } \begin{pmatrix} 1 & 1 & 3 & 3 \\ & 3 & 5 & 5 \\ & & 5 & 5 \\ & & & 7 \end{pmatrix},$$

Then by Equation (5) *the canonical divisor class* $K_{\operatorname{wGr}(2,5)} = \mathcal{O}(-19)$. *The weighted complete intersection of* $\operatorname{wGr}(2,5)$ *with two forms of degree 3 and two forms of degree 5;*

$$X = \operatorname{wGr}(2,5) \cap (f_3) \cap (g_3) \cap (f_5) \cap (g_5) \hookrightarrow \mathbb{P}_{(x_1, x_2, y_1, z_1, z_2, w_1)}(1^2, 3, 5^2, 7)$$

is a del Pezzo surface with $K_X = \mathcal{O}(-19 + (3+3+5+5)) = \mathcal{O}(-3)$. We can take X to be defined by the maximal Pfaffians of

$$\begin{pmatrix} x_1 & x_2 & f_3 & g_3 \\ & y_1 & f_5 & g_5 \\ & & z_1 & z_2 \\ & & & w_1 \end{pmatrix}, \tag{11}$$

where f_3, g_3, f_5 and g_5 are general weighted homogeneous forms in given variables and they remove the variables of the corresponding degrees from the ambient $\mathrm{wGr}(2,5)$. The coordinate point corresponding to w_1 lies on X. From the equations we have x_1, x_2 and y_1 as tangent variables and z_1, z_2 as local variables. Therefore it is an orbifold point of type $\frac{1}{7}(5,5) = \frac{1}{7}(1,1)$. The locus $X \cap \mathbb{P}(5,5)$ is locally a quadric in $\mathbb{P}^1$ which defines two points. By similar application of implicit function theorem we can show that each is an orbifold point of type $\frac{1}{5}(1,2)$. The restriction of X to weight 3 locus is an empty set, so X contains no further orbifold points. To show the quasismoothness on the base locus we use the computer algebra and write down equations for X. For example, if we choose

$$f_3 = 3x_1^3 + 3x_2^3, \quad f_5 = x_2^5 + x_1^2 y_1 + x_2^2 y_1 + z_1 + z_2,$$
$$g_3 = x_2^3 + y_1, \quad g_5 = x_1^5 + 2x_1^2 y_1 + 3x_2^2 y_1 + 3z_2$$

then the Pfaffians of (11) gives a quasismooth surface. Thus X is an orbifold del Pezzo surface of Fano index 3 with singular points; $2 \times \frac{1}{5}(1,2)$ and $\frac{1}{7}(1,1)$.

As we mentioned in Section 3.4 that we prove the existence of given orbifold del Pezzo surface by theoretical singularity analysis. Then only those which are quasismooth, wellformed and having correct basket of singularities appear in tables of examples. There are in total 8 candidate examples which fails to be quasismooth and we discuss one of them below in detail. No candidate example fails for not being wellformed.

Example 3. *(Non working candidate) A computer search also gives a candidate complete intersection orbifold del Pezzo surface of Fano index 2 given by*

$$X_{6,30} \hookrightarrow \mathbb{P}_{(x,y,z,t,u)}(1,3,9,10,15).$$

Then $F_6 = f(x,y)$ (since other variables have weight higher than 6) and

$$F_{30} = x^{30} + x^{27}y + yz^3 + \cdots$$

are the defining equations of X. The coordinate point $p = (0,0,1,0,0)$ lies on X as no pure power of z appear in F_{30}. Now we can not find two tangent variables to z in the equations of X which implies that the rank of the Jacobian matrix of X at p is equal to 1 which is less than its codimension, so X is not quasismooth at p. Thus, X is a del Pezzo surface which is not quasismooth and does not appear in the following tables.

Concluding Remark: One can use this approach to construct and classify orbifold del Pezzo surfaces with any quotient singularity in a given fixed format, under certain bounds. Moreover, we can also construct examples with rigid orbifold points of type $\frac{1}{r}(1,a)$ for $r \geq 11$ but as the weights higher the computer search output becomes slower due to the nature of algorithm. Therefore, we restrict ourself to the cases with $r \leq 10$.

Funding: This research was funded by the Deanship of Scientific Research, King Fahd University of Petroleum and Minerals via a grant number SB. 191029.

Acknowledgments: I would like to thank Erik Paemurru for pointing me to [27] for computations of Picard rank in the hypersurface case. I am grateful to an anonymous referee for their feedback which improved the exposition of the paper significantly.

Conflicts of Interest: The author declares no conflict of interest.

Appendix A. Table of Examples

Notations in Tables

- The column X represents a del Pezzo surface and the corresponding weighted projective space containing X; the subscripts give the equation degrees of X. The column I lists the Fano index of X.
- The next two columns contain the anti-canonical degree $-K_X^2$ and the first plurigenus $h^0(-K_X)$. If $h^0(-K_X) = 0$ the X is not of class TG.
- $e(X)$ denotes the topological Euler characteristics of X, $\rho(X)$ is the rank of Picard group of X, and $e_{\text{orb}}(X)$ denotes the orbifold Euler number of X. $\rho(X)$ is only listed in Table A1 of hypersurfaces and $e_{\text{orb}}(X)$ only in Table A2 of complete intersections, as discussed in Section 3.3.
- The column $\mathcal{B}$ represents the basket of singular points of X.
- In Tables A3 and A4, the last column represents the matrix of weights, which provides weights of ambient weighted projective space containing wGr(2,5) or weighted $\mathbb{P}^2 \times \mathbb{P}^2$ variety.
- We provide references to those examples which appeared in [6,8], primarily in a toric setting.

Table A1. Hypersurfaces in $w\mathbb{P}^3$.

S.No	X	I	$-K_X^2$	$h^0(-K_X)$	$e(X)$	$\rho(X)$	Basket $\mathcal{B}$
1	$X_3 \subset \mathbb{P}(1^4)$	1	3	4	9	7	
2	$X_4 \subset \mathbb{P}(1^3,2)$	1	2	3	10	8	
3	$X_6 \subset \mathbb{P}(1^2,2,3)$	1	1	2	11	9	
4	$X_{10} \subset \mathbb{P}(1,2,3,5)$	1	1/3	1	11	9	$\frac{1}{3}(1,1)$ [6]
5	$X_{12} \subset \mathbb{P}(2,3^2,5)$	1	2/15	0	10	8	$4 \times \frac{1}{3}(1,1), \frac{1}{5}(1,1)$
6	$X_{15} \subset \mathbb{P}(1,3,5,7)$	1	1/7	1	11	9	$\frac{1}{7}(1,2)$
7	$X_{15} \subset \mathbb{P}(3^2,5^2)$	1	1/15	0	11	9	$5 \times \frac{1}{3}(1,1), 3 \times \frac{1}{5}(1,1)$
8	$X_{16} \subset \mathbb{P}(1,3,5,8)$	1	2/15	1	12	10	$\frac{1}{3}(1,1), \frac{1}{5}(1,1)$
9	$X_{18} \subset \mathbb{P}(2,3,5,9)$	1	1/15	0	9	7	$2 \times \frac{1}{3}(1,1), \frac{1}{5}(1,2)$
10	$X_{20} \subset \mathbb{P}(2,5^2,9)$	1	2/45	0	10	8	$4 \times \frac{1}{5}(1,2), \frac{1}{9}(1,1)$
11	$X_{28} \subset \mathbb{P}(3,5,7,14)$	1	2/105	0	8	6	$\frac{1}{3}(1,1), \frac{1}{5}(1,2), 2 \times \frac{1}{7}(1,2)$
12	$X_2 \subset \mathbb{P}(1^4)$	2	8	9	4	2	
13	$X_4 \subset \mathbb{P}(1^3,3)$	2	16/3	6	6	4	$\frac{1}{3}(1,1)$ [6]
14	$X_6 \subset \mathbb{P}(1^3,5)$	2	24/5	6	8	6	$\frac{1}{5}(1,1)$ [8]
15	$X_6 \subset \mathbb{P}(1^2,3^2)$	2	8/3	3	8	6	$2 \times \frac{1}{3}(1,1)$ [6]
16	$X_7 \subset \mathbb{P}(1^3,6)$	2	14/3	6	9	7	$\frac{1}{6}(1,1)$ [8]
17	$X_8 \subset \mathbb{P}(1^2,3,5)$	2	32/15	3	10	8	$\frac{1}{3}(1,1), \frac{1}{5}(1,1)$
18	$X_8 \subset \mathbb{P}(1^3,7)$	2	32/7	6	10	8	$\frac{1}{7}(1,1)$
19	$X_9 \subset \mathbb{P}(1^2,3,6)$	2	2	3	11	9	$\frac{1}{3}(1,1), \frac{1}{6}(1,1)$ [8]
20	$X_9 \subset \mathbb{P}(1^3,8)$	2	9/2	6	11	9	$\frac{1}{8}(1,1)$
21	$X_{10} \subset \mathbb{P}(1^2,3,7)$	2	40/21	3	12	10	$\frac{1}{3}(1,1), \frac{1}{7}(1,1)$
22	$X_{10} \subset \mathbb{P}(1^3,9)$	2	40/9	6	12	10	$\frac{1}{9}(1,1)$
23	$X_{10} \subset \mathbb{P}(1^2,5^2)$	2	8/5	3	12	10	$2 \times \frac{1}{5}(1,1)$ [8]
24	$X_{11} \subset \mathbb{P}(1^3,10)$	2	22/5	6	13	11	$\frac{1}{10}(1,1)$
25	$X_{11} \subset \mathbb{P}(1^2,5,6)$	2	22/15	3	13	11	$\frac{1}{5}(1,1), \frac{1}{6}(1,1)$
26	$X_{11} \subset \mathbb{P}(1^2,3,8)$	2	11/6	3	13	11	$\frac{1}{3}(1,1), \frac{1}{8}(1,1)$
27	$X_{12} \subset \mathbb{P}(1^2,5,7)$	2	48/35	3	14	12	$\frac{1}{5}(1,1), \frac{1}{7}(1,1)$

Table A1. *Cont.*

S.No	X	I	$-K_X^2$	$h^0(-K_X)$	$e(X)$	$\rho(X)$	Basket $\mathcal{B}$
28	$X_{12} \subset \mathbb{P}(1^2,6^2)$	2	4/3	3	14	12	$2 \times \frac{1}{6}(1,1)[8]$
29	$X_{12} \subset \mathbb{P}(1^2,3,9)$	2	16/9	3	14	12	$\frac{1}{3}(1,1), \frac{1}{9}(1,1)$
30	$X_{13} \subset \mathbb{P}(1^2,5,8)$	2	13/10	3	15	13	$\frac{1}{5}(1,1), \frac{1}{8}(1,1)$
31	$X_{13} \subset \mathbb{P}(1^2,3,10)$	2	26/15	3	15	13	$\frac{1}{3}(1,1), \frac{1}{10}(1,1)$
32	$X_{13} \subset \mathbb{P}(1^2,6,7)$	2	26/21	3	15	13	$\frac{1}{6}(1,1), \frac{1}{7}(1,1)$
33	$X_{14} \subset \mathbb{P}(1^2,6,8)$	2	7/6	3	16	14	$\frac{1}{6}(1,1), \frac{1}{8}(1,1)$
34	$X_{14} \subset \mathbb{P}(1^2,7^2)$	2	8/7	3	16	14	$2 \times \frac{1}{7}(1,1)$
35	$X_{14} \subset \mathbb{P}(1^2,5,9)$	2	56/45	3	16	14	$\frac{1}{5}(1,1), \frac{1}{9}(1,1)$
36	$X_{15} \subset \mathbb{P}(1^2,6,9)$	2	10/9	3	17	15	$\frac{1}{6}(1,1), \frac{1}{9}(1,1)$
37	$X_{15} \subset \mathbb{P}(1^2,5,10)$	2	6/5	3	17	15	$\frac{1}{5}(1,1), \frac{1}{10}(1,1)$
38	$X_{15} \subset \mathbb{P}(1,3,6,7)$	2	10/21	1	11	9	$2 \times \frac{1}{3}(1,1), \frac{1}{6}(1,1), \frac{1}{7}(1,2)$
39	$X_{15} \subset \mathbb{P}(1^2,7,8)$	2	15/14	3	17	15	$\frac{1}{7}(1,1), \frac{1}{8}(1,1)$
40	$X_{16} \subset \mathbb{P}(1^2,6,10)$	2	16/15	3	18	16	$\frac{1}{6}(1,1), \frac{1}{10}(1,1)$
41	$X_{16} \subset \mathbb{P}(1^2,7,9)$	2	64/63	3	18	16	$\frac{1}{7}(1,1), \frac{1}{9}(1,1)$
42	$X_{16} \subset \mathbb{P}(1^2,8^2)$	2	1	3	18	16	$2 \times \frac{1}{8}(1,1)$
43	$X_{17} \subset \mathbb{P}(1,3,7,8)$	2	17/42	1	11	9	$\frac{1}{3}(1,1), \frac{1}{7}(1,1), \frac{1}{8}(1,5)$
44	$X_{17} \subset \mathbb{P}(1^2,7,10)$	2	34/35	3	19	17	$\frac{1}{7}(1,1), \frac{1}{10}(1,1)$
45	$X_{17} \subset \mathbb{P}(1^2,8,9)$	2	17/18	3	19	17	$\frac{1}{8}(1,1), \frac{1}{9}(1,1)$
46	$X_{18} \subset \mathbb{P}(1^2,8,10)$	2	9/10	3	20	18	$\frac{1}{8}(1,1), \frac{1}{10}(1,1)$
47	$X_{18} \subset \mathbb{P}(1^2,9^2)$	2	8/9	3	20	18	$2 \times \frac{1}{9}(1,1)$
48	$X_{19} \subset \mathbb{P}(1^2,9,10)$	2	38/45	3	21	19	$\frac{1}{9}(1,1), \frac{1}{10}(1,1)$
49	$X_{20} \subset \mathbb{P}(1^2,10^2)$	2	4/5	3	22	20	$2 \times \frac{1}{10}(1,1)$
50	$X_{21} \subset \mathbb{P}(3,6,7^2)$	2	2/21	0	9	7	$3 \times \frac{1}{3}(1,1), \frac{1}{6}(1,1), 3 \times \frac{1}{7}(1,2)$
51	$X_{21} \subset \mathbb{P}(1,3,9,10)$	2	14/45	1	13	11	$2 \times \frac{1}{3}(1,1), \frac{1}{9}(1,1), \frac{1}{10}(1,3))$
52	$X_{22} \subset \mathbb{P}(1,5,7,11)$	2	8/35	1	10	8	$\frac{1}{5}(1,1), \frac{1}{7}(1,3)$
53	$X_{24} \subset \mathbb{P}(3,7,8^2)$	2	1/14	0	7	5	$\frac{1}{7}(1,1), 3 \times \frac{1}{8}(1,5)$
54	$X_{30} \subset \mathbb{P}(3,9,10^2)$	2	2/45	0	9	7	$3 \times \frac{1}{3}(1,1), \frac{1}{9}(1,1), 3 \times \frac{1}{10}(1,3)$
55	$X_{36} \subset \mathbb{P}(1,7,12,18)$	2	2/21	1	11	9	$\frac{1}{6}(1,1), \frac{1}{7}(1,3)$
56	$X_6 \subset \mathbb{P}(1^2,2,5)$	3	27/5	6	5	3	$\frac{1}{5}(1,2)$
57	$X_8 \subset \mathbb{P}(1^2,2,7)$	3	36/7	6	6	4	$\frac{1}{7}(1,2)$
58	$X_{10} \subset \mathbb{P}(1,2,5^2)$	3	9/5	2	7	5	$2 \times \frac{1}{5}(1,2)$
59	$X_{12} \subset \mathbb{P}(1,2,5,7)$	3	54/35	2	8	6	$\frac{1}{5}(1,2), \frac{1}{7}(1,2)$
60	$X_{14} \subset \mathbb{P}(1,2,7^2)$	3	9/7	2	9	7	$2 \times \frac{1}{7}(1,2)$
61	$X_{15} \subset \mathbb{P}(1,5^2,7)$	3	27/35	1	9	3	$3 \times \frac{1}{5}(1,2), \frac{1}{7}(1,1)$
62	$X_6 \subset \mathbb{P}(1^2,3,5)$	4	32/5	7	4	2	$\frac{1}{5}(1,2)$
63	$X_{10} \subset \mathbb{P}(1,3,5^2)$	4	32/15	2	6	4	$\frac{1}{3}(1,1), 2 \times \frac{1}{5}(1,2)$
64	$X_{12} \subset \mathbb{P}(1,3,5,7)$	4	64/35	2	6	4	$\frac{1}{5}(1,2), \frac{1}{7}(1,3)$
65	$X_{15} \subset \mathbb{P}(1,3,5,10)$	4	8/5	2	7	5	$\frac{1}{5}(1,2), \frac{1}{10}(1,3)$
66	$X_{15} \subset \mathbb{P}(3,5^2,6)$	4	8/15	0	7	5	$2 \times \frac{1}{3}(1,1), 3 \times \frac{1}{5}(1,2), \frac{1}{6}(1,1)$
67	$X_{21} \subset \mathbb{P}(1,7^2,10)$	4	24/35	1	9	7	$3 \times \frac{1}{7}(1,3), \frac{1}{10}(1,1)$
68	$X_8 \subset \mathbb{P}(1,2,3,7)$	5	100/21	5	4	2	$\frac{1}{3}(1,1), \frac{1}{7}(1,3)$
69	$X_8 \subset \mathbb{P}(1^2,4,7)$	5	50/7	8	4	2	$\frac{1}{7}(1,2)$
70	$X_{12} \subset \mathbb{P}(1,3,4,9)$	5	25/9	3	5	3	$\frac{1}{3}(1,1), \frac{1}{9}(1,4)$
71	$X_{14} \subset \mathbb{P}(2,3,7^2)$	5	25/21	1	5	3	$\frac{1}{3}(1,1), 2 \times \frac{1}{7}(1,3)$
72	$X_{16} \subset \mathbb{P}(1,4,7,9)$	5	100/63	2	6	4	$\frac{1}{7}(1,2), \frac{1}{9}(1,4)$

Table A1. *Cont.*

S.No	X	I	$-K_X^2$	$h^0(-K_X)$	$e(X)$	$\rho(X)$	Basket $\mathcal{B}$
73	$X_{21} \subset \mathbb{P}(3,7^2,9)$	5	25/63	0	7	5	$2 \times \frac{1}{3}(1,1), 3 \times \frac{1}{7}(1,3), \frac{1}{9}(1,1)$
74	$X_{15} \subset \mathbb{P}(1,5,7,8)$	6	27/14	2	5	3	$\frac{1}{7}(1,3), \frac{1}{8}(1,5)$
75	$X_{16} \subset \mathbb{P}(1,5,8^2)$	6	9/5	2	6	4	$\frac{1}{5}(1,1), 2 \times \frac{1}{8}(1,5)$
76	$X_{10} \subset \mathbb{P}(1,2,5,9)$	7	49/9	6	3	1	$\frac{1}{9}(1,4)$
77	$X_{12} \subset \mathbb{P}(2,3,5,9)$	7	98/45	2	4	2	$\frac{1}{3}(1,1), \frac{1}{5}(1,2), \frac{1}{9}(1,4)$
78	$X_8 \subset \mathbb{P}(1,3,5,7)$	8	512/105	5	4	2	$\frac{1}{3}(1,1), \frac{1}{5}(1,2), \frac{1}{7}(1,2)$
79	$X_{14} \subset \mathbb{P}(1,5,7,9)$	8	128/45	3	4	2	$\frac{1}{5}(1,2), \frac{1}{9}(1,4)$
80	$X_{15} \subset \mathbb{P}(1,6,7,9)$	8	160/63	3	5	3	$\frac{1}{6}(1,1), \frac{1}{7}(1,3), \frac{1}{9}(1,4)$
81	$X_{15} \subset \mathbb{P}(1,5,7,10)$	8	96/35	3	5	3	$\frac{1}{5}(1,2), \frac{1}{7}(1,2), \frac{1}{10}(1,3))$

Table A2. Codimension 2 Complete Intersections.

S.No	X	I	$-K_X^2$	$h^0(-K_X)$	$e(X)$	$\rho(X)$	Basket $\mathcal{B}$
82	$X_{2,2} \subset \mathbb{P}(1^5)$	1	4	5	8	8	
83	$X_{4^2} \subset \mathbb{P}(1^2,2^2,3)$	1	4/3	2	10	28/3	$\frac{1}{3}(1,1)\,[6]$
84	$X_{4,6} \subset \mathbb{P}(1,2^2,3^2)$	1	2/3	1	10	26/3	$2 \times \frac{1}{3}(1,1)[6]$
85	$X_{6^2} \subset \mathbb{P}(1^2,3^2,5)$	1	4/5	2	12	56/5	$\frac{1}{5}(1,1)\,[8]$
86	$X_{6^2} \subset \mathbb{P}(2^2,3^3)$	1	1/3	0	9	19/3	$4 \times \frac{1}{3}(1,1)\,[6]$
87	$X_{6,7} \subset \mathbb{P}(1,2,3^2,5)$	1	7/15	1	11	133/15	$2 \times \frac{1}{3}(1,1), \frac{1}{5}(1,1)$
88	$X_{6,8} \subset \mathbb{P}(1,2,3,4,5)$	1	2/5	1	10	46/5	$\frac{1}{5}(1,2)$
89	$X_{8^2} \subset \mathbb{P}(1^2,4^2,7)$	1	4/7	2	14	92/7	$\frac{1}{7}(1,1)$
90	$X_{6,10} \subset \mathbb{P}(1,3^2,5^2)$	1	4/15	1	12	136/15	$2 \times \frac{1}{3}(1,1), 2 \times \frac{1}{5}(1,1)$
91	$X_{8,10} \subset \mathbb{P}(2,3,4,5^2)$	1	2/15	0	8	86/15	$\frac{1}{3}(1,1), 2 \times \frac{1}{5}(1,2)$
92	$X_{9,10} \subset \mathbb{P}(2,3^2,5,7)$	1	1/7	0	9	43/7	$3 \times \frac{1}{3}(1,1), \frac{1}{7}(1,2)$
93	$X_{10^2} \subset \mathbb{P}(1^2,5^2,9)$	1	4/9	2	16	136/9	$\frac{1}{9}(1,1)$
94	$X_{10,11} \subset \mathbb{P}(1,2,5^2,9)$	1	11/45	1	13	473/45	$2 \times \frac{1}{5}(1,2), \frac{1}{9}(1,1)$
95	$X_{10,12} \subset \mathbb{P}(3^2,5^2,7)$	1	8/105	0	10	512/105	$4 \times \frac{1}{3}(1,1), 2 \times \frac{1}{5}(1,1), \frac{1}{7}(1,2)$
96	$X_{10,12} \subset \mathbb{P}(2,3,5,6,7)$	1	2/21	0	8	122/21	$2 \times \frac{1}{3}(1,1), \frac{1}{7}(1,3)$
97	$X_{6,8} \subset \mathbb{P}(1,3^2,4,5)$	2	16/15	1	8	88/15	$2 \times \frac{1}{3}(1,1), \frac{1}{5}(1,2)$
98	$X_{8,10} \subset \mathbb{P}(1,3,4,5,7)$	2	16/21	1	8	136/21	$\frac{1}{3}(1,1), \frac{1}{7}(1,3)$
99	$X_{8,12} \subset \mathbb{P}(1,3,5,6,7)$	2	64/105	1	10	736/105	$2 \times \frac{1}{3}(1,1), \frac{1}{5}(1,1), \frac{1}{7}(1,2)$
100	$X_{10,12} \subset \mathbb{P}(3,4,5^2,7)$	2	8/35	0	6	124/35	$2 \times \frac{1}{5}(1,2), \frac{1}{7}(1,3)$
101	$X_{9,14} \subset \mathbb{P}(1,3,6,7,8)$	2	1/2	1	10	61/8	$\frac{1}{3}(1,1), \frac{1}{6}(1,1), \frac{1}{8}(1,5)$
102	$X_{12,14} \subset \mathbb{P}(3,4,5,7,9)$	2	8/45	0	6	164/45	$\frac{1}{3}(1,1), \frac{1}{5}(1,2), \frac{1}{9}(1,4)$
103	$X_{14,15} \subset \mathbb{P}(3,6,7^2,8)$	2	5/42	0	8	545/168	$2 \times \frac{1}{3}(1,1), \frac{1}{6}(1,1), 2 \times \frac{1}{7}(1,2), \frac{1}{8}(1,5)$
104	$X_{11,18} \subset \mathbb{P}(1,3,8,9,10)$	2	11/30	1	12	1067/120	$2 \times \frac{1}{3}(1,1), \frac{1}{8}(1,1), \frac{1}{10}(1,3)$
105	$X_{12,14} \subset \mathbb{P}(4,5,6,7^2)$	3	9/35	0	5	87/35	$\frac{1}{5}(1,2), 2 \times \frac{1}{7}(1,3)$
106	$X_{10,12} \subset \mathbb{P}(3,5^2,6,7)$	4	64/105	0	6	232/105	$2 \times \frac{1}{3}(1,1), 2 \times \frac{1}{5}(1,2), \frac{1}{7}(1,2)$

Table A3. Codimension 3 Pfaffians.

S.No	X	I	$-K^2$	$h^0(-K)$	Basket $\mathcal{B}$	Weight Matrix
107	$X_{2,2,2,2,2} \subset \mathbb{P}(1^6)$	1	5	6		1 1 1 1 1 1 1 1 1 1
108	$X_{3,3,4,4,4} \subset \mathbb{P}(1^3, 2^2, 3)$	1	7/3	3	$\frac{1}{3}(1,1)\,[6]$	1 1 2 2 1 2 2 2 2 3
109	$X_{4,4,6,6,6} \subset \mathbb{P}(1^3, 3^2, 5)$	1	9/5	3	$\frac{1}{5}(1,1)\,[8]$	1 1 3 3 1 3 3 3 3 5
110	$X_{4,5,6,6,7} \subset \mathbb{P}(1^2, 2, 3^2, 5)$	1	17/15	2	$\frac{1}{3}(1,1), \frac{1}{5}(1,1)$	1 1 2 3 2 3 4 3 4 5
111	$X_{5,5,8,8,8} \subset \mathbb{P}(1^3, 4^2, 7)$	1	11/7	3	$\frac{1}{7}(1,1)$	1 1 4 4 1 4 4 4 4 7
112	$X_{6,7,8,9,10} \subset \mathbb{P}(1, 2, 3^2, 5, 7)$	1	10/21	1	$\frac{1}{3}(1,1), \frac{1}{7}(1,4)$	1 2 3 4 3 4 5 5 6 7
113	$X_{6,6,10,10,10} \subset \mathbb{P}(1^3, 5^2, 9)$	1	13/9	3	$\frac{1}{9}(1,1)$	1 1 5 5 1 5 5 5 5 9
114	$X_{7,8,8,9,10} \subset \mathbb{P}(2, 3^2, 4, 5^2)$	1	1/5	0	$3 \times \frac{1}{3}(1,1), \frac{1}{5}(1,2), \frac{1}{5}(1,1)$	2 3 3 4 4 4 5 5 6 6
115	$X_{6,7,10,10,11} \subset \mathbb{P}(1^2, 2, 5^2, 9)$	1	38/45	2	$\frac{1}{5}(1,2), \frac{1}{9}(1,1)$	1 1 4 5 2 5 6 5 6 9
116	$X_{6,8,10,10,12} \subset \mathbb{P}(1, 3^2, 5^2, 7)$	1	29/105	1	$\frac{1}{3}(1,1), \frac{1}{5}(1,1), \frac{1}{7}(1,4)$	1 1 3 5 3 5 7 5 7 9
117	$X_{10,10,12,12,14} \subset \mathbb{P}(3^2, 5^2, 7^2)$	1	3/35	0	$3 \times \frac{1}{3}(1,1), \frac{1}{5}(1,1), 2 \times \frac{1}{7}(1,4)$	3 3 5 5 5 7 7 7 7 9

Table A3. Cont.

S.No	X	I	$-K^2$	$h^0(-K)$	Basket $\mathcal{B}$	Weight Matrix
118	$X_{11,12,12,15,16}$ $\subset \mathbb{P}(2,5^2,6,7,9)$	1	23/315	0	$3 \times \frac{1}{5}(1,2), \frac{1}{7}(1,3), \frac{1}{9}(1,1)$	2 5 5 6 6 6 7 9 10 10
119	$X_{4,7,8,8,9}$ $\subset \mathbb{P}(1^2,2,3,6,7)$	2	22/7	4	$\frac{1}{3}(1,1), \frac{1}{6}(1,1), \frac{1}{7}(1,2)$	1 1 2 5 2 3 6 3 6 7
120	$X_{4,8,9,9,10}$ $\subset \mathbb{P}(1^2,2,3,7,8)$	2	43/14	4	$\frac{1}{7}(1,1), \frac{1}{8}(1,5)$	1 1 2 6 2 3 7 3 7 8
121	$X_{4,10,11,11,12}$ $\subset \mathbb{P}(1^2,2,3,9,10)$	2	134/45	4	$\frac{1}{3}(1,1), \frac{1}{9}(1,1), \frac{1}{10}(1,3)$	1 1 2 8 2 3 9 3 9 10
122	$X_{8,9,12,13,14}$ $\subset \mathbb{P}(1,3,5,6,7,8)$	2	19/30	1	$\frac{1}{3}(1,1), \frac{1}{5}(1,1), \frac{1}{8}(1,5)$	1 2 5 6 3 6 7 7 8 11
123	$X_{12,12,14,15,15}$ $\subset \mathbb{P}(4,5^2,7^2,8)$	2	11/70	0	$2 \times \frac{1}{5}(1,2), 2 \times \frac{1}{7}(1,3), \frac{1}{8}(1,1)$	4 5 7 7 5 7 7 8 8 10
124	$X_{14,14,15,15,16}$ $\subset \mathbb{P}(3,6,7^2,8^2)$	2	1/7	0	$\frac{1}{3}(1,1), \frac{1}{6}(1,1), \frac{1}{7}(1,2), 2 \times \frac{1}{8}(1,5)$	6 6 7 7 7 8 8 8 8 9
125	$X_{16,17,17,18,18}$ $\subset \mathbb{P}(3,7,8^2,9,10)$	2	11/105	0	$\frac{1}{3}(1,1), \frac{1}{7}(1,1), 2 \times \frac{1}{8}(1,5), \frac{1}{10}(1,3)$	7 8 8 9 8 8 9 9 10 10
126	$X_{6,6,8,8,10}$ $\subset \mathbb{P}(1^2,3,5^2,7)$	3	153/35	4	$2 \times \frac{1}{5}(1,2), \frac{1}{7}(1,1)$	1 1 3 3 3 5 5 5 5 7
127	$X_{8,8,11,11,14}$ $\subset \mathbb{P}(1^2,4,7^2,10)$	4	184/35	6	$2 \times \frac{1}{7}(1,3), \frac{1}{10}(1,1)$	1 1 4 4 4 7 7 7 7 10

Table A4. Codimension 4 $\mathbb{P}^2 \times \mathbb{P}^2$.

S.No	X	I	$-K^2$	$h^0(-K)$	Basket $\mathcal{B}$	Weight Matrix
128	X_{2^9} $\subset \mathbb{P}(1^7)$	1	6	7		1 1 1 1 1 1 1 1 1
129	$X_{2,3^4,4^4}$ $\subset \mathbb{P}(1^4,2^2,3)$	1	10/3	4	$\frac{1}{3}(1,1)$ [6]	1 1 2 1 1 2 2 2 3
130	$X_{2,4^4,6^4}$ $\subset \mathbb{P}(1^4,3^2,5)$	1	14/5	4	$\frac{1}{5}(1,1)$ [8]	1 1 3 1 1 3 3 3 5
131	$X_{2,5^4,8^4}$ $\subset \mathbb{P}(1^4,4^2,7)$	1	18/7	4	$\frac{1}{7}(1,1)$	1 1 4 1 1 4 4 4 7
132	$X_{4,5^2,6^3,7^2,8}$ $\subset \mathbb{P}(1,2^2,3^3,5)$	1	4/5	1	$3 \times \frac{1}{3}(1,1), \frac{1}{5}(1,1)$	1 2 3 2 3 4 3 4 5
133	$X_{2,6^4,10^4}$ $\subset \mathbb{P}(1^4,5^2,9)$	1	22/9	4	$\frac{1}{9}(1,1)$	1 1 5 1 1 5 5 5 9
134	$X_{5,6^2,7^2,8^2,9,10}$ $\subset \mathbb{P}(1,2,3^2,4,5^2)$	1	8/15	1	$\frac{1}{3}(1,1), \frac{1}{5}(1,2), \frac{1}{5}(1,1)$	1 2 3 3 4 5 4 5 6
135	$X_{4,7^2,8^2,10,11^2,12}$ $\subset \mathbb{P}(1,2^2,3,5^2,9)$	1	26/45	1	$\frac{1}{3}(1,1), 2 \times \frac{1}{5}(1,2), \frac{1}{9}(1,1)$	1 2 5 2 3 6 5 6 9
136	$X_{6,8^2,10^3,12^2,14}$ $\subset \mathbb{P}(1,3^2,5^2,7^2)$	1	2/7	1	$2 \times \frac{1}{7}(1,4)$	1 3 5 3 5 7 5 7 9
137	$X_{7,8,10,11^2,12^2,15,16}$ $\subset \mathbb{P}(1,2,5^2,6,7,9)$	1	86/315	1	$\frac{1}{5}(1,2), \frac{1}{7}(1,3), \frac{1}{9}(1,1)$	1 2 5 5 6 9 6 7 10
138	$X_{10,11^2,12^3,13^2,14}$ $\subset \mathbb{P}(3,4,5^2,6,7^2)$	1	3/35	0	$2 \times \frac{1}{5}(1,2), 2 \times \frac{1}{7}(1,4)$	4 5 6 5 6 7 6 7 8
139	$X_{8,9,11,12^3,13,15,16}$ $\subset \mathbb{P}(2,3,5^2,6,7,9)$	1	38/315	0	$\frac{1}{3}(1,1), 3 \times \frac{1}{5}(1,2), \frac{1}{7}(1,4), \frac{1}{9}(1,1)$	2 3 6 5 6 9 6 7 10
140	$X_{4,8^2,9^2,12,13^2,14}$ $\subset \mathbb{P}(1,2,3,6^2,7^2)$	2	20/21	2	$\frac{1}{3}(1,1), 2 \times \frac{1}{6}(1,1), 2 \times \frac{1}{7}(1,2)$	1 2 6 2 3 7 6 7 11
141	$X_{4,8,9^2,10,13,14^2,15}$ $\subset \mathbb{P}(1,2,3,6,7^2,8)$	2	37/42	2	$\frac{1}{6}(1,1), \frac{1}{7}(1,1), \frac{1}{7}(1,2), \frac{1}{8}(1,5)$	1 2 6 2 3 7 7 8 12
142	$X_{4,8,9,11,12,15,16^2,17}$ $\subset \mathbb{P}(1,2,3,6,7,9,10)$	2	248/315	2	$\frac{1}{3}(1,1), \frac{1}{6}(1,1), \frac{1}{7}(1,2), \frac{1}{9}(1,1), \frac{1}{10}(1,3)$	1 2 6 2 3 7 9 10 14
143	$X_{4,9,10,11,12,16,17^2,18}$ $\subset \mathbb{P}(1,2,3,7,8,9,10)$	2	451/630	2	$\frac{1}{7}(1,1), \frac{1}{8}(1,5), \frac{1}{9}(1,1), \frac{1}{10}(1,3)$	1 2 7 2 3 8 9 10 15
144	$X_{4,11^2,12^2,18,19^2,20}$ $\subset \mathbb{P}(1,2,3,9^2,10^2)$	2	28/45	2	$\frac{1}{3}(1,1), 2 \times \frac{1}{9}(1,1), 2 \times \frac{1}{10}(1,3)$	1 2 9 2 3 10 9 10 17
145	$X_{14,15^2,16^3,17^2,18}$ $\subset \mathbb{P}(3,6,7^2,8,9,10)$	2	16/105	0	$3 \times \frac{1}{3}(1,1), \frac{1}{6}(1,1), 2 \times \frac{1}{7}(1,2), \frac{1}{10}(1,3)$	6 7 8 7 8 9 8 9 10
146	$X_{6,8^2,10^3,12^2,14}$ $\subset \mathbb{P}(1,3^2,5,7^2,9)$	5	250/63	4	$2 \times \frac{1}{3}(1,1), 2 \times \frac{1}{7}(1,3), \frac{1}{9}(1,1)$	1 3 5 3 5 7 5 7 9
147	$X_{14,15^2,16^3,17^2,18}$ $\subset \mathbb{P}(6,7^2,8,9^2,10)$	8	256/315	1	$\frac{1}{6}(1,1), 2 \times \frac{1}{7}(1,3), 2 \times \frac{1}{9}(1,4), \frac{1}{10}(1,1)$	6 7 8 7 8 9 8 9 10

References

1. Cheltsov, I.; Park, J.; Shramov, C. Exceptional Del Pezzo hypersurfaces. *J. Geom. Anal.* **2010**, *20*, 787–816. [CrossRef]
2. Cheltsov, I.; Shramov, C. Del Pezzo zoo. *Exp. Math.* **2013**, *22*, 313–326. [CrossRef]
3. Coates, T.; Corti, A.; Galkin, S.; Golyshev, V.; Kasprzyk, A. Mirror symmetry and Fano manifolds. In Proceedings of the European Congress of Mathematics, Krakow, Poland, 2–7 July 2012; European Mathematical Society Publishing House: Zurich, Switzerland, 2012, 824p; ISBN 978-3-03719-120. [CrossRef]
4. Akhtar, M.; Coates, T.; Corti, A.; Heuberger, L.; Kasprzyk, A.; Oneto, A.; Petracci, A.; Prince, T.; Tveiten, K. Mirror symmetry and the classification of orbifold Del Pezzo surfaces. *Proc. Am. Math. Soc.* **2016**, *144*, 513–527. [CrossRef]
5. Kasprzyk, A.; Nill, B.; Prince, T. Minimality and mutation-equivalence of polygons. In *Forum of Mathematics, Sigma*; Cambridge University Press: Cambridge, UK, 2017; Volume 5.
6. Corti, A.; Heuberger, L. Del Pezzo surfaces with $\frac{1}{3}(1,1)$ points. *Manuscripta Math.* **2017**, *153*, 71–118. [CrossRef]
7. Cavey, D.; Prince, T. Del Pezzo surfaces with a single $\frac{1}{k}(1,1)$ singularity. *J. Math. Soc. Jpn.* **2020**, *72*, 465–505. [CrossRef]
8. Cavey, D.; Kutas, E. Classification of Minimal Polygons with Specified Singularity Content. *arXiv* **2017**, arXiv:1703.05266.
9. Miura, T. Classification of del Pezzo surfaces with $\frac{1}{3}(1,1)$ and $\frac{1}{4}(1,1)$ singularities. *arXiv* **2019**, arXiv:1903.00679.
10. Reid, M.; Suzuki, K. Cascades of projections from log del Pezzo surfaces. In *Number Theory and Algebraic Geometry*; London Mathematical Society Lecture Note Series; Reid, M., Skorobogatov, A., Eds.; Cambridge University Press: Cambridge, UK, 2004; pp. 227–250. [CrossRef]
11. Corti, A.; Reid, M. Weighted Grassmannians. In *Algebraic Geometry*; Beltrametti, M.C., Catanese, F., Ciliberto, C., Lanteri, A., Pedrini, C., Eds.; de Gruyter: Berlin, Germany, 2002; pp. 141–163.
12. Qureshi, M.I.; Szendrői, B. Constructing projective varieties in weighted flag varieties. *Bull. Lond. Math. Soc.* **2011**, *43*, 786–798. [CrossRef]
13. Qureshi, M.I.; Szendrői, B. Calabi-Yau threefolds in weighted flag varieties. *Adv. High Energy Phys.* **2012**, *2012*, 547317. [CrossRef]
14. Brown, G.; Kasprzyk, A.M.; Qureshi, M.I. Fano 3-folds in $\mathbb{P}^2 \times \mathbb{P}^2$ format, Tom and Jerry. *Eur. J. Math.* **2018**, *4*, 51–72. [CrossRef]
15. Brown, G.; Kasprzyk, A.M.; Zhu, L. Gorenstein formats, canonical and Calabi-Yau threefolds. *Exp. Math.* **2019**, 1–19. [CrossRef]
16. Qureshi, M.I. Computing isolated orbifolds in weighted flag varieties. *J. Symb. Comput.* **2017**, *79 Pt 2*, 457–474. [CrossRef]
17. Mayanskiy, E. Weighted complete intersection del Pezzo surfaces. *arXiv* **2016**, arXiv:1608.02049.
18. Paemurru, E. Del Pezzo surfaces in weighted projective spaces. *Proc. Edinb. Math. Soc.* **2018**, *61*, 545–572. [CrossRef]
19. Qureshi, M.I. Biregular models of log del Pezzo surfaces with rigid singularities. *Math. Comput.* **2019**, *88*, 2497–2521. [CrossRef]
20. Atiyah, M.F.; Macdonald, I.G. *Introduction to Commutative Algebra*; Reading, Mass.-London-Don Mills, Ont.; Addison-Wesley Publishing Co.: Boston, MA, USA, 1969; p. ix+128.
21. Kollár, J.; Shepherd-Barron, N.I. Threefolds and deformations of surface singularities. *Invent. Math.* **1988**, *91*, 299–338.
22. Akhtar, M.; Kasprzyk, A. Singularity content. *arXiv* **2014**, arXiv:1401.5458.
23. Szendrői, B. *On Weighted Homogeneous Varieties*; 2005. Unpublished manuscript.
24. Buckley, A.; Reid, M.; Zhou, S. Ice cream and orbifold Riemann–Roch. *Izv. Math.* **2013**, *77*, 461–486. [CrossRef]
25. Blache, R. Chern classes and Hirzebruch–Riemann–Roch theorem for coherent sheaves on complex-projective orbifolds with isolated singularities. *Math. Z.* **1996**, *222*, 7–57. [CrossRef]
26. Brown, G.; Fatighenti, E. Hodge numbers and deformations of Fano 3-folds. *arXiv* **2017**, arXiv:1707.00653.

27. Dolgachev, I. Weighted projective spaces. In *Group Actions and Vector Fields*; Lec Note in Mathematics; Springer: Berlin/Heidelberg, Germany, 1981; Volume 956, pp. 34–71.
28. Hwang, D. On the orbifold Euler characteristic of log del Pezzo surfaces of rank one. *J. Korean Math. Soc.* **2014**, *51*, 867–879. [CrossRef]
29. Iano-Fletcher, A.R. Working with weighted complete intersections. In *Explicit Birational Geometry of 3-Folds*; London Math. Soc. Lecture Note Ser; CUP: London, UK, 2000; Volume 281, pp. 101–173.
30. Bosma, W.; Cannon, J.; Playoust, C. The Magma algebra system. I. The user language. *J. Symb. Comput.* **1997**, *24*, 235–265. Computational algebra and number theory (London, 1993). [CrossRef]
31. Brown, G.; Zucconi, F. Graded rings of rank 2 Sarkisov links. *Nagoya Math. J.* **2010**, *197*, 1–44. [CrossRef]

© 2020 by the author. Licensee MDPI, Basel, Switzerland. This article is an open access article distributed under the terms and conditions of the Creative Commons Attribution (CC BY) license (http://creativecommons.org/licenses/by/4.0/).

MDPI
St. Alban-Anlage 66
4052 Basel
Switzerland
Tel. +41 61 683 77 34
Fax +41 61 302 89 18
www.mdpi.com

Mathematics Editorial Office
E-mail: mathematics@mdpi.com
www.mdpi.com/journal/mathematics

www.ingramcontent.com/pod-product-compliance
Lightning Source LLC
LaVergne TN
LVHW070704100526
838202LV00013B/1029